河南省“十四五”普通高等教育规划教材
煤炭高等教育“十四五”规划教材
河南省研究生教育优质课程项目（HNYJS2018KC10）资助

工程热力学与传热学

双语

第三版

主　编　王志军　袁东升　宋文婷
副主编　程　磊　张瑞林　邓权威　赵文斌

Engineering Thermodynamics and Heat Transfer

中国矿业大学出版社
·徐州·

内 容 简 介

本书主要介绍了工程热力学与传热学的基本概念、基本定律、水蒸气及湿空气的热力性质，导热、对流、辐射三种传热方式的原理与计算方法。全书共分 11 章，每章后面附有思考与练习题。

本书为新形态教材，可作为高等院校安全工程、采矿工程、消防工程及相关专业的教材或教学参考书，也可供有关工程技术人员参考。

图书在版编目(CIP)数据

工程热力学与传热学 ：汉英对照 / 王志军，袁东升，宋文婷主编. —3 版. —徐州 ：中国矿业大学出版社，2022.10(2024.12 重印)

ISBN 978-7-5646-5349-1

Ⅰ. ①工… Ⅱ. ①王… ②袁… ③宋… Ⅲ. ①工程热力学—教材—汉、英②工程传热学—教材—汉、英 Ⅳ. ①TK123②TK124

中国版本图书馆 CIP 数据核字(2022)第 151810 号

书　　名　工程热力学与传热学
　　　　　　Engineering Thermodynamics and Heat Transfer
主　　编　王志军　袁东升　宋文婷
责任编辑　王美柱　满建康
出版发行　中国矿业大学出版社有限责任公司
　　　　　　(江苏省徐州市解放南路　邮编 221008)
营销热线　(0516)83885370　83884103
出版服务　(0516)83995789　83884920
网　　址　http://www.cumtp.com　**E-mail**:cumtpvip@cumtp.com
印　　刷　江苏淮阴新华印务有限公司
开　　本　787 mm×1092 mm　1/16　**印张** 19.25　**字数** 566 千字
版次印次　2022 年 10 月第 3 版　2024 年 12 月第 3 次印刷
定　　价　35.00 元
(图书出现印装质量问题，本社负责调换)

第三版前言

本教材第一版于2015年8月问世，受到众多院校及同行们的支持和肯定。作为本教材的编写者，我们也是教材的使用者。自第一版教材出版之日起，我们一直在不断检查、审视、完善这部教材。针对一些同行专家在使用第一版教材过程中发现的问题，2018年1月我们完成了第二版教材的修订。第二版教材获评第三届煤炭行业优秀教材一等奖(2020年)、河南省首届教材建设(高等教育类)二等奖(2021年)。根据第二版教材使用反馈情况，2020年下半年我们启动第三版教材编写工作。第三版教材于2020年11月入选河南省“十四五”普通高等教育规划教材(河南省教育厅教高〔2020〕469号)建设行列，2021年被列入煤炭高等教育“十四五”规划教材建设项目。

在第二版教材的基础上，本次修订我们重点做了以下修改：在每章开篇处设置“本章导学”“学习目标”“重点与难点”模块，以便于学生对该章知识内容的理解和总体把握，明确该章内容在课程知识体系中的地位，方便学生掌握课程知识的脉络结构；在每章末尾处增加“本章小结”，归纳总结该章主要内容，指出内容之间的内在逻辑，便于学习和掌握该章的知识；在部分章节末尾处设置“知识拓展”模块(通过扫描二维码获取)，便于学生了解更多相关知识，其中一些知识属于课程思政教育内容，在开阔知识视野的同时还可以加强思政教育；此外，还增设了部分章节内容、重新选用了典型例题及课后习题、优化了英文对照翻译、重新精绘了图片、增补了附录内容。

本版教材由河南理工大学王志军、袁东升、宋文婷担任主编，由程磊、张瑞林、邓权威、赵文斌担任副主编。本次修订附录由河南理工大学程磊和张健编写，其余各章节编写分工与第一版教材相同。

除编写人员参加修订外，使用过本教材的老师也提出了宝贵的修改意见和建议，在此特表示衷心的感谢。本教材得到河南省研究生教育优质课程项目(HNYJS2018KC10)的资助，在此一并表示感谢。此次虽然修订了较多内容，但由于编者水平所限，书中可能仍有不能令人满意之处，恳请广大读者提出意见、建议，以使本书更趋完善。

编 者

2022年7月

第一版前言

工程热力学与传热学也简称为热工学，主要研究热功转换、热能合理利用和热量传递规律。随着我国社会经济的飞速发展，能源需求量快速增长，能源资源紧缺以及高效洁净转换利用技术缺乏已成为制约我国经济发展的关键问题。因此，节能降耗、保护环境是每一个工程技术人员的责任。能源的开发和利用在很大程度上是热能的开发和利用，热工学几乎是渗透到现代工程技术各个领域的一门技术基础课，是各专业实际应用中不可缺少的重要组成部分，掌握合理用能的基本知识和理论是新世纪复合型人才应具备的重要素质。

本教材将“工程热力学”与“传热学”合二为一，抓住能量转换与传递这一主线索进行阐述，注重科学性与教学性的结合，注重理论联系实际，以便学生学完本课程后能熟练地运用热力学与传热学的基本理论与基本分析方法，进一步学好后续课程，解决实际工程技术问题。

我国教育部于2001年制定了推动大学双语教学的文件。近年来，双语教学已经逐步成为我国高等教育教学改革的一个热点，各高校纷纷参与双语教学的改革实践。“工程热力学与传热学”目前已经有较完善的基本理论，其涉及的知识的应用又是当前国际能源领域的学者和科研技术人员研究的热门课题。对“工程热力学与传热学”实施双语教学，有利于在知识传授方面深入浅出地介绍本学科领域的知识，又能在语言环境上营造出氛围，使学生能够较为轻松地学习，有利于培养学生的综合素质。

本书由河南理工大学王志军、袁东升、宋文婷担任主编，由王志军拟定全书内容提纲并负责统稿、定稿。全书共分11章，各章节编写分工如下：第1、3章由河南理工大学王志军编写；第2章由河南理工大学袁东升编写；第4章由山东科技大学赵文斌与河南理工大学程磊编写；第5章由河南理工大学王志军与程磊编写；第6、7章由河南理工大学宋文婷编写；第8、9章由河南理工大学邓权威编写；第10章由河南理工大学宋文婷与王志军编写；第11章由河南理工大学程磊与河南工程学院张瑞林编写；附录由河南理工大学程磊编写。

本书编写过程中参考了相关教材、专业书籍等文献资料，在此谨向被引用的文献作者和引用但未注明的文献作者一并表示衷心感谢！同时，向中国矿业大学出版社的领导和编辑们为本书的出版给予的大力支持和帮助表示诚挚的谢意！

由于编者水平所限，书中不足之处在所难免，恳请读者及使用本书的师生批评指正。

编　者

2015年7月

目　录

第1章 绪　论

本章导学

从学科上讲，工程热力学与传热学（简称热工学）是研究有关热能利用的能量传递和能量转换的一门学科。能源是人类社会赖以生存和发展的重要物质基础。能源的利用不仅关系着经济的可持续发展，而且直接影响人类生存的生态环境。热能的利用是能量利用的主要形式，学习相关的热能知识十分必要。热工学几乎是渗透到现代工程技术各个领域的一门技术基础课，是各专业实际应用中不可缺少的重要组成部分，掌握合理用能的基本知识和理论是新世纪复合型人才应具备的重要素质。

学习目标

(1) 能源与热能利用：了解自然界的能源及其主要分类方法，能源利用的主要形式。掌握热能利用的意义和分类。

(2) 工程热力学与传热学的研究内容及研究方法：掌握本学科的研究对象和研究目的；掌握本学科研究的主要内容；了解本学科研究的主要方法，为学好本课程奠定基础。

(3) 了解工程热力学的发展简史。

重点与难点

(1) 热能利用的意义和分类。

(2) 工程热力学与传热学之间的关系。

1.1 能源与热能利用

为人类生产与日常生活提供各种能量和动力的自然资源称为能源，能源是人类赖以生存和发展所必需的燃料和动力来源，是发展生产和提高人类生活水平的重要物质基础。自然界中所有的活动都涉及物质和能量的相互作用。人类对各种能源的不断开发和利用大大地推动了人类社会生产力的发展。纵观世界各国的国民经济发展状况，能源消费水平一定程度上反映出社会生产力的发展水平。

风能、水力能、燃料的化学能、太阳能和地热能等能源，是自然界中以自然形态存在可直接开发利用的能源，称为一次能源。这些能源，有些可以以机械能的形式直接被利用，有些需要经过加工转化后才能利用。由一次能源直接或间接转化而来的能源称为二次能源，其中主要是热能、机械能和电能。因此,能量的利用过程,实质是能量的传递和转换过程。各种能源及其转换和利用情况大致如图 1-1 所示。

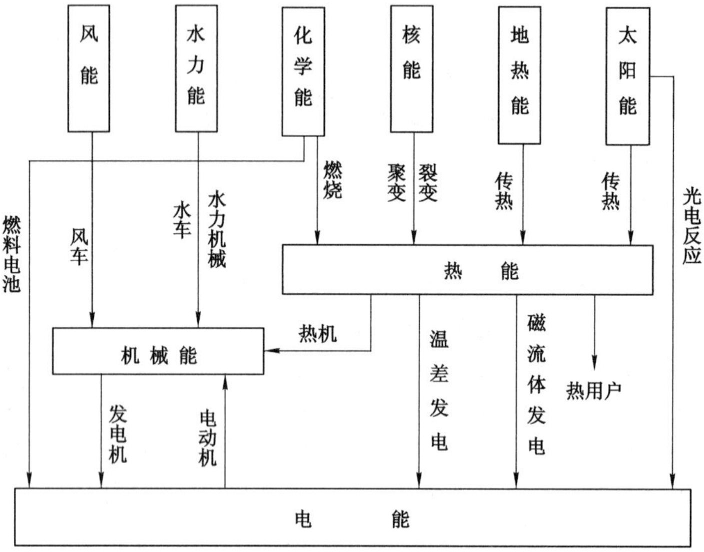

图 1-1　能源的转换和利用情况

目前,本着可持续发展的思想,国内外均在提倡风能、太阳能、潮汐能等清洁能源的应用。就目前国际国内技术水平而言,在今后相当长的时期内,主要能源仍将以煤炭、石油及天然气等矿物燃料为主,并将加快核能利用的步伐。由图 1-1 可知,除了水力能和风能是机械能外,其余一次能源都直接或间接向人类提供热能形式的能量,且占有相当大的比例,据统计,经热能这个环节而被利用的能量在世界上占 85%以上。热能是一次能源利用过程中重要的转换形式,从某种意义上讲,能源的开发和利用就是热能的开发和利用。因此,热能的开发及高效利用是能源可持续应用的最重要的部分,对于人类社会的发展有着重要的意义。

热能的利用可分为直接利用和间接利用两种。热能的直接利用是指直接用热能加热物体,热能的形式不发生变化。如工业生产中的冶炼、加热、蒸煮、干燥及分馏等,再如日常生活中的热水供应及采暖等。工业中直接利用热能的设备很多,如各种工业炉窑、工业锅炉、加热器、冷却器、蒸发器和冷凝器等。热能的间接利用是指把热能通过各种类型的发动机(热机)及发电机转变为机械能或电能,如蒸汽动力装置、燃气动力装置、火箭发动机、内燃机等都能实现热能向机械能或电能的转换。热能的间接利用是热能利用的重要方式,这种间接利用方式是极其重要的,是人类文明及生产发展的物质基础。人类发明蒸汽机引起了第一次工业革命,使手工作坊式生产走向大规模的工业生产,对人类改造自然、发展生产起了重大作用。

在热能的间接利用中,主要存在热能转化为机械能或电能过程中的有效程度的问题。如在热力发电中,最简单的装置,热能有效利用率只有 25%左右,先进的大型装置也只能达到约 40%～45%,最高接近 47%,仍有一半以上的热能无法利用。这部分无法利用的热能称为废热。再如交通运输中的汽车、火车、飞机及轮船,热能的有效利用率更低。这些装置排放到大气中的废气带有大量有害物质,已经给人类赖以生存的环境造成了严重的污染。因此,如何在动力装置中提高热能的有效利用率并消除污染,是能源科技工作者的首要任务。目前正在研究中的大型热能动力装置,如能按理想工况进行运转,有可能将热能的利用率提高到55%。提高热能动力装置的热效率是节约能源的主要途径之一。

热能直接利用所消耗的燃料也占有较大的比例,节约燃料的消耗同样是十分重要的。而直接利用热能的设备也存在换热效率问题。因此,如何提高换热设备的换热效率是当今重要的研究课题之一。

国民经济的发展离不开燃料热能的间接利用和直接利用,然而地球上的矿物燃料资源毕竟有限,如何提高热能的有效利用率,减少燃料的消耗量,不仅是我国科技界的重大课题,也是一个世界性的学术问题。因此,对物质热力性质、热能转换及热量传递规律的研究,具有十分重要的意义。

1.2 工程热力学与传热学的研究内容及研究方法

热能间接利用所涉及的热能和机械能的转换属于工程热力学范畴。热能直接利用所涉及的研究热量传递规律的学科属传热学。工程热力学与传热学基本理论组合形成了热能工程学的理论基础,简称热工学,它是研究热能利用的基本原理和规律,以及热能利用的能量传递和能量转换,以提高热能利用经济性(即节能)为主要目的的一门学科。

工程热力学主要研究热能和机械能及其他形式的能量之间相互转换的规律。热能转变为机械能必须借助某种传递能量的载能物质——工质和设备(热机)。为实现工质在热机中通过状态的变化,源源不断地把热能转变成机械能对外输出并提高转换效率,必须掌握热能和机械能转换的基本理论,了解工质的性质以及能量转换过程中工质参数变化的特性。所以,工程热力学的主要内容包括:基本概念和基本定律——热力学第一定律和第二定律,工质在热能动力设备中吸热、膨胀做功等状态变化过程的特点及热量和功量的计算分析,常用工质的热力性质。

传热学主要研究热量传递的规律。热量是指在温差的作用下传递的能量。热力学第二定律指出,热量可以自发地由高温热源传给低温热源。这就是说,有温差就会有热量传递(简称传热)。由于在自然界以及在人们的日常生活和生产实践中,温差几乎无处不在,所以热量传递是普遍存在的物理现象,对人们的生活和生产实践具有广泛而深刻的影响。因此,认识传热的规律,掌握控制与优化热量传递的方法和技术措施,对国民经济建设以及改善人民生活具有重要的意义。传热学的主要内容就是对传热的基本方式和实际的传热过程进行分析,研究物体内部或物体与物体之间的热能传递机理,从而找出提高传热效率或减少热损失的途径以及确定常见简单物体内温度分布的规律。

工程热力学与传热学采用宏观的研究方法,其特点是根据工程热力学的基本定律和传热学的热量传递规律,运用严密的逻辑推理,对物体的宏观现象进行分析研究,而不涉及物质的微观结构和微观粒子的运动情况。这种抽出共性,抓主要矛盾的处理问题方法,在工程热力学与传热学的理论分析中特别有用,不仅不脱离实际,而且能反映客观事物的本质。

本章小结

（1）热能利用是能源利用的主要形式

人类利用的主要能源有水力能、风能、地热能、太阳能、燃料的化学能和原子核能。在这些能源中，除水力能和风能是机械能外，其余都直接或间接向人类提供热能形式的能量，且占了相当大的比例。因此，从某种意义上讲，能源的开发和利用就是热能的开发和利用。

（2）热能利用的两种形式

热能的利用形式可分为直接利用和间接利用两种。热能的直接利用是指直接用热能加热物体，热能的形式不发生变化。热能的间接利用是指把热能转换为机械能（或进而再转换为电能）以满足人类生产和生活对动力的需要。在热能的间接利用中，热能的能量形式发生了转换。

（3）工程热力学与传热学的主要内容

热能间接利用所涉及的热能和机械能的转换属工程热力学范畴。它的主要内容有热力学基本定律、工质热力性质和工质热力过程等理论基础。

热能直接利用所涉及的研究热量传递规律的学科属传热学。它的主要内容是对三种不同热传递方式进行研究。热量传递的三种基本方式为热传导、热对流和热辐射。

应用热工理论对实际工程中的工程热力过程、热力循环、装置和换热设备进行分析研究，以提高它们的能量利用经济性，是本课程的另一主要内容。

（4）工程热力学与传热学的研究方法

工程热力学与传热学采用宏观和唯象的方法进行研究，即把物质视为连续体，用宏观物理量去描述物质的性质、行为，以大量观察和实验总结出来的基本定律为依据，通过推理、演绎得出具有高可靠性和普遍适用性的结论、公式，以解决热量传递和热能转换的问题。

（5）热能利用简史和热工理论基础的学科发展

人类对热能的直接利用可以追溯到远古的钻木取火和对火的利用，至今生产和生活中的各种加热、烘干、冶炼等设备都是直接利用热能的设备。

18 世纪瓦特发明的蒸汽机实现了热能向机械能的能量转换，开创了热能间接利用的新纪元，带来了第一次工业革命。从此，热能的间接利用得到了广泛深入的发展，转换技术和装置不断创新。

18 世纪的第一次工业革命使生产力得到空前发展，同时导致了涉及研究热能利用的热工理论基础的两大支柱——工程热力学和传热学的出现和不断发展。

知识拓展：工程热力学的发展简史

Chapter 1 Preface

1.1 Energy Resources and Utilization of Thermal Energy

In nature, all kinds of activities concern mass and energy interactions. Human being can not survive and develop without fuels. The exploration and utilization of various kinds of energy resources prompt the development of social productivities of human being. And the energy consumption condition indicates the situation of productivities to some extent.

There are large amount of energy resources, such as wind-power resources, water-power resources, chemical energy stored in fuels, solar energy, geothermal energy and nuclear energy, etc. At present, wind resources, solar energy and tidy energy technologies are encouraged to be used at home and abroad to maintain sustainable development. However, in a very long term, the mineral fuels such as coal, crude oil and natural gases will be in domination at domestic. And the exploration of nuclear energy will be strengthened. Fig. 1-1 tells us the energy conversion from various kinds of energies into thermal energy is a very important step before they are finally used. Therefore, it is very important to use the thermal energy emitted by combustion of fuels or by fission or fusion of nuclear energy efficiently to achieve sustainable energy utilization.

There are two ways to use the thermal energy emitted by fuel combustion. One is using thermal energy directly. For example, during the process of melting, heating, drying and fractionating, etc., thermal energy is used directly. There are many facilities, in which thermal energies are used directly, such as, furnaces, boilers, heaters, condensers, evaporators and chillers, and so on. The other way is indirect utilization, that is, to transform thermal energy into mechanical energy or electricity through turbines and power generators. For example, equipment in steam power plants, gas-turbine, engine of rocket, internal combustion engines, etc. , can be used to accomplish energy conversion. Most of thermal energy is used indirectly. Since the first industrial revolutionary aroused by the invention of steam engines, the productivities have developed greatly.

An important issue is how to efficiently transform thermal energy into mechanical energy or electricity by indirect use of thermal energy. For instance, only 25% of thermal energy can be transformed into power in simple thermal power plants. Even in large installations, the energy transformation efficiency can only reach about 40% to 45%. This part of thermal energy which is not used effectively is called waste heat. Another example, the effectiveness of

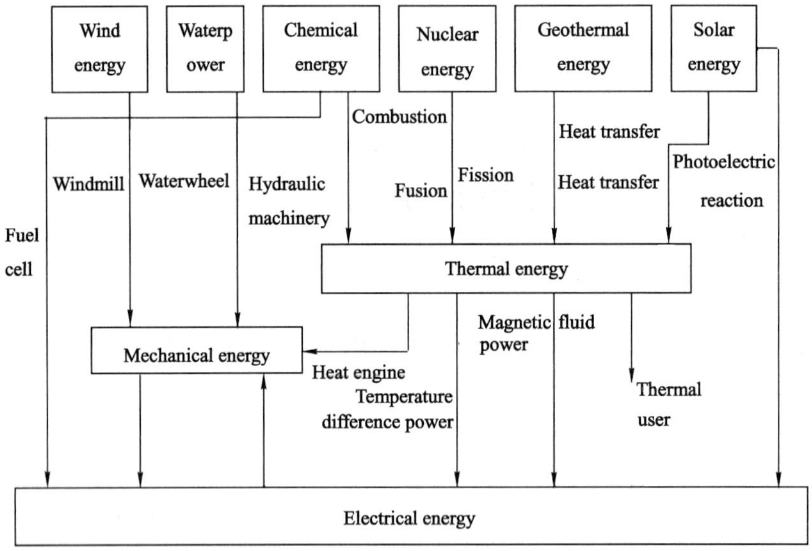

Fig. 1-1 Utilization of Thermal Energy

thermal energy utilization is even less in vehicles, locomotives, air-crafts and steamboats, etc. The exhaust gases from these equipments include a large amount of harmful substances. This has brought severe pollution on the environment. Thus, how to improve thermal efficiency of these equipments is the most important thing, to which we should pay attention.

The amount of fuels burned when thermal energy is used directly is also very huge. Thus, it is very important to save fuels. Thus, how to effectively accomplish heat transfer in heat exchangers is another key topic we are facing. This topic will be introduced in the branch of Heat Transfer.

Economy can not develop without direct and indirect use of thermal energy. However, the amount of mineral fuels is limited. Thus, how to improve thermal efficiency and how to save energy are critical research topics worldwide. Therefore, it is necessary to study thermodynamic properties of substances and the rules which govern thermal energy transformation and transfer.

1.2 Main Contents and Study Methods of Engineering Thermodynamics & Heat Transfer

It is indispensible to converse thermal energy into mechanical energy without some installations and working medium. The working medium can flow through the installations and output expansion work through volume changes. Therefore, the generally used working mediums are gaseous substances. In the course of Thermodynamics, it is to study thermodynamic systems, thermal equilibrium states, thermodynamic processes, thermodynamic cycle and working medium to better the performance of heat engines, refrigerators and heat pumps and to improve the thermal efficiency and energy utilization effectiveness. Therefore, the characteristics of thermodynamic processes will be studied based on the fundamental laws of thermodynamics. And thermo-physical properties of gases and fluids will be introduced as well as the law governs phase change processes, such as evaporation and condensation. Chemical reaction processes, such as combustion, melting and resolving are involved in practice. These may concern some basic knowledge of Chemical Thermodynamics. Thus, the following three parts will be introduced in Thermodynamics: ① Two fundamental laws of thermodynamics—the First Law and the Second Law of Thermodynamics; ② Thermodynamic properties of working mediums commonly used; ③ Thermodynamic processes, cycles and the ways to improve thermal efficiency.

Heat transfer describes the exchange of thermal energy, between physical systems depending on the temperature and pressure, by dissipating heat. Systems which are not isolated may decrease in entropy. Most objects emit infrared thermal radiation near room temperature. The fundamental modes of heat transfer are conduction or diffusion, convection, advection and radiation.

The exchange of kinetic energy of particles through the boundary between two systems which are at different temperatures from each other or from their surroundings. Heat transfer always occurs from a region of high temperature to another region of lower temperature. Heat transfer changes the internal energy of both systems involved according to the First Law of Thermodynamics. The Second Law of Thermodynamics defines the concept of thermodynamic entropy, by measurable heat transfer.

第 2 章　热力学基本概念

本章导学

在工程热力学中，要用到一组基本概念，这些基本概念是工程热力学独特研究方法的基础。对于这些基本概念，一开始就必须予以重视，正确理解它们的含义。然后，随着课程的展开，逐步学会熟练地利用它们来分析和解决问题。本章先对工程热力学的研究对象和研究方法，以及一些主要的基本概念和术语作简要阐述，并以此为基础，在以后的章节中逐步扩展、深化。

学习目标

（1）掌握工程热力学的一些基本术语和概念：热力系、平衡状态、准平衡过程、可逆过程、循环等。

（2）掌握状态参数的特征、分类和基本状态参数 p、v、T 的定义和单位。

（3）了解工程热力学分析问题的特点、方法和步骤。

重点与难点

（1）平衡状态、准平衡过程与可逆过程。

（2）状态量与过程量。

2.1　热力系统

2.1.1　热力系统含义

分析研究任何现象，首先应明确研究对象。工程热力学中，为了研究问题的需要，以人为划定的一个或多个任意几何面所围成的空间作为研究对象，这种空间内物质的总和称为热力系统，简称热力系或系统。热力系以外的物质称为外界，热力系与外界的交界称为边界。边界可以是真实的，也可以是假想的；可以是固定的，也可以是移动的。如图 2-1 所示的气缸活塞机构，若取气缸中的气体为热力系，则气体和气缸壁间构成的边界是真实的，气体和活塞内表面构成的边界为移动的。如图 2-2 所示的汽轮机，若取汽轮机中的蒸汽为热力系，蒸汽和汽轮机之间就是实际边界，而进、出口处则为假想的边界。

对一个具体的问题，热力系统的选择方式可以有很多种，应该视具体情况而定。确定系统以后，该如何来进行分析和研究呢？在物理学或者力学中，主要研究系统与外界之间所存

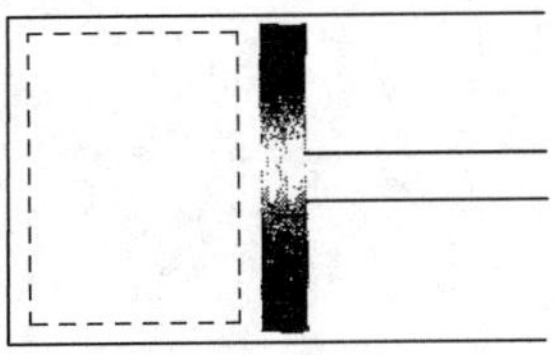

图 2-1　闭口系

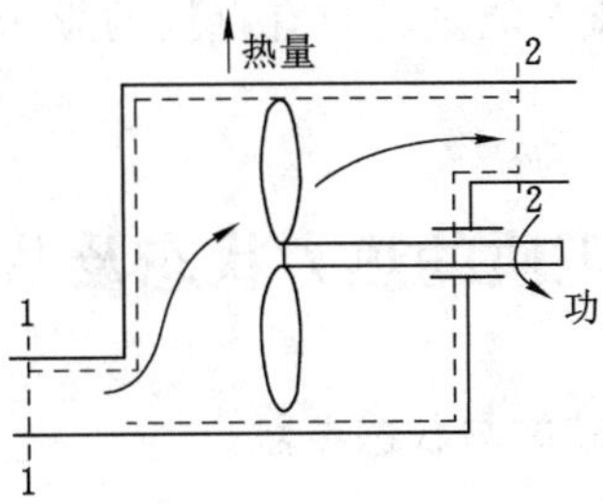

图 2-2　开口系

在的力的作用。而在热力学中，则主要研究系统和外界之间的能量及物质的相互作用。

2.1.2　热力系统分类

根据热力系统与外界之间的能量和物质交换情况，热力系统可分为各种不同的类型。

(1) 闭口系：指与外界之间没有物质交换的热力系统。系统中物质的质量不变，有时也称为控制质量系统，如图 2-1 所示。

(2) 开口系：指与外界有物质交换的热力系统，有质量的流入或流出。如图 2-2 所示，运行中的汽轮机就可视为开口系，在运行过程中，有蒸汽不断地流进流出。由于开口系是一个划定的空间范围，所以开口系又称控制容积系统。

(3) 绝热系：指与外界没有热量交换的热力系统。在热力学中“绝热”是一个理想化的概念，实际上即使对系统的边界采用很好的绝热保温措施，也难免与外界有一定的热量交换。但当系统与外界交换的热量与其他形式的能量交换相比足够小时，即可以认为是“绝热”的。

(4) 孤立系：指与外界不发生任何相互作用的热力系统。既没有任何形式的能量交换，也不发生物质的交换。显然，孤立系也是一个理想化的概念。有时为了便于问题的研究，可以把系统连同与其进行能量和物质交换的相关物体或空间视为一体，构成一个孤立系。孤立系内部各个子系统间存在各种相互作用，而孤立系与外界之间则无任何相互作用。

(5) 简单可压缩系：热力系由可压缩流体构成，并且在边界上只与外界有一种功的交换。工程热力学讨论的大部分系统都是简单可压缩系。

此外，也可按热力系统内部状况的不同，将热力系统分为均匀系、非均匀系，单元系、多元系等类型。

对热力系统进行分析时，应该注意，系统与外界发生的各种相互作用均穿越边界，对相互作用(质量、热量或功量交换)的分析应当从边界着手进行。正确选取系统对于热力学分析具有重要的意义。选取不当，就难以进行热力学分析或无法得到正确的结论。

2.1.3　工质

能量转换必须通过物质来实现，物质是实现能量转换不可缺少的内部条件。用来实现能量相互转换的媒介物称为工质。例如，内燃机以燃气为工质，蒸汽动力装置以水蒸气为工质，等等。

从理论上讲，气、液、固三态物质都可作为工质，但工程热力学所研究的能量转换一般都是通过工质体积变化来实现的，而对体积变化最灵敏的是气(汽)态物质。因此，在工程热力学中，主要选择气(汽)态物质为工质。

工质对能量转换有直接的影响，故对不同工质性质的研究也是工程热力学的主要内容之一。

2.2 工质的热力状态及基本状态参数

2.2.1 状态与状态参数

系统与外界之间能够进行能量交换（传热或做功）的根本原因，在于两者之间的热力状态存在差异。例如，锅炉中的热量传递是由于燃料燃烧生成的高温烟气与汽锅内汽水之间存在温度差；又如热力发动机中能量的转换是由于热力发动机中的高温高压工质与外界环境的温度、压力有很大的差别。这种温度、压力上的差异标志着工质物理特性数值的不同。我们把工质在热力变化过程中的某一瞬间所呈现的宏观物理状态称为工质的热力状态，简称为状态。热力状态反映工质大量分子热运动的平均特性，描述工质状态特性的宏观物理量称为工质的状态参数。

状态参数是状态的函数，热力状态一定，描述状态的状态参数也确定。描述一种状态的状态参数是唯一确定的一组数值。只要其中的一个参数发生变化，则系统状态就不同。

状态参数是点函数，这是状态参数的一个重要的基本特性。如图 2-3 所示，当系统从初态 1 经历不同的途径变化至终态 2 时，状态参数的变化量仅与初、终状态有关，而与状态变化所经历的途径无关，即

$$\Delta x_{1-2} = x_2 - x_1 = \int_1^2 \mathrm{d}x \tag{2-1}$$

式中 x——工质某一状态参数。

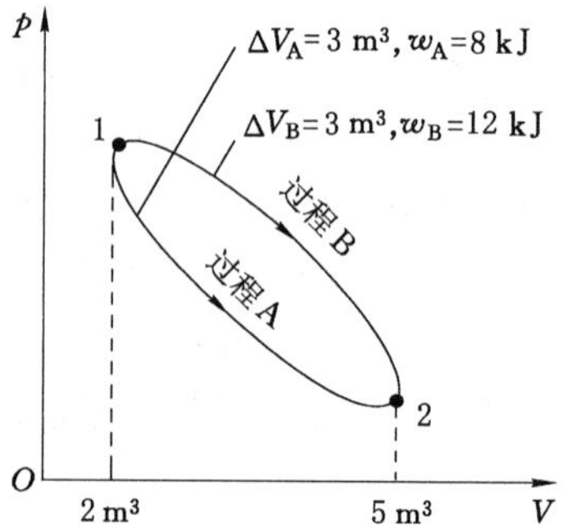

图 2-3 状态

热力系统经过一系列变化之后又回到初始状态时，其状态参数的变化量为零，即状态参数的循环积分为零：

$$\oint \mathrm{d}x = 0 \tag{2-2}$$

状态参数分为基本状态参数和导出状态参数。基本状态参数可以直接或间接用仪表测量，如温度（T）、压力（p）、比容（v）或密度（ρ）等。导出状态参数则无法直接测量得到，而只能由基本状态参数间接计算得出，如内能（u）、焓（h）、熵（s）等。下面首先介绍基本状态参数，其他的状态参数在以后各章中逐步介绍。

2.2.2　基本状态参数

2.2.2.1　温度

温度是物体冷热程度的标志。从微观上看，温度标志物质分子热运动的激烈程度。温度概念的建立以及温度的测量是以热力学第零定律（或称热平衡定律）为依据的。热力学第零定律表述为：如果两个物体分别与第三个物体处于热平衡，则这两个物体彼此也处于热平衡。其中，第三个物体可用作温度计，即当温度计与被测的物体达到热平衡时，温度计所指示的温度就是被测物体的温度。

进行温度测量需要有温度的数值表示方法，即需要建立温度的标尺或温标。国际单位制（SI）采用热力学温标作为基本温标，用这种温标确定的温度称为热力学温度，用符号 T 表示，单位是开尔文，符号为 K。热力学温标取水的三相点（纯水的固、液、汽三相平衡共存的状态）为基准点，并定义其温度为 273.16 K。因此，1 K 等于水的三相点热力学温度的1/273.16。

与热力学温标并用的还有热力学摄氏温标，简称摄氏温标。用这种温标确定的温度称为摄氏温度，以符号 t 表示，单位为 ℃，并定义 $t=T-273.15$，由此可知，摄氏温标与热力学温标仅起点不同。摄氏温度 0 ℃相当于热力学温度 273.15 K。显然，水的三相点温度为 0.01 ℃。

2.2.2.2　压力

（1）压力及其单位

垂直作用于器壁单位面积上的力称为压力（即压强），用符号 p 表示，即

$$p=\frac{F}{A} \tag{2-3}$$

式中，F 为垂直作用于面积 A 上的力。分子运动学说把气体的压力看作大量气体分子撞击器壁的平均效果。

SI 规定压力单位为帕斯卡（Pa），1 Pa=1 N/m^2。因 Pa 这个单位太小，常用 kPa（千帕）和 MPa（兆帕）作为压力的单位，并有 1 MPa=10^3 kPa=10^6 Pa。

工程上还有其他压力单位，如 bar（巴）、atm（标准大气压）、mmHg（毫米汞柱）、mmH$_2$O（毫米水柱）、at（工程大气压）等，它们与 Pa 之间的换算关系为：

1 bar=10^5 Pa，1 atm=1.013×10^5 Pa，1 mmHg=133.3 Pa，

1 mmH$_2$O=9.8 Pa，1 at=9.8×10^4 Pa

（2）绝对压力、表压力和真空度

工质的真实压力称为绝对压力，用 p 表示，一般用如图 2-4(a)、图 2-4(b)所示的测量微小压力的 U 形管压力计或如图 2-4(c)所示的弹簧管式压力计测量。压力计本身处于某种环境（通常是大气环境）压力下，因此压力计所测得的压力是工质的真实压力（即绝对压力）与环境压力之差，称为表压力，用 p_g表示；或称为真空度，用 p_v表示。

绝对压力 p、表压力 p_g、真空度 p_v和环境压力 p_b之间的关系如图 2-5 所示，并表述如下：

当绝对压力 p 高于环境压力 p_b时，则 $p=p_b+p_g$；

当绝对压力 p 低于环境压力 p_b时，则 $p=p_b-p_v$。

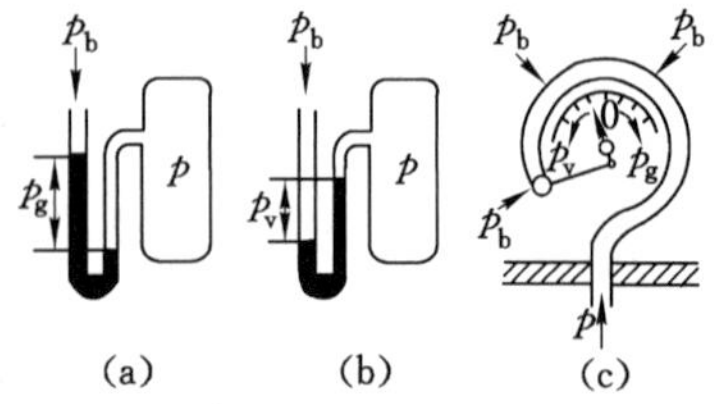

图 2-4　压力计示意图

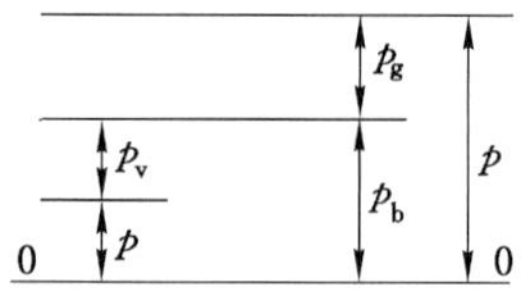

图 2-5　各压力间的关系

环境压力随测量时间、地点变化而不同，可用压力计测定。即使绝对压力不变，表压力和真空度也会因环境压力变化而变化。因此，作为工质状态参数的压力是绝对压力。本书中如不注明是“相对压力或表压力”，都应理解为“绝对压力”，注意只有绝对压力才是状态参数。

2.2.2.3　比容

单位质量的工质所占有的体积称为比容，用符号 v 表示，单位为 m^3/kg。如果质量为 m 的工质占有的体积为 V，则工质的比容 $v=V/m$。单位体积工质的质量称为密度，用符号 ρ 表示，单位为 kg/m^3。比容与密度互为倒数，即 $\rho v=1$。显然，比容和密度都是说明工质在某一状态下分子疏密程度的物理量，二者互不独立，通常以比容作为状态参数。

2.2.3　强度性参数与广延性参数

描述系统状态特性的各种参数，按其与物质数量的关系，可分为两类：

(1) 强度性参数：系统中单元体的某强度性参数值与整个系统的该强度性参数值相同，与质量多少无关，没有可加性，如温度 T、压力 p。当强度性参数不相等时，便会发生能量的传递，如在温差作用下发生热量传递，在压差作用下发生功的传递。可见，强度性参数在热力过程中起着推动力作用，称为广义力或势。一切实际热力过程都是在某种势差推动下进行的。

(2) 广延性参数：整个系统的某广延性参数值等于系统中各单元体该广延性参数值之和，它们与系统中质量多少有关，具有可加性，如系统的质量 m、体积 V、内能 U、焓 H 和熵 S 等。在热力过程中，广延性参数的变化起着类似力学中位移的作用，称为广义位移。

广延性参数用大写字母表示(质量除外)，如 V,U,H,S 等。广延性参数除以系统的总质量，即得单位质量的广延性参数，也称比参数，用小写字母表示，如比容 v、比内能 u、比焓 h、比熵 s 等。习惯上，除比容外，常将“比”字省略，简称为内能、焓、熵等。比参数没有可加性。

2.3　平衡状态、状态公理及状态方程

2.3.1　平衡状态

在没有外界作用的情况下，工质(或系统)的宏观性质不随时间而变化的状态称为热力平衡状态，简称平衡状态。处于平衡状态的热力系，各处具有均匀一致的温度、压力等状态参数。如果系统内部各处状态参数不同，如内部存在温差，则将产生从高温处向低温处的热传递，系统的状态随之变化，这种状态称为非平衡状态。如果没有外界的影响，非平衡状态最终将过渡到平衡状态，即最终系统内部的温差会消失而达到平衡，这种平衡称为热平衡。温差是推动系统发生状态变化的一种动力，也称之为“热不平衡势”。

热力学平衡的条件有多种，除热平衡(温度相等)之外，还有力平衡。若系统内部或者系统与外界之间存在不平衡力的作用，则系统状态也会发生改变，直到系统内部或者系统与外界之间达到力平衡。此外，系统的热力学平衡还包括相平衡、化学平衡等条件。当系统发生相变、化学反应、扩散或溶解等变化时，系统状态同样随之改变。力、化学势与温度一样是系统状态改变的动力参数。只有在热力系统同时满足热平衡、力平衡和化学平衡时，系统才处于热力平衡状态。这时，系统各部分具有均匀一致的状态参数。

用状态参数描述系统状态特性，只有在平衡状态下才有可能，否则系统各部分状态不同就不可能用确定的参数值描述整个系统的特性，平衡状态的概念正是基于这一点提出来的，是工程热力学的基本概念。

2.3.2　状态公理

描述系统特性的参数有许多，它们之间有内在的联系。当某些参数确定后，系统平衡状态便完全确定，所有其他状态参数也随之有确定的值。那么，在一定的限制条件下，确定系统平衡状态的独立参数究竟需要几个呢？实践经验表明，对于纯物质系统，与外界发生任何一种形式的能量传递都会引起系统状态的变化，且各种能量传递形式可单独进行，也可同时进行，于是归纳出一条状态公理，即

$$\text{纯物质系统平衡状态的独立参数个数}=n+1$$

式中，n 表示传递可逆功的形式种类数，而加 1 表示能量传递中的热量传递。例如，对除热量传递外只有容积功(膨胀功或压缩功)传递的简单可压缩系，$n=1$，于是确定系统平衡状态的独立参数个数为 $1+1=2$。对简单可压缩系，已知任意两个独立的状态参数就可以确定其所处状态，系统所有的其他状态参数都可表示为这两个独立参数的函数。

2.3.3　状态方程

由状态公理可知，对于简单可压缩系，在工质的基本状态参数 p,v,T 中，只要其中任意两个确定，另一个也随之确定。如 $p=f_1(T,v)$，$T=f_2(p,v)$，$v=f_3(p,T)$，或表示成隐函数的形式 $F(p,v,T)=0$。这种表示状态参数之间关系的方程称为状态方程。状态方程的具体形式取决于工质的性质。例如，理想气体的状态方程为 $pv=RT$。

两个独立参数可确定系统的一个状态，因此以任意两个独立参数作为坐标，构成平面直角坐标系，图上的一点即表示一个确定的平衡状态，该图也称为热力状态图。工程上经常应用的热力状态图有 p-v 图和 T-s 图，如图 2-6 所示。

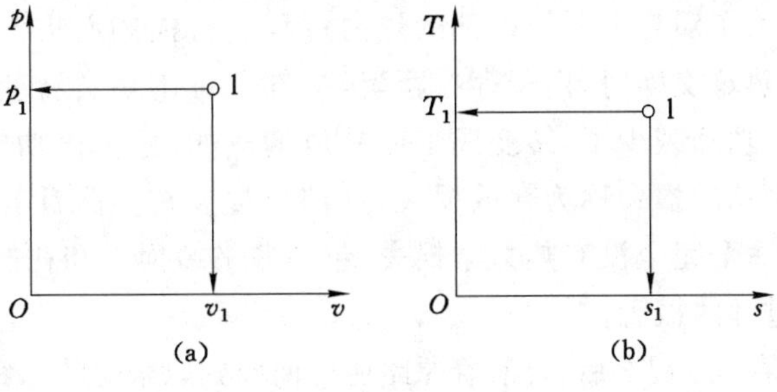

图 2-6　热力状态图

只有在平衡状态下，热力系才有确定的宏观热力状态参数，因此热力状态图上的每一点

都对应热力系的一个平衡状态，非平衡状态无法在热力状态图上表示。

2.4　准平衡过程与可逆过程

热能和机械能的相互转换必须通过工质的状态变化才能实现。系统由一个状态到达另一个状态的变化过程称为热力过程，简称过程。当系统发生状态变化时，必然破坏原有的平衡，才可能到达另一个平衡态。在两个平衡态之间存在着一系列非平衡状态。由于非平衡状态不能用宏观参数来描述，为此要定义一些理想化的过程。

2.4.1　准平衡过程(准静态过程)

在无限小势差(如温差、压差)的推动下，由一系列连续的平衡状态组成的过程称为准平衡过程，也称为准静态过程。而有限势差推动下的过程为非平衡过程。准平衡过程要求推动过程进行的势差无限小，这样保证系统在任意时刻皆无限接近平衡状态。准平衡过程由于经历了一系列非常接近平衡状态的状态，因此可以在状态参数坐标图上用状态点描述状态，用连续的实线表示过程。而非平衡过程由于所经历的不平衡状态没有确定的状态参数，因而不能表示在状态参数坐标图上。

要实现无限小不平衡势差推动下的准平衡过程，从理论上讲要无限缓慢。实际过程都是在有限势差作用下以有限的速度和在有限时间内进行的非平衡过程。理论上，在没有外界作用条件下，一个系统从非平衡态达到完全平衡态需要很长时间，但是从非平衡态趋近平衡态所需时间往往不很长，这段时间叫作弛豫时间。实际上，在系统内外的不平衡势(如压差、温差等)不是很大的情况下，弛豫时间非常短，可以将实际过程近似地看作准平衡过程。例如，在活塞式热力机械中，活塞运动的速度一般在 10 m/s 以内，但气体的内部压力波的传播速度等于声速，通常为每秒数百米。相对而言，活塞运动的速度很慢，这类情况就可按准平衡过程处理。因此，大多数过程可以认为是准平衡过程。以后讨论的大部分热力过程都能假定是准平衡过程。

建立准平衡过程概念的优点是：可以用确定的状态参数变化描述过程，可以在状态参数坐标图上表示过程，可以用状态方程进行计算，可以计算热力过程中系统和外界的功量和热量交换。

2.4.2　可逆过程

准平衡过程只是为了对热力系统内部的热力过程进行描述而提出的，但是当研究涉及系统与外界的功量和热量交换时，由于存在摩阻，一部分可用功会转变为热，虽然能量的总量没有变化，但是可用功却减少了，转变成了低品位的热能，这种由功转变为热的现象称为耗散效应，而造成可用功的损失称为耗散损失，这部分损失在实际计算中很难确定。因此，理论分析时可以设想一个完全没有热力学损失(包括非平衡损失和耗散损失)的理想热力过程。为此，引出可逆过程的概念。

如果系统完成了某一过程之后，再沿着原路逆行而回到原来的状态，外界也随之回到原来的状态而不留下任何变化，则这一过程称为可逆过程。否则就是不可逆过程。实现可逆过程需具备两个条件：一是过程是准平衡过程，即没有势差(或势差无限小)，因为有限势差的存在必然导致不可逆，如传热没有温差，做膨胀功没有压差，等等。二是过程没有耗散效应，如机械运动没有摩擦，导电没有电阻，等等。因此，可逆过程就是无耗散效应的准平衡过程。

实际中，只要过程中包含下列因素之一就是不可逆过程，如：温差传热、自由膨胀、混合过程、节流过程、摩擦生热、黏性流体、阻尼振动、电阻热效应、燃烧过程、非弹性变形以及磁滞损耗等。

可逆过程是理想化的过程，并不存在。实际过程只是近似可逆过程。那么，为什么要提出这一概念呢？这是由于将实际的过程理想化为可逆过程来处理，可给工程分析和计算带来很大的方便，且所得结论可作为实际过程能量转换效果的比较标准，对实际应用有重要的指导作用，所以可逆过程是工程热力学中一个最基本的概念。在后面的讨论中，除了特别指明以外，均将各种热力过程当作可逆过程来对待。

2.5　热力循环

要使工质连续不断地做功，单有一个膨胀过程是不可能的，当它与环境压力达到平衡时，便不能再继续膨胀做功了。为了使工质能周而复始地做功，就必须使膨胀后的工质回到初始状态，如此反复地循环。工质从某一初态出发，经历一系列状态变化，最终又回到初始状态的全部过程称为热力循环，简称循环。全部由可逆过程组成的循环称为可逆循环；若循环中有部分过程或全部过程是不可逆的，则该循环为不可逆循环。

根据循环所产生的效果不同，可以把循环分为正循环和逆循环。将热能转变为机械能的循环称为正循环，所有热力发动机都是按正循环工作的，所以正循环也称为动力循环或热机循环；而将热量从低温热源传递到高温热源的循环称为逆循环，与正循环相反，逆循环消耗外界提供的功量，如制冷循环。如图 2-7(a)所示，1—2—3—4—1 为正循环，为了对外输出有效功量，循环的膨胀功应大于压缩功，所以在状态参数坐标图上正循环都是按顺时针方向进行的。图 2-7(b)中 1—4—3—2—1 为逆循环，逆循环需要消耗功，故在状态参数坐标图上是按逆时针方向进行的。

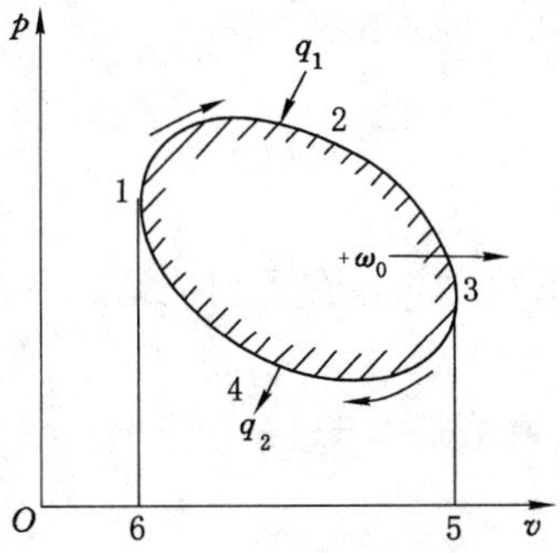

(a) 正循环

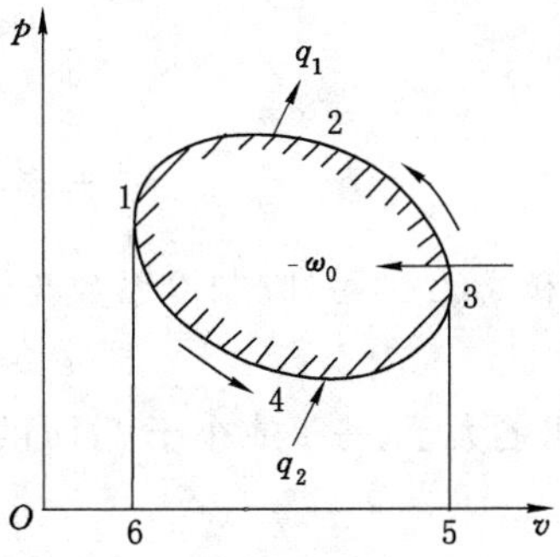

(b) 逆循环

图 2-7　任意循环在 p-v 图上的表示

2.5.1　正循环

设有 1 kg 工质在气缸中进行一个正循环 1—2—3—4—1。过程 1—2—3 表示膨胀过程，所做膨胀功在 p-v 图上为面积 123561。为使工质回到初态，必须对工质进行压缩，此时所消耗的压缩功为面积 341653。正循环所做净功 ω_0 为膨胀功与压缩功之差,即循环所包围的面积 12341(正值)。

对正循环 1—2—3—4—1,在膨胀过程 1—2—3 中工质从热源吸热 q_1,在压缩过程 3—4—1 中工质向冷源放热 q_2。由于在循环过程中,工质回到初态,工质的状态没有变化,因此,工质内部所具有的能量也没有变化。循环过程中工质从热源吸收的热量 q_1 与向冷源放出的热量 q_2 的差值,必然等于循环 1—2—3—4—1 所做的净功 ω_0,即 $\omega_0 = q_1 - q_2$。

正循环中热转换功的经济性指标用循环热效率表示:

$$\text{循环热效率} = \frac{\text{循环中转换为功的热量}}{\text{工质从热源吸收的热量}}$$

$$\eta_t = \frac{\omega_0}{q_1} = \frac{q_1 - q_2}{q_1} = 1 - \frac{q_2}{q_1} \tag{2-4}$$

由式(2-4)可得出结论:循环热效率总是小于 1。从热源吸收的热量 q_1 只能有一部分变为净功 ω_0,在这一部分热量转换为功的同时,必然有另一部分热量 q_2 流向冷源,没有这部分热量流向冷源,热量是不可能连续不断地转变为功的。

2.5.2 逆循环

如图 2-7(b)所示,热力循环按逆时针方向进行(即循环 1—4—3—2—1)时,就成了逆循环。由 p-v 图可知,逆循环的净功为负值,即逆循环需消耗功。工程上逆循环有两种用途:如以获得制冷量为目的,称为制冷循环,这时制冷工质从冷源吸收热量 q_2(或称冷量);如以获得供热量为目的,则称为供热循环,这时工质将从冷源吸收的热量 q_2,连同循环中消耗的净功 ω_0,一并向较高温度的供热系统供给热量 q_1($q_1 = q_2 + \omega_0$)。逆循环的经济指标采用工作系数(coefficient of performance,COP)表示。分别有制冷系数 ε_1(或 COP_R)和供热系数 ε_2(或 COP_H)。

制冷系数:

$$\varepsilon_1 = \frac{q_2}{\omega_0} = \frac{q_2}{q_1 - q_2} \tag{2-5}$$

供热系数:

$$\varepsilon_2 = \frac{q_1}{\omega_0} = \frac{q_1}{q_1 - q_2} \tag{2-6}$$

由式(2-5)和式(2-6)可知,制冷系数与供热系数之间存在下列关系:

$$\varepsilon_2 = 1 + \varepsilon_1 \tag{2-7}$$

制冷系数可能大于、等于或小于 1,而供热系数总是大于 1。

本 章 小 结

本章讨论了热能转换所涉及的基本概念和术语。这是学习两大基本定律和其他后续内容的基础。其中,一些重要概念如平衡状态、准平衡过程和可逆过程体现了热力学研究问题和处理问题的方法。

热力系统,是根据某种研究目的人为划定的研究对象。按照热力系统和外界的物质和能量交换情况进行分类,常用的热力系统有开口系、闭口系、绝热系和孤立系等。

工质，是实现能量转换的媒介物质。

热力系统某一瞬间所呈现的宏观物理状态称为热力状态。用于描述工质所处状态的宏观物理量称为状态参数。基本状态参数有压力、温度和比容。

平衡状态具有确定的状态参数，这是平衡状态的特点。准平衡过程是实际过程进行得足够缓慢的极限情况。实现准平衡过程的条件是推动过程进行的不平衡势差无限小，即 $\Delta p \to 0$，$\Delta T \to 0$。可逆过程与准平衡过程的差别就在于无耗散损失。一个可逆过程必须同时也是一个准静态过程，但准静态过程则不一定可逆。

要实现连续的能量转换，就必须实施热力循环。根据循环所产生的效果不同，可以把循环分为正循环和逆循环。无论是正循环还是逆循环，其经济性都可以表示为：

$$\text{经济性指标} = \frac{\text{收益}}{\text{代价}}$$

思考与练习题

(1) 开口系与外界有物质交换，而物质与能量又不可分割，所以开口系一定不是绝热系，这种观点对吗？为什么？

(2) 温度计测温的基本原理是什么？

(3) 某工质热力状态保持一定，试问，测定该工质的压力表读值能否发生变化？为什么？

(4) 热力学中引入平衡状态的概念，解决了热力分析中的什么问题？

(5) 铁棒一端浸入冰水混合物中，另一端浸入沸水中。经过一段时间，铁棒各点温度保持恒定，试问，铁棒是否处于平衡状态？

(6) 准平衡过程是如何处理“平衡状态”与“热力变化过程”的矛盾的？

(7) 温度高的物体比温度低的物体具有更多的热量，这种说法对吗？

(8) 如果当地大气压为 10^5 Pa，试完成下列计算：① 表压力为 1.5 MPa 时的绝对压力；② 真空表读值为 4 kPa 时的绝对压力；③ 绝对压力为 90 kPa 时的真空值；④ 绝对压力为 1 MPa时的表压力。

(9) 某烟囱高 40 m，烟囱内烟气平均密度为 0.75 kg/m^3。若当地的大气压力为 0.09 MPa，温度为 15 ℃，空气密度在烟囱高度范围内为常数，试求烟囱底部的绝对压力及真空度。

(10) 某蒸汽动力厂，发电量 $P=5\times10^4$ kW，锅炉耗煤量 $M=1.9\times10^4$ kg/h，煤的发热量 $Q_L=3\times10^4$ kJ/kg，试求该动力厂的热效率。

知识拓展：可逆过程

Chapter 2 Basic Concept of Thermodynamics

Thermodynamics is a science about energy utilization and conversion. There are many basic concepts and special terminologies. We start this chapter with a discussion on some basic concepts such as system, state, property, equilibrium, process and cycles.

2.1 Thermodynamic System

2.1.1 Thermodynamic System Implication

A thermodynamic system, or simply a system, is defined as a quantity of matter or a region in space chosen for study. The purpose of choosing system is to make certain what the analytic targets are. This means that the system can be an object, a set of objects, or part of an object, or a given region. The mass or region outside the system is called the surroundings. The surface that separates the system from its surroundings is called the boundary. Note that the boundary is the contact surface shared by both the system and its surroundings.

Considering a piston-cylinder device shown in Fig. 2-1, there is a gas of fixed mass enclosed in the cylinder. As it is heated, the piston may move. There is no mass crossing the boundary. This means that the system works with a fixed amount of mass. We choose the fixed mass in space for study. But the volume is variable. That is, the boundary of a system can be fixed or movable.

Fig. 2-2 shows a turbine. There may exist heat transfer through boundaries. We can choose the volume occupied by the turbine for study. The inner surfaces of the turbine form the real part of the boundary. And the entrance and exit areas form the imaginary part, since there are no physical surfaces there. The fluid enters the system from section 1-1, and exits from 2-2. And the mass in the system may be variable. The boundary of a system can be physical or imaginary.

In general, any arbitrary region in space can be selected as a control volume. There are no concrete rules for the selection of control volumes, but the proper choice certainly makes the analysis much easier. After the system is properly chosen, we will begin to study the interactions between the system and its surroundings, including the energy and mass interactions.

2.1.2 Categories of Thermodynamic Systems

Based on the different interactions, systems can be classified as following.

(1) Closed system: a closed system consists of a fixed amount of mass, and mass cannot cross its boundary. And it is also known as a control mass. The volume of a closed system does not have to be fixed. Everything outside the boundary is the surroundings. There is no mass can enter or leave the system. But energy, in the form of heat or work, can cross the boundary, shown in Fig. 2-1.

(2) Open system: an open system is a properly selected region in space. Substances flowing through such device as a compressor, turbine, or nozzle are encountered in engineering. These devices are also known as control volumes. There is substance flowing in and out of the system. The boundaries of a control volume are called a control surface, and they can be real or imaginary. Both mass and energy can cross the boundary of a control volume, shown in Fig. 2-2.

(3) Adiabatic system: if there is no heat entering and leaving a system, the system is called adiabatic system. It is an idealized system. In engineering analysis, if the heat transferred compared with other forms of energy exchange is sufficiently small, it can be regard as an adiabatic system.

(4) Isolated system: if there are no any interactions between a system and its surroundings, the system is called isolated system. No mass and energy is allowed to cross the boundary. An isolated system must be a closed system. Apparently, it is an idealized system too. Sometime for the convenience of solving problems, we choose all objects in the system and surroundings as a new system. This system is an isolated system. Within the system, each subsystem can have interactions with each other, but this new system has no interactions with its surroundings.

(5) Simple compressible system: a system is called a simple compressible system, when all other forms of work can be ignored besides interactions of heat exchange and expansion work. That is, there are no electrical, magnetic, gravitational, motion and surface tension effects involved.

In engineering analysis, the system under study must be defined carefully. In most cases, the systems investigated are quite simple and obvious. It may seem like a tedious and unnecessary task to define the system. In other cases, however, the system under study may be rather involved, and a proper choice of the system may greatly simplify the analysis.

2.1.3 Working Fluid

Energy transformations are always fulfilled by some substances. The substance is called "working fluid" or "working medium". For example, in combustion engines, heat is transformed into work by virtual of gas expansion and the working fluids are gases. In steam power plant, steam (water) is the working fluid.

In general, all matters can be working fluids. And heat is always transformed into work by means of expansion of the working fluid. Because gas (or vapor) can change state sensitively and rapidly, the working fluids involved in practical applications are mainly gases (or vapors).

As working fluid has an immediate impact on energy conversion, the characteristics of the working fluid are also important contents to study in thermodynamics.

2.2 Thermodynamic States and Basic Properties

2.2.1 States and States Properties

The thermodynamic state, or simply state is the condition of a system. A system not undergoing any change gives a set of properties that completely describes the condition of that system. A property is a characteristic that helps to describe the state of a system. At this point, all the properties can be measured or calculated throughout the entire system. The state of a system is described by its properties.

State properties are point functions. At a given state, all the properties of a system have fixed values. This is an important basic characteristic of state properties. They are determined by the state only, but not related to how a system reaches this state. During any change from the same initial state 1 to final state 2, the changes in a certain state property are the same, as shown in Fig. 2-3. It is only related to the state 1 and 2, but regardless of the path followed. Namely:

$$\Delta x_{1-2} = x_2 - x_1 = \int_1^2 \mathrm{d}x \tag{2-1}$$

When a thermodynamic system goes through a series of processes, and returns to its initial state at the end of the process, the total change in its state properties is zero:

$$\oint \mathrm{d}x = 0 \tag{2-2}$$

State properties are divided into basic state properties and derived properties. The basic state properties can be measured directly or indirectly, such as pressure p, temperature T and specific volume v. The derived properties can be calculated in terms of the basic state properties, such as internal energy u, enthalpy h, entropy s, and so on.

2.2.2 Basic State Properties

2.2.2.1 Temperature

Based on our physiological sensations, we say the temperature is very high in summer, and it is low in winter. But we cannot assign numerical values to temperatures based on our sensations alone. It is not easy to give an exact definition for it. In an isolated system, when a body is brought into contact with another one that is at a different temperature, heat is transferred from the body at higher temperature to the other at lower temperature until both bodies attain the same temperature. At that point, the heat transfers tops, and the two bodies are said to have reached thermal equilibrium. Therefore, temperature is the property.

"The Zeroth Law of Thermodynamics" states that if two bodies are in thermal equilibrium with a third body, they are also in thermal equilibrium with each other as well. It serves as the basis for the validity of temperature measurement. When the test system and the thermometer which has been numerical calibrated are in thermal equilibrium, temperature reading of the thermometer is equivalent to the test system. Temperature is the property describing the characteristic of the heat balance.

Temperature scales are used to record temperature readings. To establish a temperature scale, it needs to take advantage of a certain physical properties of a substance and to choose reference points and sub-degree methods. All temperature scales are based on some easily reproducible states, such as the freezing and boiling points of water, which are also called the ice point and the steam point, respectively. Temperature scale in the SI is the Kelvin scale, also called thermodynamic temperature scale. It is independent of the properties of the substances that are used to measure temperature. The unit of thermodynamic temperature T is defined as 1/273.16 of the thermodynamic temperature of the triple point of water (the state at which all three phases of water coexist in equilibrium), which is assigned the value 273.16 K. And the temperatures on this scale are called absolute temperatures. The unit is the Kelvin, designated by K.

The temperature scale used synchronously in the SI today is the Celsius scale and its unit is the Celsius, designated by ℃. The temperature on this scale is denoted by t. It is related to the Kelvin scale temperature T by $t=T-273.15$.

2.2.2.2 Pressure

(1) Pressure and units

Pressure p is defined as the force exerted by a fluid on per unit area. That is:

$$p = \frac{F}{A} \tag{2-3}$$

Where F stands for the force perpendicular to the surface, N; A stands for the area of a surface, m^2.

There is a certain amount of gas within the system, a large number of gas molecules randomly move and continuously hit the wall. For a particular molecule, this collision is intermittent, but for the whole system with a large number of molecules, the collision is simultaneous and continuous, pressure behaves a constant at macro-scale. Pressure at any point in a fluid is the same in all directions.

The pressure unit is Newton per square meter (N/m^2), which is called Pascal (Pa). 1 Pa = 1 N/m^2. But the Pascal is too small for pressure encountered in engineering practice. Therefore, its multiples kilopascal (kPa) and Mega-Pascal (MPa) are commonly used. And 1 MPa $=10^3$ kPa $=10^6$ Pa.

Moreover, the other four pressure units commonly used in practice are bar, standard atmosphere (atm), engineering atmosphere (at), mmH_2O, and mmHg. The conversion of the different pressure units refers to as following:

$$1\ \text{bar}=10^5\ \text{Pa},\ 1\ \text{atm}=1.013\times10^5\ \text{Pa},$$

$$1\ \text{mmHg}=133.3\ \text{Pa},\ 1\ \text{mmH}_2\text{O}=9.8\ \text{Pa},\ 1\ \text{at}=9.8\times10^4\ \text{Pa}$$

(2) Absolute pressure, gage pressure and vacuum pressure

Pressure is measured with pressure gage (gauge). There are two types of pressure gage used commonly (Fig. 2-4). They are the Bourdon-tube pressure gage and the U-tube manometer. However, they are calibrated to read zero in the atmosphere, and so they indicate the difference between the absolute pressure and the local atmospheric pressure. The actual pressure at a given position is called the absolute pressure, and their difference is called the gage pressure. Pressure below atmospheric pressure is called vacuum pressure and is measured by vacuum gage that indicates the difference between the atmospheric pressure and the absolute pressure. Absolutes, gage, and vacuum pressures are all positive quantities and are related to each other as following, as illustrated in Fig. 2-5.

If $p>p_b$, then $p=p_b+p_g$ for pressure higher than the local atmospheric pressure.

If $p<p_b$, then $p=p_b-p_v$ for pressure lower than the local atmospheric pressure.

where p_b, p_g and p_v are the local atmospheric pressure, the gage pressure and the vacuum pressure respectively.

The atmospheric pressure changes not only with elevation but also with weather conditions. If atmospheric pressure is at different condition, the gage pressure is different even though the absolute pressure of the system is consistent. Therefore, the gage pressure is not a state property. Only absolute pressure is a state property.

2.2.2.3 Specific Volume

Specific volume is defined as the volume per unit mass. It has the unit of m^3/kg. $v=V/m$. Density is the reciprocal of specific volume and is defined as the mass per unit volume, kg/m^3. $\rho v=1$.

In general, the density of a substance depends on its pressure and temperature. The density of most gases is proportional to the pressure and inversely proportional to the temperature. The densities of liquids are essentially constant and are often approximated to be incompressible.

2.2.3 Intensive Properties and Extensive Properties

In accordance with the mass of a system, properties are considered to be either intensive or extensive.

Intensive properties are those independent of the mass of a system, such as temperature, pressure, and density. Extensive properties are those whose values depend on the size or extent of system, such as total volume, internal energy, enthalpy, and entropy. Extensive properties can he added. Extensive properties per unit mass are called specific properties, such as specific volume (v), specific internal energy (u), specific enthalpy (h) and specific entropy (s).

2.3 Equilibrium State, State Postulate and State Equation

2.3.1 Equilibrium State

Thermodynamics mainly deals with equilibrium states. The word equilibrium implies a state of balance. At an equilibrium state, there are no unbalanced potentials or driving forces within the system. For example, if the temperature is not the same throughout the entire system, the system is not in thermal equilibrium, then heat will flow from the high temperature part to other low temperature parts until the temperature turns to be the same throughout the entire system. Then the system is in thermal equilibrium.

There are many types of equilibrium, and a system is not in thermodynamic equilibrium unless all the relevant types of equilibrium are satisfied. Besides the thermal equilibrium mechanical equilibrium is related to pressure. And a system is in mechanical equilibrium if there is no change in pressure at any point of the system with time. A system involving two phases is in phase equilibrium when the mass of each phase reaches equilibrium level and keeps the level there. Chemical equilibrium means the chemical composition of a system does not change with time, that is, no chemical reaction occurs. A system will not be in equilibrium unless all the relevant equilibriums are satisfied.

When the system is in equilibrium, the state properties are the same throughout the entire system, including the pressure, temperature, etc. The state can be described by its properties.

2.3.2 The State Postulate

When a system is in equilibrium, the state can be described by its properties. Do we need to specify all the properties in order to fix a state? Actually, the properties are not isolated. Once a sufficient number of properties are specified, the rest of the properties are certain automatically. That is, specifying a certain number of properties is sufficient to fix a state.

The state postulate indicates the state of a simple compressible system is completely specified by two independent intensive properties. Two properties are independent if one property can be varied while the other one is held constant.

2.3.3 Equation of State

The state postulate indicates that the state of a simple compressible system can be described by only specifying two independent properties. Temperature and pressure, for example, are always independent properties for a single phase system, and together they can fix the state of a system and other properties. $p=f_1(T,v)$, $T=f_2(p,v)$, and $v=f_3(p,T)$. It can also be expressed in the following form: $F(p,v,T)=0$. These relations are called the equations of state.

As an equilibrium state of a system can be specified by any two independent properties, we may choose two independent properties as two coordinates to constitute a Cartesian coordinates system which is called the state diagram. Some familiar diagrams include p-v diagram and T-s diagram, as shown in Fig. 2-6.

2.4 Quasi-Equilibrium Process and Reversible Process

When the state of a system changes from one equilibrium state to another, it undergoes a process. The series of states through which a system passes during a process is called the path of the process. If there are unbalanced potentials existing in an equilibrium state, the equilibrium is destroyed. The state of the system will change until it reaches another equilibrium state. There are a series of unbalanced states between two equilibrium states. Because unbalanced state can not be described by certain state properties, thus some ideal processes will be defined.

2.4.1 Quasi-Equilibrium Process (Quasi-Static Process)

When a process proceeds in such a manner that a system remains infinitesimally close to an equilibrium state at all times, it is called a quasi-equilibrium, or quasi-static process.

In order to fulfill a quasi-equilibrium process, it is necessary to maintain mechanical equilibrium within a system, not only internal equilibrium, but also in equilibrium with the surroundings. The conclusion can be extended to heat transfer and chemical reactions, etc.

A quasi-equilibrium process can be viewed as a sufficiently slow process that allows the system to adjust itself internally so that the properties in one part of the system do not change any faster than those at other parts. A working producing devices operating on quasi-equilibrium processes can deliver the most work. Although a quasi-equilibrium process is an idealized process and can not represent actual process completely. However, in engineering practice, many actual processes can closely approximate it. They can be modeled as quasi-equilibrium with negligible error. As quasi-equilibrium state is easy to be analyzed, it is chosen to serve as a base case with which actual processes can compare.

2.4.2 Reversible Process

Reversible process is defined as a process that can be reversed without leaving any trace on the surroundings. Processes that are not reversible are called irreversible processes. For a reversible process, not only the system but also the surroundings can return to their original states if the process is reversed.

Firstly, reversible process must be a quasi-equilibrium process. Otherwise the driving force will be irreversibility in the process. Secondly, there is no any irreversible dissipation effect (The work loss, which is converted into heat by friction, resistance, magnetism, etc., is called dissipation loss) involved in reversible process.

Actually, irreversible process can be discriminated easily. The process is irreversible if only it involves the following factors, such as temperature difference, unrestrained expansion, mixing of fluids, throttling process, electric resistance, inelastic deformation of solids, and chemical reactions, etc. A reversible process involves none of these.

Reversible processes are idealized processes, and they can be approached but never reached in reality. You may be wondering, then, why we are bothering with such fictitious processes. As reversible processes are easy to analyze, they play an important role in thermodynamics theory and in practical applications. Reversible process is a basic concept in thermodynamics.

2.5 Thermodynamic Cycle

A thermodynamic cycle occurs when a system undergoes several processes and returns to its initial state at the end of the process. That is, the initial and final states are identical for a cycle. In engineering, it is necessary to go through a cycle to obtain work continuously.

According to the function of a cycle, it can be categorized as power cycle and refrigeration cycle.

2.5.1 Power Cycle

The devices or systems used to produce a net power output are often called engines. Heat engines are designed for the purpose of converting thermal energy to work. And the thermodynamic cycles they operate on are called power cycles.

For heat engines, the working fluid receives heat, q_1, from a high-temperature source, and then expands to deliver work. To complete the cycle, there must be a compressible process. In a cycle, only part of the heat absorbed can be converted to the work. The work obtained is called net work ω_0. The other part of the heat absorbed, q_2, will be rejected to a low-temperature sink. Thus, the net work of a system is also equal to the net heat transfer to the system $\omega_0 = q_1 - q_2$.

That is, the net work output during a cycle is equal to the net heat input, only part of the heat transferred to the heat engine is converted into work and the other part is rejected as heat in order to complete the cycle.

The fraction of the heat input that is converted to the net work output is a measure of the performance of a heat engine and is called the thermal efficiency η_t. It indicates how well an energy conversion is, and can be expressed as:

$$\text{Thermal efficiency} = \frac{\text{Net work output}}{\text{Total heat inout}}$$

or

$$\eta_t = \frac{\omega_0}{q_1} = \frac{q_1 - q_2}{q_1} = 1 - \frac{q_2}{q_1} \tag{2-4}$$

The thermal efficiency is usually used to evaluate the thermal economy of a power cycle.

2.5.2 Refrigeration Cycle

The cycle on which a device operates reversed from power cycle is called refrigeration cycle. Refrigerators and heat pumps are these kings of devices. They consume some work and transfer heat from a lower temperature region to a high temperature one. $q_1 = q_2 + \omega_0$, where q_2 is the magnitude of the heat removed from the refrigerated space at temperature T_2; q_1 is the magnitude of the heat rejected to the warmer space at temperature T_1; ω_0 is the net work input to the cycle and it is a necessary term.

Both refrigerators and heat pumps operate on refrigeration cycles, while they just differ in their functions. The function of a refrigerator is to maintain the refrigerated space at a low temperature by removing heat from it. Discharging the heat to a higher-temperature medium is merely a necessary part of the operation, but not the purpose. The function of a heat pump, however, is to maintain a heated space at a high temperature. This is accomplished by absorbing heat from a low-temperature source, such as from well water or cold outdoor air in winter, and supplying the heat to a warmer indoor environment.

The performance of refrigerators and heat pumps is expressed in terms of the coefficient of performance (COP), defined as:

$$\mathrm{COP}=\frac{\text{Desired output}}{\text{Required input}}$$

The performance of refrigerators can be expressed as:

$$\varepsilon_1 = \frac{q_2}{\omega_0} = \frac{q_2}{q_1 - q_2} \tag{2-5}$$

The performance of heat pumps can be expressed as:

$$\varepsilon_2 = \frac{q_1}{\omega_0} = \frac{q_1}{q_1 - q_2} \tag{2-6}$$

A comparison of Eq. (2-5) and Eq. (2-6) reveals that:

$$\varepsilon_2 = 1 + \varepsilon_1 \tag{2-7}$$

It indicates that COP_R (ε_1) can be less, greater or equal to 1, however, COP_H (ε_2) is always greater than 1.

第 3 章　热力学第一定律

本章导学

热力学第一定律是工程热力学的基本定律之一，是能量转换与守恒定律在涉及热现象的能量转换过程中的应用。热力学第一定律的能量转换及守恒的实质是不难理解的，问题的复杂性在于应用。能量的转换有赖于物质的状态变化，同时能量有质的差异，因此要把转移中的能量和储存于物系中的能量分开。迁移能又分为热量和功量，储存能也有内部储存能（内能）与外部储存能的区别。热能转变为机械能要靠物系容积的改变，但实际表现出来的并不总是膨胀功，因而应用中还要引入其他不同形式的功。另外，实际中的能量转移过程都是瞬变过程，所以又要抽象为与时间无关的稳流过程。由于以上种种原因，同一性质的能量方程将以不同形式、不同内容出现。本章对热力学第一定律的实质、数学描述及在工程中的应用逐一进行讨论。

学习目标

深入理解热力学第一定律的实质，熟练掌握热力学第一定律及其表达式。掌握能量、储存能、迁移能（功量和热量）的概念。掌握膨胀功、流动功、轴功、技术功、热量的概念及计算方法。熟练掌握在 p-v 或 T-s 图上表达膨胀功、技术功、热量的方法。熟悉焓的概念及计算式。

重点与难点

（1）焓的概念。

（2）功、稳定流动过程中几种功的关系。

（3）能量方程的应用。

3.1　热力学第一定律的实质

能量转换与守恒定律是自然界中最普遍、最基本的定律之一，它可表述为：自然界的一切物质都具有能量，能量既不能被创造，也不会凭空消失，而只能从一种形式转换成另一种形式，或从一个系统转移到另一个系统，其总量保持恒定。将能量守恒定律应用于涉及热现象的能量转换过程即为热力学第一定律，可表述为：热能和机械能在转移或转换时，能量的总量必定守恒。对于任何热力系统中的任意过程，能量守恒定律的表达式为：

进入系统的总能量－离开系统的总能量＝系统总储存能的变化量

热力学第一定律可以确定热力过程中热力系统与外界进行能量交换时，各种形态能量在数量上的守恒关系。根据该定律可以断定，不消耗任何能量而连续做功的所谓“第一类永动机”是不可能实现的。

根据热力学第一定律建立起来的能量方程，在各种热力过程的分析和计算中有广泛的应用。分析热力过程，选取热力系统十分重要，同一现象选取不同的热力系统，系统与外界之间的能量关系也不同，由此建立起来的能量方程亦各不相同。

3.2 热力系统的储存能

自然界中能量的形式与物质的运动形态有关，物质处于不同的运动形态，便有不同的能量形式。系统的储存能包括两部分：一部分与系统内工质的分子结构及微观运动形式有关，由系统本身（内部）的状态决定，称为内能，也称热力学能；另一部分取决于系统工质与外力场的相互作用（如重力势能）及以外界为参考坐标的系统宏观运动所具有的能量（宏观动能），这两种能量统称为外部储存能。

3.2.1 内能

内能是因物质分子微观运动而具有的能量，包括内动能、内位能和内化学能、原子核能。

（1）内动能

内动能是由分子的热运动而形成的能量，包括分子的移动动能、转动动能以及分子中原子的振动动能。因此，内能是温度的函数。物质的温度越高，分子内动能越大。

（2）内位能

内位能是由分子间存在相互作用力而形成的能量，取决于分子间的距离，与气体的比容有关。因此，内能也是比容的函数。

（3）内化学能和原子能

内化学能和原子能是为维持稳定的分子结构和原子核结构而具有的能量。热力过程不涉及化学反应和原子核反应，工质的内化学能和原子能保持不变。因此，工程热力学中，内能的变化只包括内动能和内位能的变化。

用 U 表示质量为 m (kg)的工质的内能，单位是 J 或 kJ。用 u 表示 1 kg 质量工质的内能，称为比内能，单位是 J/kg 或 kJ/kg。内能的大小只与系统所处的状态有关，是温度和比容的函数，即 $u=f(T,v)$。又因为 p,v,T 三者之间存在着一定关系，所以内能也可以写成 $u=f(T,p)$ 或 $u=f(v,p)$，可见，内能也是状态参数。

3.2.2 外部储存能

（1）宏观动能

质量为 m 的物体相对系统外的参考坐标以速度 c 运动时，该物体具有的宏观动能为：

$$E_k = \frac{1}{2}mc^2$$

（2）重力势能

重力场中质量为 m 的物体相对系统外的参考坐标系的高度为 z 时，具有的重力势能为：

$$E_p = mgz$$

式中　g——重力加速度。

由于 c，z 是独立于热力系统内部状态的外参数，因此将系统的宏观动能和重力势能称为外部储存能。

3.2.3　系统的总储存能

热力系统的总储存能 E 为内能与外部储存能之和，单位是 J 或 kJ。

$$E = U + E_k + E_p \tag{3-1a}$$

或

$$E = U + \frac{1}{2}mc^2 + mgz \tag{3-1b}$$

能量是广延性参数，1 kg 质量物体的总储存能记作 e，也称为比总储存能，表示为：

$$e = u + \frac{1}{2}c^2 + gz \tag{3-1c}$$

当热力系统静止，且高度为零时，其宏观动能和重力势能为零，$E=U$ 或 $e=u$，此时系统总储存能的变化就是内能的变化，即

$$\Delta E = \Delta U \text{ 或 } \Delta e = \Delta u$$

3.3　系统与外界传递的能量

系统与外界传递的能量是指系统与外界热力源（热源、功源、质源）或其他有关物体之间进行的能量传递。能量从一个物体传递到另一个物体，可有两种方式：一种是做功，另一种是传热。

3.3.1　热量

在热力学中，热量指热力系统与外界之间仅仅由于温差作用而通过边界传递的能量。当系统与外界之间达到热平衡时，热量传递随之停止。热量一旦通过边界传入或传出系统，就变成系统或外界的储存能的一部分，即内能，有时习惯上称为热能或内能，如图 3-1 所示。内能与热量之间有原则的区别，热量是与过程特性有关的过程量，是过程路径的函数，而内能是状态量，取决于热力状态。因此，不能说系统具有多少热量，而只能说系统具有多少能量。

热量用 Q 或 q 表示，单位为 J 或 J/kg。由于热量是过程量，因此微元热力过程中传递的热量用 δQ 或 δq 表示。穿越系统边界的热量有方向性。通常规定：系统吸热，热量为正；系统对外放热，热量为负。

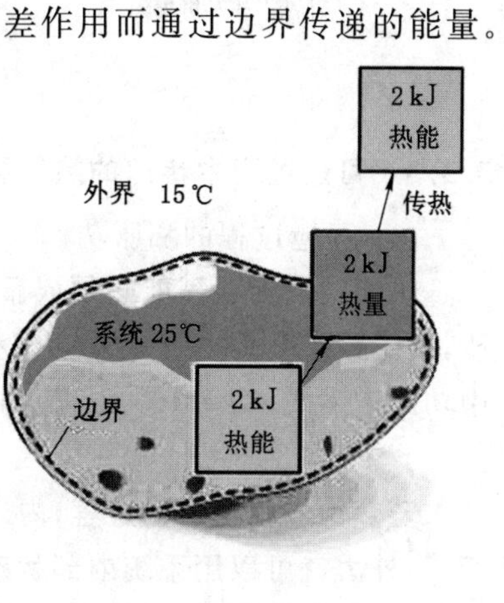

图 3-1　热量传递过程

3.3.2　功量

在热力学中，功是除温差以外的其他不平衡势差所引起的系统与外界传递的能量。由

于外界功源有各种不同形式，如电、磁、机械装置等，相应的功也有各种不同的形式，如电功、磁功、机械拉伸功、弹性变形功、表面张力功和膨胀功、轴功等。

工程热力学主要研究热能与机械能的转换，而膨胀功是热转换为功的必要途径。另外，热工设备的机械功往往通过机械轴传递，因此，我们最感兴趣的是膨胀功和轴功。

（1）膨胀功（也称容积功）：在压差作用下，由于系统工质容积发生变化而传递的机械功称为膨胀功。系统容积变化是做膨胀功的必要条件，而不是充分条件，即工质膨胀过程也可以没有功的输出。例如，在绝热刚性容器中，用隔板将容器分为两部分，一部分存有气体，另一部分为真空，当隔板抽去后，气体绝热自由膨胀，压力降低，比容增大，但没有功的输出，这是典型的不可逆过程。另外，做膨胀功除工质的容积变化外，还应当有功的传递和接收机构。

膨胀功是与过程特性有关的过程量，一旦过程结束，系统与外界之间功的传递就停止。工程热力学中规定：系统膨胀，对外做功为正；系统压缩，外界对系统做功为负。

（2）轴功：系统通过机械轴的旋转与外界传递的机械功称为轴功，如图 3-2(a)所示。如图 3-2(b)所示，外界功源向刚性绝热闭口系输入轴功，该轴功通过耗散效应转换成热量，被系统吸收，增加了系统的内能。但是，由于密闭刚性容器中的工质不能膨胀，热量不可能自动地转换为机械功，因此，刚性闭口系无法靠加热来连续地向外界输出轴功。如图 3-2(c)所示，开口系可与外界传递轴功 W_s（输入或输出）。工程上许多动力机械，如汽轮机、内燃机、风机、压气机等都靠机械轴传递功。通常规定，系统输出的轴功为正，对系统输入的轴功为负。

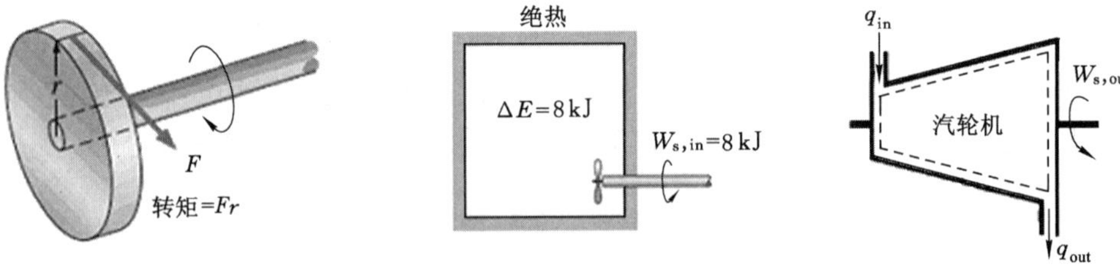

(a) 定义轴功的模型　　(b) 输入闭口系的轴功　　(c) 开口系的轴功

图 3-2　轴功示意图

3.3.3　可逆过程中传递的热量和膨胀功

（1）可逆过程的膨胀功

热能转换为机械能必须依靠工质的膨胀。如图 3-3 所示，气缸内有质量为 m (kg)的工质，当工质克服外力 F 推动活塞移动微小距离 dS 时，工质将对外做微小膨胀功。按物理学中功的定义式：功＝力×距离，则有：

$$\delta W = F\mathrm{d}S$$

若热力过程是可逆过程，内外没有势差，作用在活塞上的外力与工质作用在活塞上的力相等，外力就可以用系统内部状态参数来表示，即

$$F = pA$$

式中，A 是活塞的截面积。

得到工质在微元热力过程中所做的膨胀功：

$$\delta W = F\mathrm{d}S = pA\mathrm{d}S = p\mathrm{d}V \tag{3-2a}$$

对于由状态 1 到 2 的可逆过程，所做的膨胀功为：

$$W = \int_{V_1}^{V_2} p\mathrm{d}V \tag{3-2b}$$

单位质量工质所做的膨胀功称为比膨胀功，即

$$\delta w = p\mathrm{d}v \tag{3-2c}$$

或

$$w = \int_{v_1}^{v_2} p\mathrm{d}v \tag{3-2d}$$

式(3-2)为可逆过程膨胀功的计算公式。显然，具体计算时，除了工质的初、终态以外，还必须知道工质在状态变化过程中的压力和比容的变化规律。

p-v 图上可直接显示出过程中工质做功的多少，因此也称 p-v 图为示功图。膨胀功在 p-v 图上的大小就是过程曲线与坐标轴 v 轴之间所包围的面积，在图 3-3 中为 1—2—n—m 斜线区域的面积。

膨胀功的多少不仅取决于工质的初、终状态，还与过程特性有关。在初、终状态相同的情况下，如果过程经过的途径不同，则膨胀功的大小也不相同，如图 3-4 所示。这表明膨胀功是过程量而不是状态量，因而不能说“系统在某状态下具有多少功量”。用数学语言表达，微元功 δw 不是全微分，“δ”表示微小量，而不是微小增量“d”，故它的积分式：

$$\omega = \int_1^2 \delta\omega \neq \omega_2 - \omega_1$$

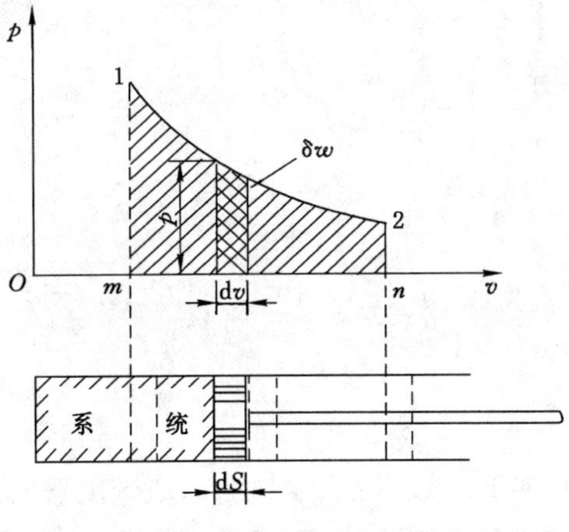

图 3-3　膨胀功在 p-v 图上的表示

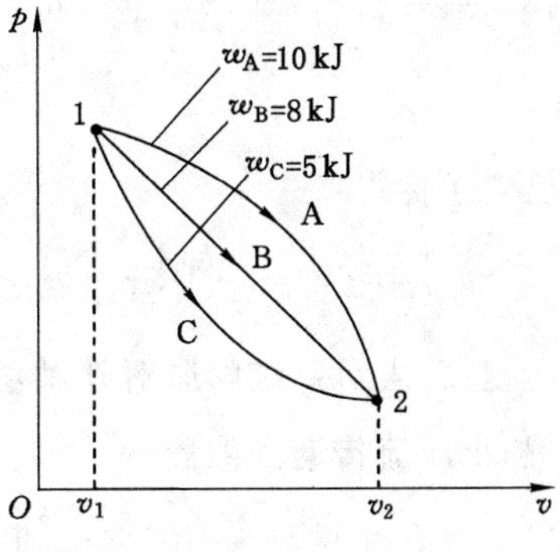

图 3-4　不同过程时的膨胀功

(2) 可逆过程的热量

热量和功量都是与过程特征有关的量，它们必然具有某些共性。那么，可逆过程的热量计算式是否也类似于功量的呢？在可逆过程中，引入熵的概念，对比功的计算式，热量也可用如下数学表达式计算：

$$\delta q = T\mathrm{d}s \tag{3-3a}$$

式中，s 称为比熵，单位为 J/(kg・K)。比熵的定义式为：

$$\mathrm{d}s = \frac{\delta q}{T} \tag{3-3b}$$

对质量为 m (kg)的工质而言，其熵为：

$$dS = \frac{\delta Q}{T} \tag{3-4a}$$

这里仅作为基本概念给出了熵的定义，其物理意义将在热力学第二定律中作进一步讨论。

对于从状态 1 到状态 2 的可逆过程，工质与外界交换的热量可用式(3-4b)计算：

$$q = \int_1^2 T\mathrm{d}s \quad 或 \quad Q = \int_1^2 T\mathrm{d}S \tag{3-4b}$$

T-s 图上可以直接显示出一个热力过程中换热量的多少，如图 3-5 所示，区域 1—2—s_2—s_1 的面积表示可逆过程传递的热量，故 T-s 图又称为示热图。从图中分析可知，初、终态相同但中间途径不同的各种过程，其传递热量也不相同，这说明热量也是过程量，它与过程特性有关。

根据熵的变化，可以很容易地判断一个可逆过程中系统与外界之间热量交换的方向：若 $\mathrm{d}s>0$，则 $q>0$，系统吸热；若 $\mathrm{d}s<0$，则 $q<0$，系统放热；若 $\mathrm{d}s=0$，则 $q=0$，系统绝热，因此可逆绝热过程又称为定熵过程。

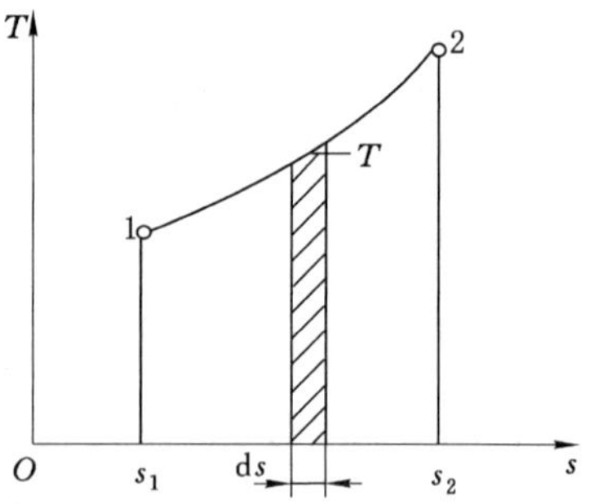

图 3-5　示热图

【例 3-1】 一定质量的气体处于封闭气缸内，经历一个可逆膨胀过程，该过程中压力与体积之间按 $pV=$ 常数的关系变化。已知初态 $p_1 = 0.5$ MPa，$V_1 = 0.4\ \mathrm{m^3}$，终态 $V_2 = 0.8\ \mathrm{m^3}$，求气体所做的膨胀功。

【解】 取气缸内气体为系统，对于可逆过程，有：

$$W = \int_{V_1}^{V_2} p\mathrm{d}V$$

由已知条件得：

$$W = \int_{V_1}^{V_2} \frac{p_1 V_1}{V}\mathrm{d}V = p_1 V_1 \ln\frac{V_2}{V_1} = 0.5\times10^6\times0.4\times\ln\frac{0.8}{0.4}\times10^{-3} = 138.63\ (\mathrm{kJ})$$

功为正值，表示系统膨胀对外做功。

3.3.4　随物质流传递的能量

开口系与外界有物质的交换。随物质的交换，如下两部分能量会被带入或带出系统。

(1) 流动工质本身具有的储存能：流动工质本身具有内能、宏观动能和重力势能，随工质流进或流出控制体而带入或带出控制体。即

$$E = U + \frac{1}{2}mc^2 + mgz$$

或者

$$e = u + \frac{1}{2}c^2 + gz$$

(2) 流动功：工质流入具有一定压力的系统时，后面的流体需要推开前面的流体而前进，这样后面的流体必须对前面的流体做推动功。

为获得推动功的计算式，假设在如图 3-6 所示的控制体中有一体积为 V 的流体单元。后面的流体将推动该流体单元进入控制体，即可认为存在一个假想的活塞。流体单元足够小以至于具有相同的状态特征。流体压力为 p，流体单元断面积为 A，通过假想活塞作用于

流体单元的力 $F=pA$。

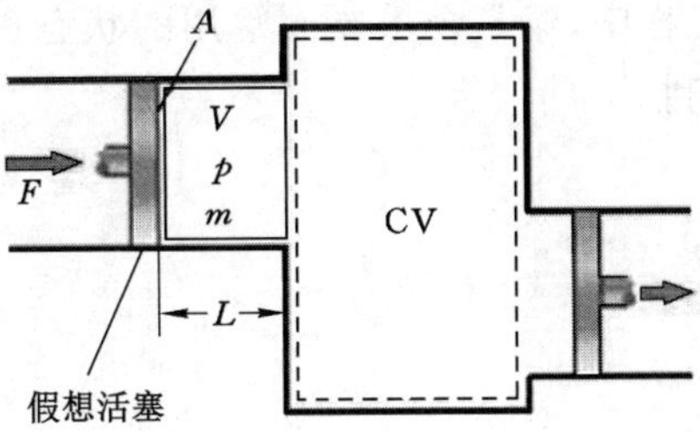

图 3-6　流动功示意图

活塞被推动 L 的距离才可将流体单元推入控制体，该过程需要做的推动功为：

$$W_f = FL = pAL = pV \tag{3-5}$$

流体进入与流出控制体的推动功的表达式是相同的。推动功是流体克服某种作用力发生宏观位置移动所消耗的功。在移动过程中流体仅发生位置变化，进入控制体过程中热力状态不发生变化。在流动过程中，流体穿过边界进、出开口系时与外界交换的推动功的差值称为流动功，即 $W_f=\Delta(pV)$。因此，流动功可视为流动过程中系统与外界因物质的进、出而传递的机械功。工质进出控制体所交换的流动功可用进、出口界面的状态参数计算，即

$$W_f = p_2V_2 - p_1V_1 \tag{3-6a}$$

或

$$w_f = p_2v_2 - p_1v_1 \tag{3-6b}$$

流动功是一种特殊的功，其数值取决于控制体进、出口界面处工质的热力状态，与热力过程无关。流动功是为推动流体通过控制体界面而传递的机械功，它是维持流体正常流动所必须传递的能量。

3.3.5　焓及其物理意义

如上所述，流动工质传递的总能量应包括物质流本身的储存能加上流动功，即

$$E_{总} = U + \frac{1}{2}mc^2 + mgz + pV$$

或

$$e_{总} = u + \frac{1}{2}c^2 + gz + pv$$

其中，u 和 pv 取决于工质的热力状态。为简化计算，将流动工质传递的总能量中取决于工质的热力状态的那部分能量写在一起，引入一个新的物理量，称为焓，定义式为：

$$H = U + pV \tag{3-7a}$$

或

$$h = u + pv \tag{3-7b}$$

由于 u 和 p，v 都是工质的状态参数，所以焓也是工质的状态参数。对于流动工质，焓为内能和流动功的代数和。焓具有能量意义，表示流动工质向流动前方传递的总能量(共四项)中取决于热力状态的那部分能量。若工质的动能和势能可以忽略，则焓代表随流动工质传递的总能量。对于不流动工质，因 pv 不是流动功，焓只是一个复合状态参数，没有明确的物理意义。

在热工设备中，工质总是不断地从一处流到另一处，随着工质流动而转移的能量就是焓而不是内能。因此，在热力工程中，焓是个重要而常用的状态参数，它的引入给热工问题的分析和计算带来了很大的便利。

3.4 闭口系能量方程

3.4.1 闭口系能量方程表达式

闭口系与外界没有物质交换，与外界之间只有热量和功量两种形式的能量交换。对闭口系进行能量分析时，通常采用控制质量法，即取一定质量的物质作为研究对象。对闭口系涉及的许多热力过程而言，系统总能量中的宏观动能和重力势能一般均不发生变化。因此，热力过程中系统总能量的变化，等于系统内能的变化，即 $\Delta E = \Delta U = U_2 - U_1$ 或 $\Delta e = \Delta u = u_2 - u_1$。

如图 3-7 所示，取气缸中工质为系统。在热力过程中，系统从外界热源吸收热量 Q，对外做膨胀功 W，系统总储存能变化了 ΔU。根据能量守恒定律，建立闭口系的能量守恒方程：

$$Q - W = \Delta U \quad 或 \quad Q = W + \Delta U \tag{3-8a}$$

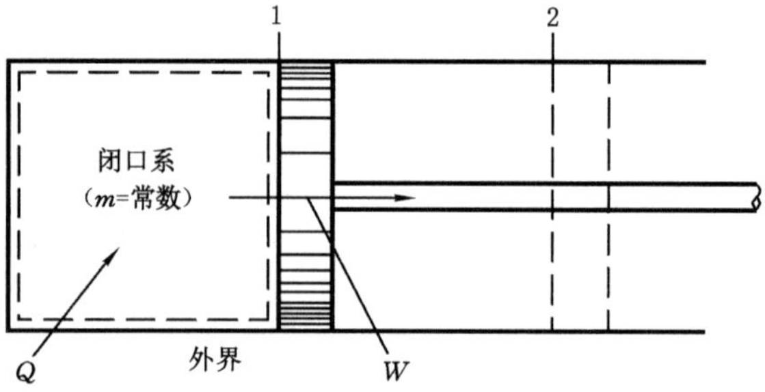

图 3-7 闭口系能量转换

对于单位质量工质，能量方程为：

$$q = \Delta u + w \tag{3-8b}$$

对于微元热力过程：

$$\delta Q = \mathrm{d}U + \delta W \quad 或 \quad \delta q = \mathrm{d}u + \delta w \tag{3-8c}$$

以上表达式均为闭口系的能量方程，也称为热力学第一定律的基本表达式。其表示加给系统一定量的热量，一部分用于改变系统的内能，另一部分用于对外做膨胀功（热转换为功）。由于建立以上方程的唯一依据是能量守恒定律，因而它们适用于闭口系任何工质的各种热力过程，无论过程是否可逆。

能量方程表达式是代数方程，如果是外界对系统做功，或系统对外放热，系统内能减少，则方程式各项为负值。

对于可逆过程，由于 $\delta w = p\mathrm{d}v$ 或 $w = \int_1^2 p\mathrm{d}v$，于是有：

$$\delta q = \mathrm{d}u + p\mathrm{d}v \quad 或 \quad q = \Delta u + \int_1^2 p\mathrm{d}v \tag{3-9}$$

式(3-9)仅适用于可逆过程。

应当指出，由于热能转换为机械能必须通过工质膨胀才能实现，因此，闭口系能量方程反映了热功转换的实质，是热力学第一定律的基本表达式。虽然式(3-8)和式(3-9)是基于闭口系推导得出的，但其热量、内能和膨胀功三者之间的关系也适用于开口系。

3.4.2　热力学第一定律在循环过程中的应用

在动力循环或制冷循环中，工质在设备内部周而复始地工作着，与外界没有物质交换，故属闭口系。如图 3-8 所示，工质沿 1—2—3—4—1 过程完成一个循环，系统终态与初态相同，即 $U_{终}=U_{始}$，$\Delta U=0$，因此，循环过程的热量与功量在数量上相等，即

$$Q=W \quad 或 \quad \oint \delta Q=\oint \delta W \tag{3-10a}$$

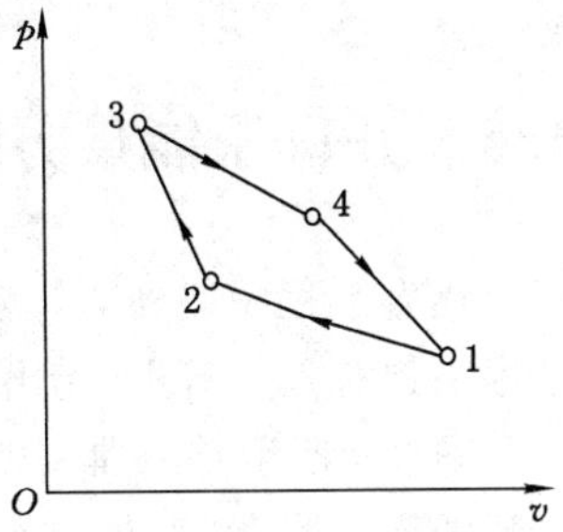

图 3-8　热力循环

也可以对循环中的单个过程逐一进行分析，设循环工质质量为 1 kg，对每一过程，按照热力学第一定律建立能量方程：

$$q_{12}=u_2-u_1+w_{12},\ q_{23}=u_3-u_2+w_{23},$$
$$q_{34}=u_4-u_3+w_{34},\ q_{41}=u_1-u_4+w_{41}$$

对于整个循环：
$$\sum \Delta u=0 \quad 或 \quad \oint \mathrm{d}u=0$$

因而
$$q_{12}+q_{23}+q_{34}+q_{41}=w_{12}+w_{23}+w_{34}+w_{41}$$

即

$$\oint \delta q=\oint \delta w \tag{3-10b}$$

式(3-10)表明，工质经历一个循环回到初始状态后，它在整个循环过程中从外界得到的净热量应等于对外做的净功。式(3-10)称为闭口系循环过程的热力学第一定律表达式。由此可见，循环工作的热力发动机向外界不断地输出机械功必须要消耗一定的热能，不消耗能量而能够不断地对外做功的机器(所谓第一类永动机)是不可能制造出来的。

【例 3-2】　定量工质，经历一个由四个过程组成的循环，试填充表 3-1 中所缺数据，并判断该循环是正循环还是逆循环。

表 3-1

过程	Q/kJ	W/kJ	ΔU/kJ
1—2	1 390	0	
2—3	0		−395
3—4	−1 000	0	
4—1	0		

【解】　根据式(3-8)计算出：

$$\Delta U_{12}=Q_{12}-W_{12}=1\ 390\ (\mathrm{kJ}),\ W_{23}=Q_{23}-\Delta U_{23}=395\ (\mathrm{kJ}),$$
$$\Delta U_{34}=Q_{34}-W_{34}=-1\ 000\ (\mathrm{kJ})$$

因为 $\oint \mathrm{d}u=0$，所以：

$$\Delta U_{41}=-(\Delta U_{12}+\Delta U_{23}+\Delta U_{34})=5\ (\mathrm{kJ})$$

故：
$$W_{41}=Q_{41}-\Delta U_{41}=-5\ (\mathrm{kJ})$$

由于循环中$\oint\delta q=\oint\delta w=390\ \mathrm{kJ}>0$，即吸热对外膨胀做功，可断定该循环为正循环。

3.5　开口系能量方程

热力工程中遇到的许多设备，如汽轮机、压气机、风机、锅炉、换热器及空调机等，在工作过程中都有工质流进、流出，都是开口系，通常选取控制体进行分析。

图 3-9 所示是一典型的开口系。系统与外界之间有热量、质量和轴功的交换。工程中遇到的实际过程，系统与外界的质量交换与能量交换并不都是恒定的，有时随时间发生变化。所以控制体内既有能量变化，又有质量变化，在分析时必须同时考虑控制体内的质量变化和能量变化。按质量守恒原理：

进入控制体的质量－离开控制体的质量＝控制体中质量的增量

按能量守恒原理：

进入控制体的能量－控制体输出的能量＝控制体中储存能的增量

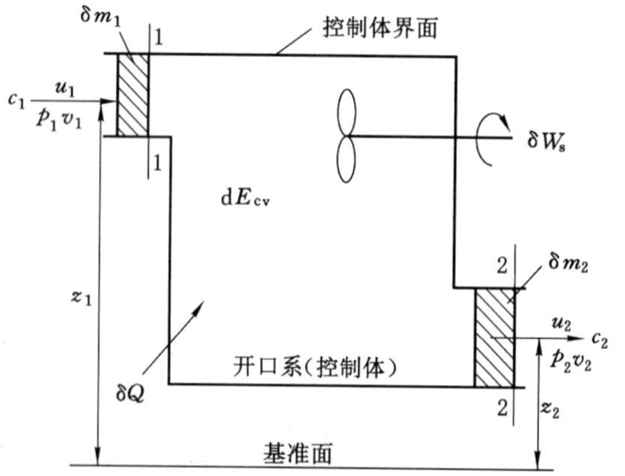

图 3-9　开口系能量转换

把控制体内质量和能量随时间而变化的过程称为不稳定流动过程，如储罐的充气或排空就是这种过程。控制体内的质量和能量不随时间变化，各点状态参数保持一定的流动过程称为稳定流动过程，或稳态稳流过程。

下面先从最普遍的不稳定流动过程着手，用热力学第一定律分析图 3-9 所示的控制体，从而导出开口系能量方程的普遍表达式。

设控制体在 τ 到$(\tau+\mathrm{d}\tau)$ 的时间内进行了一个微元热力过程。在 $\mathrm{d}\tau$ 这段时间内，由控制体入口界面 1—1 处流入的工质质量为 δm_1，出口界面 2—2 处流出的工质质量为 δm_2，控制体从热源吸热 δQ，对外做轴功 δW_s。控制体的能量收入与支出情况如下：

$$\text{进入控制体的能量}=\delta Q+\left(h_1+\frac{1}{2}c_1^2+gz_1\right)\delta m_1$$

$$\text{控制体输出的能量}=\delta W_s+\left(h_2+\frac{1}{2}c_2^2+gz_2\right)\delta m_2$$

控制体储存能的变化量：　　$dE_{cv} = E_{cv,\tau+d\tau} - E_{cv,\tau}$

根据热力学第一定律建立能量方程：

$$\delta Q + \left(h_1 + \frac{1}{2}c_1^2 + gz_1\right)\delta m_1 - \delta W_s - \left(h_2 + \frac{1}{2}c_2^2 + gz_2\right)\delta m_2 = dE_{cv}$$

整理得：

$$\delta Q = \left(h_2 + \frac{1}{2}c_2^2 + gz_2\right)\delta m_2 - \left(h_1 + \frac{1}{2}c_1^2 + gz_1\right)\delta m_1 + \delta W_s + dE_{cv} \tag{3-11a}$$

这就是开口系的普遍适用能量方程，它反映了流动过程中能量转换的一般规律。式(3-11a)是在普遍情况下推导出的，对不稳定流动和稳态稳流，可逆与不可逆过程都适用，也能适用于闭口系。

工程上常用流率：

$$\dot{m} = \lim_{\delta\tau\to 0}\left(\frac{\delta m}{\delta\tau}\right), \dot{W} = \lim_{\delta\tau\to 0}\left(\frac{\delta W}{\delta\tau}\right), \dot{Q} = \lim_{\delta\tau\to 0}\left(\frac{\delta Q}{\delta\tau}\right)$$

$$\dot{Q} = \left(h_2 + \frac{1}{2}c_2^2 + gz_2\right)\dot{m}_2 - \left(h_1 + \frac{1}{2}c_1^2 + gz_1\right)\dot{m}_1 + \dot{W}_s + dE_{cv}/d\tau \tag{3-11b}$$

对于闭口系，由于系统边界没有物质流进和流出，$\delta m_1 = \delta m_2 = 0$，而通过界面的功为膨胀功 δW，系统能量变化为 dE，于是式(3-11a)变为：$\delta Q = dE + \delta W$。又因为在闭口系中工质的动能和势能没有变化，$dE = dU$，故得：$\delta Q = dU + \delta W$。

【例 3-3】 有一储气罐从压缩空气总管充气，总管内压缩空气的参数恒定，分别为 $p_1 = 1$ MPa，$T_1 = 300$ K。储气罐与总管相连的管段上有配气阀，如图 3-10 所示。充气前，阀门关闭着而储气罐内是真空。阀门开启后，压缩空气进入罐内，直到罐内压力与总管压力相等。假设罐壁绝热，充气过程中储罐内气体状态均匀变化，求充气前后储气罐内压缩空气能量的变化。

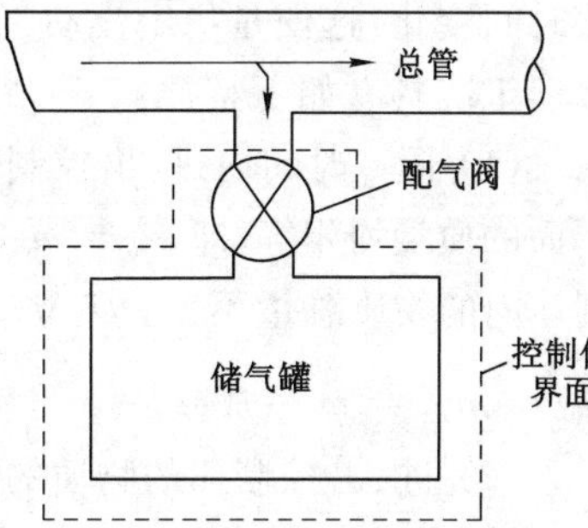

图 3-10　例 3-3 图

【解】 取储气罐为控制体，如图中虚线所示为控制体的界面。从图中及题意可知：

(1) 只有一股气流通过控制体界面流进储罐，而没有气流流出界面，故 $\delta m_2 = 0$；

(2) 储气罐壁是绝热的，即 $\delta Q = 0$；

(3) 系统与外界没有轴功传递，即 $\delta W_s = 0$；

(4) 进入系统的动能及势能忽略不计，即 $\left(\frac{1}{2}c_1^2 + gz_1\right)\delta m_1 = 0$；

(5) 控制体内动能与势能没有变化，因此，$dE_{cv} = dU_{cv} = d(mu)_{cv}$。

由式(3-11)简化后得：　　$h_1\delta m_1 = d(mu)_{cv}$

即整个充气过程中进入控制体的能量等于控制体中内能的增量。

两边积分：

$$\int_0^{m_1} h_1\delta m_1 = \int_1^2 d(mu)_{cv}$$

则：

$$h_1 m_1 = (mu)_{cv,2} - (mu)_{cv,1}$$

由于充气前储气罐内为真空，因而$(mu)_{cv,1} = 0$，$m_1 = m_{cv,2}$，上式变为：

$$h_1 = u_{cv,2}$$

即压缩气体从总管进入储气罐，其焓在充气结束后转变为储气罐内气体的内能，储气罐内工质能量的增量等于流入系统的气体带入的焓值。

假设充入的空气为理想气体，应用理想气体焓定义式 $h = u + RT$ 及理想气体内能计算式 $u = c_v T$ 和 $c_p = c_v + R$ 等关系，可得罐内空气温度 $T_{cv,2} = \frac{c_p T_1}{c_v} = \kappa T_1 = 1.4 \times 300 = 420$ (K)。

由此可见，从压缩空气总管中进入储气罐气体的焓，在充气结束后已转变为储气罐内气体的内能，因而温度由 300 K 升高为 420 K。

3.6 开口系稳定流动能量方程及其应用

3.6.1 稳定流动能量方程

实际的热工设备，通常都是在稳定工况下运行的，工质以恒定的流量连续不断地进出系统，系统内部及界面上各点工质的状态参数和宏观运动参数都保持一定，不随时间变化，这便是稳定流动工况，如图 3-11 所示。稳定流动工况具有如下特征：

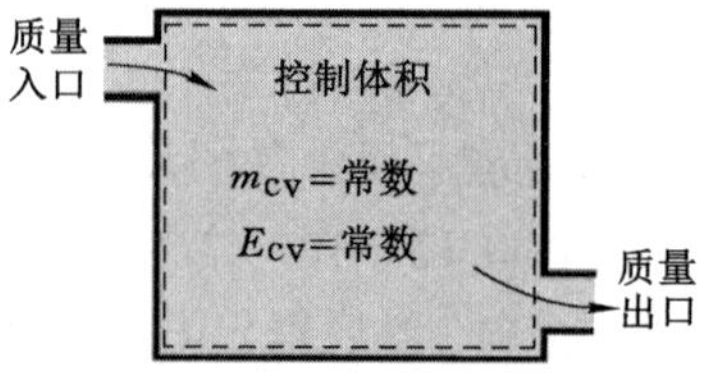

图 3-11 稳定流动系统

(1) 同一时间内进、出控制体界面及流过系统内任意断面的质量均相等，即 $m_{in} = m_{out} = m =$ 常数。也就是说，控制体内的物质总量不变，即 $\Delta m_{cv} = 0$。用质量流率表示为 $\dot{m}_{in} = \dot{m}_{out} = \dot{m} =$ 常数。

(2) 进、出控制体的物质的状态参数不随时间而改变。

(3) 单位时间内控制体与外界的功和热量交换不随时间变化。

(4) 控制体储存能保持不变，即系统的储存能增量为零，$dE_{cv} = 0$。

于是式(3-11b)可写成：

$$\dot{Q} = \left[(h_2 - h_1) + \frac{1}{2}(c_2^2 - c_1^2) + g(z_2 - z_1) \right]\dot{m} + \dot{W}_s \tag{3-12a}$$

由于 $\dot{Q} = \dot{m}q$，$\dot{W}_s = \dot{m}w_s$，对于单位质量工质：

$$q = (h_2 - h_1) + \frac{1}{2}(c_2^2 - c_1^2) + g(z_2 - z_1) + w_s = \Delta h + \frac{1}{2}\Delta c^2 + g\Delta z + w_s \tag{3-12b}$$

对于微元热力过程：

$$\delta q = dh + \frac{1}{2}dc^2 + g dz + \delta w_s \tag{3-12c}$$

式(3-12)是开口系稳定流动能量方程，是式(3-11)应用于稳定流动情况的一个特例，普遍适用于稳定流动各种热力过程，稳定流动能量方程中的动能变化 $\frac{1}{2}\Delta c^2$、势能变化 $g\Delta z$

及轴功 w_s 都属于机械能，是热力过程中可被直接用来做功的能量，统称为技术功，记为 w_t。即

$$w_t = \frac{1}{2}\Delta c^2 + g\Delta z + w_s \tag{3-13}$$

引入技术功概念后，稳定流动能量方程可写成：

$$q = \Delta h + w_t \tag{3-14}$$

根据焓的定义 $h = u + pv$，由式(3-14)得：

$$w_t = q - \Delta h = (\Delta u + w) - (\Delta u + p_2v_2 - p_1v_1) = w + p_1v_1 - p_2v_2$$

$$w = w_t + \Delta(pv) \tag{3-15}$$

式(3-15)表明，工质在稳定流动过程中所做的膨胀功是隐含的，表现为一部分消耗于维持工质流动所需的流动功 $\Delta(pv)$，一部分用于增加工质的宏观动能和重力势能，其余部分才作为热力设备输出的轴功。膨胀功是简单可压缩系热变功的源泉。

对于稳定流动的可逆过程，技术功为：

$$\delta w_t = \delta q - \mathrm{d}h = (\mathrm{d}u + p\mathrm{d}v) - \mathrm{d}(u + pv) = \mathrm{d}u + p\mathrm{d}v - \mathrm{d}u - p\mathrm{d}v - v\mathrm{d}p$$

整理得：

$$\delta w_t = -v\mathrm{d}p \tag{3-16a}$$

即可逆过程中技术功为：

$$w_t = -\int_1^2 v\mathrm{d}p \tag{3-16b}$$

因此，对于可逆的稳定流动过程，有：

$$\delta q = \mathrm{d}h - v\mathrm{d}p \tag{3-17}$$

式(3-16)和式(3-17)仅适用于可逆过程。由式(3-16a)可知，微元热力过程的技术功，在图3-12所示的 p-v 图上用斜线所示的微元面积表示。技术功 w_t 在图 3-12 所示的 p-v 图上用过程线 1—2 与纵坐标间围成的面积表示，即 w_t = 面积 12341。

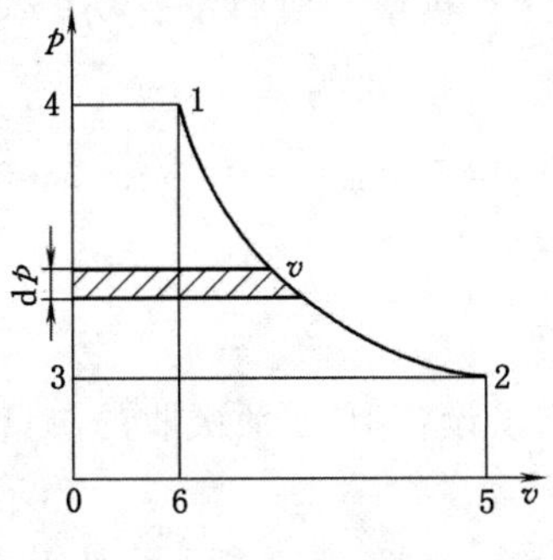

图 3-12　技术功

技术功、膨胀功及流动功之间的关系，由式(3-15)和式(3-16)及图 3-12 得出：

$$w_t = w + p_1v_1 - p_2v_2 = \text{面积 } 12561 + \text{面积 } 41604 - \text{面积 } 23052$$

显然，技术功也是过程量，其值不仅取决于初、终状态，而且与过程特性有关。在一般的工程设备中，往往可以不考虑进、出口工质动能和势能的变化，此时技术功就等于轴功。

3.6.2 稳定流动能量方程的应用

稳定流动能量方程在工程上有着广泛的应用，在不同条件下可适当将稳定流动能量方程加以简化，以便于应用。下面以典型设备为例，说明稳定流动能量方程的应用。

(1) 动力机械

将工质的热能转化为功并向外输出的设备称为动力机械，也称为热机，如蒸汽机、汽轮机、内燃机和燃气轮机等。汽轮机和燃气轮机是旋转式机械设备，蒸汽机和内燃机是往复式机械设备。

动力机械进、出口的高度差一般很小，进、出口工质的流速变化也不大，又因工质在动力机械中停留的时间很短，系统通过边界的散热量相对输出的功量所占比例很小，可以忽略，即

$$g(z_2 - z_1) \approx 0, \frac{1}{2}(c_2^2 - c_1^2) \approx 0, q \approx 0$$

将上述条件代入式(3-12b)，得到工质流经动力机械的能量方程：

$$w_s = h_1 - h_2 \tag{3-18}$$

由此得出，汽轮机、燃气轮机等动力机械所做的轴功等于工质的焓降。

(2) 压缩机械

泵、风机、压气机等为消耗机械能(轴功)提升工作流体压力的设备，统称为压缩机械。水泵将水的压力提高，风机、压气机将空气的压力提高，这类设备均是耗能设备。能量转换中的散热、动能差、势能差与消耗的功相比可忽略，能量方程为：

$$-w_s = h_2 - h_1 \tag{3-19}$$

压缩机械中绝热压缩消耗的轴功等于压缩工质焓的增量。

(3) 热交换器

锅炉、加热(或冷却)器、蒸发器、冷凝器等热交换器是完成冷、热流体交换热能的设备。工质流经热交换器时与外界有热量交换而没有功的交换，动能差与势能差也可忽略不计，即

$$w_s = 0, g\Delta z \approx 0, \frac{1}{2}\Delta c^2 \approx 0$$

故得：

$$q = h_2 - h_1 \tag{3-20}$$

因此，在锅炉、冷凝器等热交换设备中，工质吸收的热量等于焓的增量。

(4) 喷管和扩压管

喷管和扩压管都是变截面短管，如图 3-13 所示。喷管能使工质压力降低而加速，由于流速很快，喷管与外界不交换热量，不做功，势能差可以忽略，能量方程为：

$$\frac{1}{2}(c_2^2 - c_1^2) = h_1 - h_2 \tag{3-21}$$

可见，喷管中流体动能的增量等于工质的焓降。与喷管功能相反的是扩压管，它可使流体减速而压力升高，类似地，扩压管中流体动能的减少量等于焓的增量。

(5) 流体的混合(合流与分流)

几股流体汇合成一股流体的流动称为合流，如图 3-14 所示。工程中的混合式换热器、

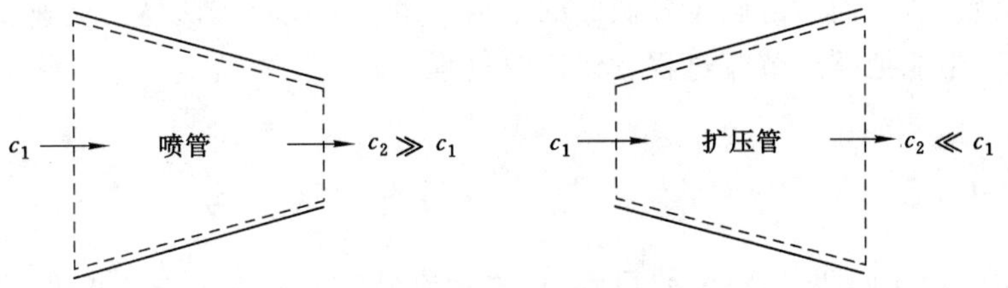

图 3-13　喷管、扩压管示意图

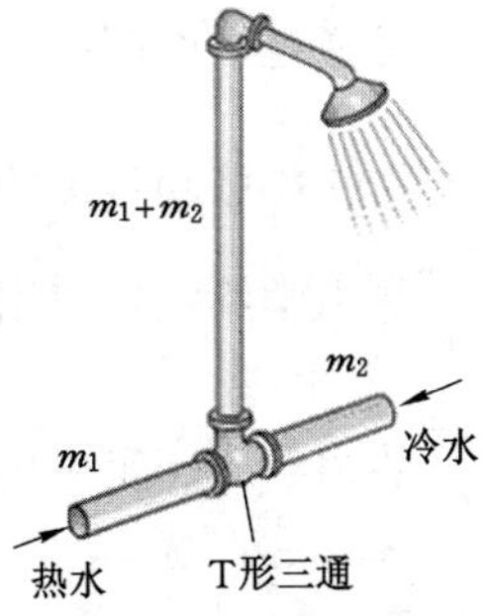

图 3-14　流体的混合（合流三通）

蒸汽动力厂的冷却塔都属于合流的例子。取混合室为控制体，混合为稳流工况，在绝热条件下进行，无功量交换，且忽略流体动能、势能变化，则控制体的能量方程式简化为：

$$\Delta H = 0 \text{ 或 } H_1 + H_2 = H_3$$

即

$$m_1 h_1 + m_2 h_2 = m_3 h_3$$

根据质量守恒定律，有：

$$m_1 + m_2 = m_3$$

终态工质的焓为：

$$h_3 = \frac{m_1 h_1 + m_2 h_2}{m_1 + m_2} \tag{3-22}$$

(6) 绝热节流

流体流经管道中的阀门或缩孔等突然变窄的断面时，由于存在阻力流体压力明显降低，这种特殊流动现象称为节流。图 3-15 所示的调节阀、多孔板等装置均可称为节流阀。

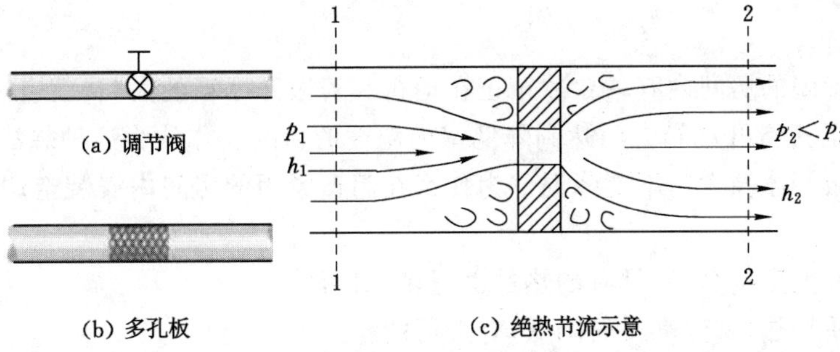

图 3-15　绝热节流

取流体节流前、后稳定段 1—1、2—2 为界面构成控制体，稳态稳流的流体快速流过狭窄断面，来不及与外界换热也没有功量的传递，可理想化称为绝热节流。若忽略流体进、出口界面的动能、势能变化，则控制体能量方程可简化为：

$$h_1 = h_2 \tag{3-23}$$

式(3-23)表明，绝热节流前、后焓值不变，即能量数量相等。但需要指出，节流过程是典型的不可逆过程，流体在节流元件附近发生强烈的扰动及涡流，流体处于极度不平衡状态，故不能用平衡态热力学方法分析节流元件附近的状况。但在距节流元件较远的地方，如截面 1—1 和 2—2 处，流体处于平衡状态，可以得出两处焓值相等。截面 1—1 和 2—2 之间流体处于不平衡状态，因而不能确定各截面的焓值。由于节流过程流体的流速变化很大，焓值并不处处相等，不能把整个节流过程看作定焓过程。

本章小结

热力学第一定律阐明了热能与机械能相互转换的数量关系。能量的转换有赖于物质的状态变化，同时能量有质的差别，正确认识储存能、内能、迁移能等各种能量是应用热力学第一定律的基础。

可逆过程的膨胀功 $w=\int_1^2 p\mathrm{d}v$。

类比膨胀功得出熵的定义式 $\mathrm{d}s=\frac{\delta q}{T}$ 和可逆过程热量的计算式 $q=\int_1^2 T\mathrm{d}s$。

热力学第一定律用于闭口系的能量方程为：

$$Q=\Delta U+W$$

热力学第一定律用于稳定流动系统的能量方程为：

$$Q=\Delta H+\frac{1}{2}m\Delta c^2+mg\Delta z+W_s$$

引入技术功的概念后稳定流动系统的能量方程为：

$$Q=\Delta H+W_t$$

可见，热能转变为机械能要靠物系体积的改变，但实际表现出来的并不总是膨胀功，引入其他几种不同形式的功，使同一性质的能量方程以不同形式、不同内容出现。

思考与练习题

(1) 门窗紧闭的房间内有一台冰箱正在运行。若敞开冰箱的大门，就有一股凉气扑面。于是有人就想通过敞开冰箱大门达到降低室内温度的目的。你认为这种想法可行吗？

(2) 既然敞开冰箱大门不能降温。为什么在门窗紧闭的房间内安装空调器后却能使温度降低呢？

(3) 物质的温度越高，所具有的热量也越多，对否？

(4) 对工质加热，其温度反而降低，有否可能？

(5) 热量和内能有什么区别？有何联系？

(6) “任何没有容积变化的过程就一定不对外做功”这种说法对吗？

(7) 气体初态 $p_1=0.5$ MPa，$v_1=172\ \mathrm{m^3/kg}$，按 $pv=$常数的规律，可逆膨胀到 $p_2=0.1$ MPa，试求膨胀功。

(8) 某气缸中有 0.5 kg 的气体，从初态 $p_1=0.7$ MPa，$v_1=0.02\ \mathrm{m^3/kg}$，可逆膨胀到终态 $v_2=0.05\ \mathrm{m^3/kg}$，各膨胀过程维持以下关系：① $p=$定值；② $pv=$定值。试计算各过程所做的膨胀功，并示意在 p-v 图上。

(9) 说明以下论断是否正确：① 气体吸热后一定膨胀，内能一定增加；② 气体膨胀时一定对外做功；③ 气体压缩时一定消耗外功。

(10) 气体在某一过程中吸热 12 kJ，同时内能增加 20 kJ。问此过程是膨胀过程还是压缩过程？对外所做的功是多少？

(11) 闭口系经历了一个由四个过程组成的循环。试填充表 3-2 中所缺数据。

表 3-2

过程	Q/kJ	W/kJ	ΔE
1—2	1 100	0	
2—3	0	100	
3—4	−950	0	
4—1	0		

(12) 某闭口系从初态 a 出发分别经历两条不同的路径到达终态 b，如图 3-16 所示。

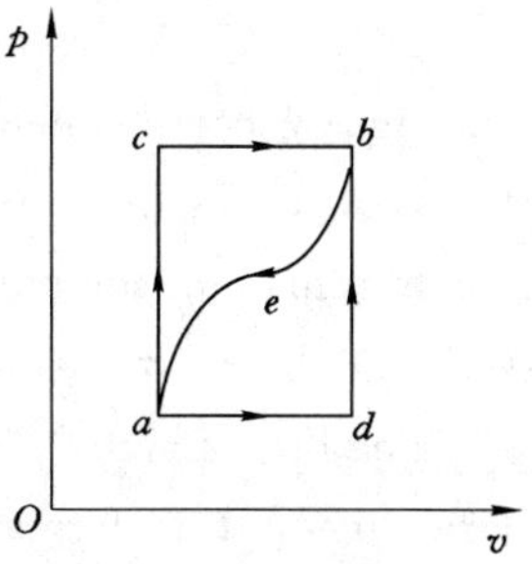

图 3-16　习题(12)图

① 从 a 经 c 至 b 时，热力系吸热 80 kJ，对外做功 30 kJ，求其内能的变化量；

② 根据问题①的结果，若从 a 经 d 至 b 时，对外做功 10 kJ，问热力系是吸热还是放热？交换的热量是多少？

③ 根据问题①的结果，若闭口系从终态 b 沿着曲线 e 返回至初态 a 的过程中外界对热力系做功 30 kJ，问热力系是吸热还是放热？交换的热量是多少？

④ 根据问题①的结果，若设状态点的内能 $U_a=0$，$U_d=20$ kJ，那么 $d\to b$ 路径的内能变化量为多少？

⑤ 根据问题①及问题②的结果，若闭口系从 a 点出发经 $a\to c\to b\to d\to a$ 回到 a 点，热力系交换的净功、净热量为多少？

(13) 膨胀功、流动功、轴功和技术功有何差别？相互有无联系？试用 p-v 图说明之。

(14) 流动功为何出现在开口系能量方程中，而不出现在闭口系能量方程中？

(15) 下列各式，适用于何种条件？

$$\delta q = \mathrm{d}u + \delta w;\ \delta q = \mathrm{d}u + p\mathrm{d}v;\ \delta q = c_v\mathrm{d}T + p\mathrm{d}v;\ \delta q = \mathrm{d}h;\ \delta q = c_p\mathrm{d}T - c\mathrm{d}p$$

(16) 某气体在压气机中被压缩，压缩前气体的参数是 $p_1=100$ kPa，$v_1=0.845\ \mathrm{m^3/kg}$，压缩后的参数是 $p_2=800$ kPa，$v_2=0.175\ \mathrm{m^3/kg}$。设在压缩过程中每千克气体的内能增加 150 kJ，同时向外界放出热量 50 kJ，压气机每分钟生产压缩气体 10 kg。试求：① 压缩过程中对每千克气体所做的压缩功；② 每生产 1 kg 压缩气体所需的轴功；③ 带动此压气机要用多大功率的电动机；④ 压缩前、后气体焓的变化。

知识拓展：第一类永动机

知识拓展：热力学第一定律的形成历程

Chapter 3 The First Law of Thermodynamics

3.1 The Essence of the First Law of Thermodynamics

Energy Conservation Principle is one of the most fundamental laws in nature. It states that energy can be neither created nor destroyed during a process; it can only change forms, or it can be transferred from one body to another. However, the total amount of energy remains constant. That is, energy is always conserved. It provides a sound basis for studying the relationships among the various forms of energy and energy interactions.

The First Law of Thermodynamics mainly states that the quantity of energy is conserved during energy transformation from thermal energy to other forms of energy. Energy balance for any system undergoing any kind of process can be expressed compactly as:

Total energy entering the system − Total energy leaving the system = Change in the total energy of the system

That is, the net change (increase or decrease) in the energy of the system during a process is equal to the difference between the total energy entering and leaving the system.

It is called the general energy balance relation. It is very helpful to understand various forms of energy and to recognize the forms of energy transfer for the successful use of this relation to solve engineering problems.

3.2 Total Energy of a System

Energy can exist in numerous forms, such as internal (sensible, latent, chemical, and nuclear), kinetic, potential, electric, and magnetic energies, and their sum constitutes the total energy E of a system. The various forms of energy that make up the total energy of a system include two groups: microscopic and macroscopic. The microscopic forms of energy are those related to the molecular structure of a system and the degree of the molecular activity, and they are independent of outside reference frames. The sum of all the microscopic forms of energy is called internal energy (or thermodynamic energy) of a system. The macroscopic forms of energy are those a system possesses as a whole with respect to some outside reference frame, such as kinetic and potential energies.

3.2.1 Internal Energy

(1) Internal kinetic energy

For a macroscopic sample of matter at rest, on a microscopic level the atoms composing the sample are in continual, random motion, such as translation, rotation, vibration. As a result, the sample of matter at rest possesses microscopic kinetic energies, which are called internal kinetic energies. The atomic kinetic energies are all positive, scalar numbers. Since this molecular motion is primarily a function of temperature, the internal energy is sometimes called "thermal energy". As temperature rises, the atomic kinetic energy increases.

(2) Internal potential energy

As the atoms move, they stretch or compress the bonds holding them together. Molecules in a gas hardly feel any force from other molecules far away. But when two molecules approach closely, the potential energy rises rapidly, causing them to repel one another and move apart again. At any instant, the sample has a microscopic, internal potential energy, which is the sum of all potential energy contributions describing the interactions between the atoms or molecules.

(3) Chemical energy and nuclear energy

Chemical energy is the energy possessed by a substance to keep the arrangement of atoms composing the molecules. It is part of internal energy. It is essentially the energy required to break chemical bonds. If a chemical reaction occurs which breaks the bonds of one type and forms bonds of another type, chemical energy may go up or down. Nuclear (atomic) energy is the energy possessed by a system due to the cohesive forces holding protons and neutrons together as the atom's nucleus.

The internal energy includes all of the kinetic energy associated with the atomic-level, random motion of the atoms of the system, and all of the potential energy associated with all possible interactions between the atoms. The internal energy is denoted as U. On a unit mass basis, it is called specific internal energy and denoted by u. It has the unit J/kg or kJ/kg.

The internal energy, u, which arises from molecular motion, depends only on the state of the system and is a function of the state of a system. It is a point function, that is, $u=f(p,T)$, or $u=f(p,v)$, or $u=f(v,T)$. It does not depend on how a system reaches that state. It has exact differentials and a small change in internal energy is denoted by dU or du using the symbol d.

3.2.2 Potential Energy and Kinetic Energy

(1) Kinetic energy

The energy that a system possesses as a result of its macroscopic motion relative to some reference frame is called kinetic energy, denoted by E_k. When all parts of a system move at the same speed, the kinetic energy can be expressed as $E_k=mc^2/2$, where c denotes the speed or velocity of the system relative to some fixed reference.

(2) Potential energy

The energy that a system possesses as a result of its elevation in the gravitational field is called potential energy, denoted as E_p, and is expressed as $E_p = mgz$, where g is the gravitational acceleration rate and z is the elevation of the center of the gravity of a system relative to some arbitrarily selected reference level.

Both c and z are independent of the state of the system. Thus, E_k and E_p are irrelevant to the state of the system.

3.2.3 Total Energy and Energy Change of a System

In the absence of electric, magnetic, and surface tension effects (i.e., for simple compressible systems), the total energy of a system consists of kinetic, potential, and internal energies and is expressed as:

$$E = U + E_k + E_p \tag{3-1a}$$

or

$$E = U + \frac{1}{2}mc^2 + mgz \tag{3-1b}$$

On a unit mass basis, the total energy is denoted as e and has the unit J/kg or kJ/kg, that is:

$$e = u + \frac{1}{2}c^2 + gz \tag{3-1c}$$

The value of a property does not change unless the state of the system changes. Therefore, the energy change of a system is zero if the state of the system does not change. Also, the change in the total energy of a simple compressible system during a process is the sum of the changes in its internal, kinetic, and potential energies and can be expressed as:

$$\Delta E = \Delta U + \Delta E_k + \Delta E_p$$

Most systems encountered in practice are stationary, that is, they do not involve any changes in their velocity or elevation during a process. It means the changes in kinetic and potential energies are zero (that is, $\Delta E_k, \Delta E_p = 0$) for stationary systems. And the total energy and the change reduces to for such systems.

$$E = U \quad \text{or} \quad e = u$$

$$\Delta E = \Delta U \quad \text{or} \quad \Delta e = \Delta u$$

3.3 Energy Transfer by Heat, Work or Mass

Mass and energy can cross the boundary to fulfill energy transfer and transformation. Generally, energy crosses the boundary in the form of heat or work.

3.3.1 Energy Transfer by Heat

Heat is defined as the form of energy that is transferred between two systems (or a system and its surroundings) by virtue of a temperature difference. That is, an energy interaction is heat only if it takes place because of a temperature difference. Once the temperature equality is established, energy transfer stops. It means that there cannot be any heat transfer between two systems that are at the same temperature.

Heat is energy in transition. It is recognized only as it crosses the boundary of a system. The body contains energy, but this energy is transferred only as it passes through the system boundary to enter its surroundings, as shown in Fig. 3-1. Once in the surroundings, the transferred heat becomes part of the internal energy of the surroundings. Thus, in thermodynamics, the term heat simply means heat transferred. The direction of heat transfer is always from a higher temperature body to a lower temperature one. Conventionally, the transfer of heat into a system is frequently referred to as heat addition and the transfer of heat out of a system as heat rejection. And heat addition is defined to be positive; heat rejection to be negative.

However, heat is a path function and the amount of heat transferred depends on the path followed. It is different from the internal energy, which is a function of state. A path function has inexact differentials and a differential amount of heat is denoted by δQ or δq. As a form of energy, heat has energy unit, kJ. The amount of heat transferred during a process between two states (states 1 and 2) is denoted just by Q. Heat transfer per unit mass of a system is denoted as q and is determined from $q=Q/m$, kJ/kg.

3.3.2 Energy Transfer by Work

Work, like heat, is a kind of energy interaction between a system and its surroundings. An energy interaction that is not caused by a temperature difference between a system and its surroundings is work. Heat is easy to recognize: its driving force is a temperature difference between the system and its surroundings. More specifically, work is the energy transfer associated with the force acting through a distance. A rising piston, a rotating shaft, and an electric wire crossing the system boundaries are all associated with work interactions.

Boundary work is an indispensable way to complete the energy transformation from thermal energy to mechanical energy. In addition, mechanical work is always transferred by virtual of a rotating of shaft. Therefore, boundary work and shaft work will be introduced in the following.

(1) Moving boundary work: the expansion and compression work is often called moving boundary work, or simply boundary work. The volume change is positive during an expansion process and negative during a compression process. Thus, the boundary work is positive during an expansion process and negative during a compression process. There are two requirements for a boundary work interaction between a system and its surroundings to occur: a force acts on the boundary and the boundary must move. Therefore, the presence of forces on the boundary without any displacement of the boundary does not constitute a work interaction. Likewise, the displacement of the boundary without any force to oppose or drive this motion, such as the expansion of a gas into an evacuated space, is not a work interaction since no energy is transferred.

Work only occurs during a process and it is a path function. As the process finishes, the work interaction stops.

(2) Shaft work: generally shaft work means the work done by a rotating machinery, which is carried by a shaft [Fig. 3-2(a)]. Usually, it is stipulated that the shaft work a system outputs is positive, and the shaft work input to a system is negative. As illustrated in Fig. 3-2(b), the paddle-wheel work is done on an adiabatic closed system through the dissipative effect such as friction. As a result, the internal energy of the system increases. However, shaft work can not be done by a closed system continuously for the heat added to the system can not be changed into mechanical work directly. But, for an open system, as mass flows in or out, it may push the shaft to output work [Fig. 3-2 (c)]. Steam turbines, pumps and compressors are devices usually encountered in practice, which transfer work through shaft.

3.3.3 Moving Boundary Work and Heat Transfer in Reversible Process

(1) Moving boundary work in reversible process

The energy conversion can be fulfilled by the state change of a working fluid. As shown in Fig. 3-3, there is some working fluid in the cylinder. Suppose the process is reversible, the system is in mechanical equilibrium. The force exerted on the system is the pressure of the system, and the piston sectional area is A, that is $F=pA$.

The piston moves a differential amount distance dS in the direction of the force. The work done can be calculated by:

$$\delta W = F\mathrm{d}S = pA\mathrm{d}S = p\mathrm{d}V \tag{3-2a}$$

In process 1-2, the total work can be obtained by integration:

$$W = \int_{V_1}^{V_2} p\mathrm{d}V \tag{3-2b}$$

The work done per unit mass of the system is denoted by w and is defined as:

$$w = \int_{V_1}^{V_2} p\mathrm{d}v \tag{3-2c}$$

The integral above is the formula to calculate moving boundary in reversible process. It can be evaluated only if we know the function relationship between p and V during the process. That is, $p=f(V)$ or $p=f(v)$ should be available.

The quasi-equilibrium expansion process described is shown on a diagram in Fig. 3-3. The differential area dA is equal to $p\mathrm{d}V$, the differential work. The total area under the process curve 1-2 on a p-v diagram is the total work. Thus, it reveals that the area under the process curve is equal, in magnitude, to the work done by a closed system during a quasi-equilibrium expansion or compression process. On the p-v diagram, it represents the boundary work done per unit mass. A gas can follow several different paths as it expands from state 1 to state 2. In general, each path will have a different area underneath it. Since this area represents the magnitude of the work, the work done will be different for every process (Fig. 3-4). As we know, work is a path function and is not a property. It depends on the path followed as well as its initial and final states. The systems do not possess work at a state. A differential amount of heat or work is represented by δQ or δW, respectively, instead of dQ or dW.

(2) Heat transferred in reversible process

Heat is defined as the form of energy that is transferred between two systems (or system to its surroundings) at different temperatures when they are brought into thermal contact. Thus, heat is a transient quantity that can appear only at the boundaries of the systems.

Since work and heat are both path functions, they must possess some common characteristics. Does heat have similar defining formula? Analogically, we can define the state property entropy. When a system undergoes a reversible process, heat transferred per unit mass of the system is defined as:

$$\delta q = T\mathrm{d}s \tag{3-3a}$$

Where s is the entropy per unit mass. The unit is J/(kg · K). Namely:

$$\mathrm{d}s = \frac{\delta q}{T} \tag{3-3b}$$

The total entropy of m kg working fluid is:

$$\mathrm{d}S = \frac{\delta Q}{T} \tag{3-4a}$$

Notice that s is a state property. The term entropy is generally used to refer to both total entropy and entropy per unit mass since the context usually clarifies which one is meant.

The amount of heat transfer is also equal to the area under the process curve on a T-s diagram during a quasi-equilibrium process (Fig. 3-5).

In this chapter, we just put forward the concept of entropy. Further discussion on entropy will be done in Chapter 5.

The heat transferred between the system and its surroundings during a reversible process from 1 to 2 is:

$$q = \int_1^2 T\mathrm{d}s, Q = \int_1^2 T\mathrm{d}S \tag{3-4b}$$

Based on the change in entropy, it is easy to judge the direction of heat transferred between a system and its surroundings:

If $\mathrm{d}s > 0$, then $q > 0$, heat is transferred to a system.

If $\mathrm{d}s < 0$, then $q < 0$, heat is transferred from a system.

If $\mathrm{d}s = 0$, then $q = 0$, the system is adiabatic. It means that a reversible adiabatic process is an isentropic process.

3.3.4 Total Energy of a Flowing Fluid and Enthalpy

Mass flow in and out of the system serves as an additional mechanism of energy transfer. When mass enters a system, the energy of the system increases because mass carries energy with it. Likewise, when some mass leaves a system, the energy contained within the system decreases because the leaving mass takes out some energy with it.

(1) Energy stored in a fluid at rest: the total energy of a simple compressible system

consists of three parts: internal, kinetic, and potential energies. That is:

$$E = U + \frac{1}{2}mc^2 + mgz \tag{3-5a}$$

On a unit mass basis, it is:

$$e = u + \frac{1}{2}c^2 + gz \tag{3-5b}$$

(2) Flow work: unlike closed systems, control volumes involve mass flow across their boundaries, and some work is required to push the mass into or out of the control volume. This work is known as the flow work, or flow energy, and it is indispensible to maintain a continuous flow through a control volume. Thus, the fluid entering or leaving a control volume possesses an additional form of energy—the flow work.

To obtain a relation for flow work, consider a fluid element of volume V as shown in Fig. 3-6. The fluid immediately upstream forces this fluid element to enter the control volume; thus, it can be regarded as an imaginary piston. The fluid element can be chosen to be sufficiently small so that it has uniform properties throughout. If the fluid pressure is p and the cross-sectional area of the fluid element is A, the force applied on the fluid element by the imaginary piston is:

$$F = pA$$

To push the entire fluid element into the control volume, this force must act through a distance L. Thus, the work done in pushing the fluid element across the boundary is:

$$W = FL = pAL = pV$$

The flow work relation is the same whether the fluid is pushed into or out of the control volume. Unlike other work quantities, flow work is expressed in terms of properties. In fact, it is the product of two properties of the fluid. For this reason, some people view it as a combination property (like enthalpy) and refer to it as flow energy, or transport energy instead of flow work.

The flow work consumed to push a flow through the open system can be expressed as

$$W_f = p_2V_2 - p_1V_1 \tag{3-6}$$

or on a unit mass basis:

$$w_f = p_2v_2 - p_1v_1$$

3.3.5 Total Energy of Flowing Fluid and Enthalpy

For simplicity, combining pv and U together will form another new concept, the enthalpy h, that is:

$$H = U + pV \tag{3-7}$$

So the energy of a flowing fluid reduces to:

$$H + \frac{1}{2}mc^2 + mgz \quad \text{or} \quad h + \frac{1}{2}c^2 + gz$$

By using the enthalpy instead of the internal energy to represent the energy of a flowing fluid, we do not need to concern about the flow work. The energy associated with pushing the fluid into or out of the control volume is automatically taken care of by enthalpy. From now on, the energy of a fluid stream flowing into or out of a control volume is represented by $H+mc^2/2+mgz$, and no reference will be made to flow work or flow energy.

For closed systems, the product pV does not represent any form of energy, the $h=u+pv$ is just viewed as a combination property.

3.4 Energy Balance for Closed Systems

So far, we have considered various forms of energy, such as heat Q, work W, and total energy E individually. And the purpose of the following section is to relate them to each other during a process.

3.4.1 Energy Balance of a Closed System

For stationary systems, the changes in kinetic and potential energies are zero (that is, $\Delta E_k, \Delta E_p = 0$), and the total energy change during a process reduces to the change in its internal energy and can be expressed as $\Delta E = \Delta U = U_2 - U_1$. Or on per unit mass basis, it can be expressed as $\Delta e = \Delta u = u_2 - u_1$.

The energy balance can be expressed as:

$$Q = \Delta U + W \tag{3-8a}$$

for a closed system as shown in Fig. 3-7.

In Eq. (3-8a), ΔU is the change in internal energy of a system; Q is the heat added to the system, and W is the work done by the system.

The equation can also be written on per unit mass basis as:

$$q = \Delta u + w \tag{3-8b}$$

These equations are called energy conservation equations. They state that for a stationary closed system, the change in energy of a system is equal to the difference between the heat added to the system and the work done by the system. They are applicable to any process in closed system, whether the process is reversible or irreversible.

The first law can also be written in differential form as:

$$\delta Q = dU + \delta W \quad \text{or} \quad \delta q = du + \delta w \tag{3-8c}$$

Where the symbol "δ" is used to denote those which are not exact differentials but dependent of path.

Work interaction between a closed system and its surrounding may involve boundary work, shaft work and electrical work, etc. However, for reversible processes, the only form of work may be involved is moving boundary work. Then, for reversible processes, we can substitute $\delta W = p dV$ into the above equations and obtain:

$$\delta Q = dU + p dV \quad \text{or} \quad \delta q = du + p dv \tag{3-9}$$

Though Eq.(3-8) and Eq.(3-9) are derived based on closed systems, the relationship among heat transfer, expansion work and thermodynamic energy is suitable for open systems. As the mass in a dosed system does not change in a process, the energy balance equation for closed systems also called control mass form of the conservation equation.

3.4.2 Application of the First Law in a Cyclic Process

For a cyclic process, as shown in Fig. 3-8, the final state is identical with the initial state. That is:

$$U_2 = U_1 \text{ and } \Delta U = 0$$

Therefore, heat and work transfers during the cyclic process are numerically equal. That is:

$$Q = W \quad \text{or} \quad \oint \delta Q = \oint \delta W \tag{3-10}$$

3.5 Energy Balance for Open Systems

Open systems are very common in engineering practice, such as turbines, compressors, fans, boilers, heat exchangers, refrigerators and heating systems, etc. A jet engine is also an open system, which takes in air and fuel, mixes and burns them, and then exhaust the hot combustion products at high speed to produce thrusts. Design of an efficient jet engine requires a careful thermodynamic analysis. In this chapter, we will develop the energy conservation equation for open systems.

A schematic of a general open system is shown in Fig. 3-9. Rather than focus on a particle of mass which moves through the engine, it is more convenient to focus on the volume occupied by the engine. Therefore, the energy balance for open system is also called control volume form of the conservation law.

3.5.1 Mass Balance for Open System

Processes involve changes within the control volume with time are called unsteady-flow, or transient flow processes. The mass and energy contents of the control volume as well as the energy interactions across the boundary do not remain constant during unsteady-flow process. Some familiar examples, such as the charging of rigid vessels from supply lines, discharging a fluid from a pressurized vessel, etc., are unsteady-flow processes. However, during a steady-flow process, no changes occur within the control volume and the state remains constant.

For the control volume shown in Fig. 3-9, net mass transfer to or from a system during a process is equal to the net change in the total mass of the system during that process and can be expressed as:

$$\sum m_{\text{in}} - \sum m_{\text{out}} = \Delta m_{\text{cv}}$$

3.5.2 Energy Balance for Open System

The energy content of a control volume changes with time during an unsteady-flow process. The magnitude of change depends on the amount of energy transfer across the system boundaries as heat or work as well as on the amount of energy transported into and out of the control volume by mass during the process. When analyzing an unsteady-flow process, we must keep track of the energy content of the control volume as well as the energies of the incoming and outgoing flow streams.

The fluid that enters or leaves has an amount of energy per unit mass:

$$e = h + \frac{1}{2}c^2 + gz$$

Thus, the energy entering the control volume is $\delta Q + (h_1 + \frac{1}{2}c_1^2 + gz_1)\delta m_1$;

The energy leaving the control volume is $\delta W_s + (h_2 + \frac{1}{2}c_2^2 + gz_2)\delta m_2$.

The energy change of the control volume is:

$$\mathrm{d}E_{cv} = E_{cv,\tau+\mathrm{d}\tau} - E_{cv,\tau}$$

The energy balance equation for an unsteady-flow process can be expressed explicitly as:

$$\delta Q = \left(h_2 + \frac{1}{2}c_2^2 + gz_2\right)\delta m_2 - \left(h_1 + \frac{1}{2}c_1^2 + gz_1\right)\delta m_1 + \delta W_s + \mathrm{d}E_{cv} \tag{3-11}$$

This is the general energy balance equation for any process in open systems. It is suitable for unsteady and steady flows as well as for processes in closed systems, reversible or irreversible.

If there is no change in the internal and potential energy of the closed system, then $\mathrm{d}E_{cv} = \mathrm{d}U$.

$$\delta Q = \mathrm{d}U + \delta W$$

3.6 Energy Balance for Steady-Flow Processes

3.6.1 Energy Balance Expression for Steady-Flow Processes

A large number of engineering devices, such as turbines, compressors, and nozzles operate for long periods of time under the same conditions once the transient startup periods complete and steady operations are established. And they are classified as steady-flow devices. Processes involving such devices can be represented reasonably by a somewhat idealized process, called the steady-flow process, during which a fluid flows through a control volume steadily. That is, the fluid properties can change from point to point within the control volume, but at any point, they remain constant during the entire process. (Remember, steady means no change with time)

During a steady-flow process, no intensive or extensive properties within the control volume change with time. Thus, the volume V, the mass m, and the total energy content E of the control volume remain constant, as shown in Fig. 3-11.

(1) The boundary work is zero for steady-flow systems (since V_{cv} = constant), and the

total mass entering the control volume must be constant and equal to the total mass leaving it ($m_{in}=m_{out}=m=$constant), that is, m_{cv} must the constant ($\Delta m_{cv}=0$).

(2) The fluid properties at an inlet or outlet remain constant during a steady-flow process, though the properties may be different at different inlets and outlets. They may even vary over the cross of an inlet or an outlet. However, all proper-constant, including the velocity an elevation, must remain with time at a fixed point at an inlet or outlet.

(3) The heat and work interactions between a steady-flow system and its surroundings do not change with time, either. Thus, the power delivered by a system and the rate of heat transferred to or from a system remain constant during a steady-flow process as well.

(4) During a steady-flow process, the total energy content entering the control volume must be equal to the total mass or energy leaving it, and thus the change in the total energy remains to be zero, $\Delta E_{cv}=0$.

Therefore, the amount of energy entering a control volume in all forms (by heat, work, or mass) must be equal to the amount of energy leaving it. Then the general energy balance reduces for a steady-flow process to be:

$$\dot{Q}=\left[(h_2-h_1)+\frac{1}{2}(c_2^2-c_1^2)+g(z_2-z_1)\right]\dot{m}+\dot{W}_s \tag{3-12a}$$

On a unit mass basis, it becomes:

$$q=(h_2-h_1)+\frac{1}{2}(c_2^2-c_1^2)+g(z_2-z_1)+w_s \tag{3-12b}$$

In which, $\mathrm{d}c^2/2+g\mathrm{d}z+\delta w_s$ belongs to mechanical energy and can be used directly in technique. Therefore, this part of energy carried by flowing fluid is also called technical work, denoted as w_t.

$$w_t=\frac{1}{2}\Delta c^2+g\Delta z+w_s \tag{3-13}$$

Thus, Eq. (3-12b) simplifies to be the following form.

$$q=\Delta h+w_t \tag{3-14}$$

In terms of the definition of enthalpy, $h=u+pv$:

$$w_t=q-\Delta h=(\Delta u+w)-(\Delta u+p_2v_2-p_1v_1)=w+(p_1v_1-p_2v_2)$$

$$w=w_t+\Delta(pv) \tag{3-15}$$

It indicates that the expansion work of the working medium in the steady flow process is implicit, showing that one part is consumed for the flow work $\Delta(pv)$ needed to maintain the flow of the working medium, one part is used to increase the macroscopic kinetic energy and gravitational potential energy of the working medium, and the rest is used as the shaft work output by the thermal equipment. Expansion work is the source of the transformation of thermal energy into work in the simple compressible system.

For a reversible steady flow process, we have:

$$\delta w_t=\delta q-\mathrm{d}h=\mathrm{d}u+p\mathrm{d}v-\mathrm{d}u-p\mathrm{d}v-v\mathrm{d}p$$

then:

$$\delta w_t = -vdp \tag{3-16a}$$

That is, the technical work output during a reversible process is:

$$w_t = -\int_1^2 vdp \tag{3-16b}$$

$$\delta q = dh - vdp \tag{3-17}$$

On p-v diagram, the area enclosed by the path of the process 1-2 and the vertical axis p is the same as the amount of technical work during the reversible process. That is $w_t =$ area 12341.

From Eq. (3-16), Eq. (3-17) and Fig. 3-12, we can find the relation among the technical work, expansion work and flow work on p-v diagram to be:

$$w_t = w + (p_1 v_1 - p_2 v_2) = \text{area } 12561 + \text{area } 41604 - \text{area } 23052$$

It means technical work is a function of process and its value is determined by the process as well as the initial and final states.

3.6.2 Some Steady-Flow Devices in Engineering Practice

Steady-flow conditions can be closely approximated by devices that are intended for continuous operation, such as turbines, pumps, boilers, condensers, and heat exchangers or power plants or refrigeration systems. They operate essentially under the same conditions for long periods of time.

In this section, some commonly encountered steady-flow devices are described, and the thermodynamic aspects of the flow through them are analyzed. The conservation of mass and the conservation of energy principles for these devices are illustrated with examples.

(1) Power machines

In steam, gas, or hydroelectric power plants, the device that drives the electric generator is the turbine. As the fluid passes through the turbine, work is done against the blades, which are attached to the shaft. As a result, the shaft rotates, and the turbine produces work.

$$g(z_2 - z_1) \approx 0, \frac{1}{2}(c_2^2 - c_1^2) \approx 0, q \approx 0$$

From Eq. (3-12b), then:

$$w_s = h_1 - h_2 \tag{3-18}$$

(2) Compressors

Compressors, as well as pumps and fans, are devices used to increase the pressure of fluids. Note that turbines produce power output, whereas compressors require power input. Work is supplied to these devices from an external source through a rotating shaft. Therefore, compressors involve work inputs. Similarly, the work input to the compressor is equal to the enthalpy increase of the compressed fluid. That is:

$$-w_s = h_2 - h_1 \tag{3-19}$$

(3) Heat exchangers

Heat exchangers are widely used in various industries. And they come in various designs. The simplest form of a heat exchanger is the tube and shell heat exchanger. It is a device where two fluids exchange heat without mixing. It is composed of two concentric pipes of different diameters. One fluid flows in the inner pipe, and the other in the annular space between the two pipes. Heat is transferred from the hot fluid to the cold one through the wall separating them. Sometimes the inner tube makes a couple of turns inside the shell to increase the heat transfer area, and thus the rate of heat transfer.

Heat exchangers typically involve no work interactions $w_s = 0$ and negligible kinetic and potential energy changes. That is, $\Delta c^2/2 \approx 0, g\Delta z \approx 0$, for each fluid stream.

$$q = h_2 - h_1 \tag{3-20}$$

(4) Nozzles and diffusers

Various nozzles and diffusers are commonly utilized in jet engines, rockets, spacecrafts, and even garden hoses. A nozzle is a device that increases the velocity of a fluid at the expense of pressure. A diffuser is a device that increases the pressure of a fluid by slowing it down. Obviously, nozzles and diffusers perform opposite tasks. The cross-sectional area of a nozzle decreases in the flow direction for subsonic flows and increases for supersonic flows. The reverse is true for diffusers. Since the fluid has high velocities. Thus it does not spend enough time in this short device. The rate of heat transfer between the flowing fluid and the surroundings is usually very small and insignificant. Then, the flow can be assumed to be adiabatic. Nozzles and diffusers typically involve no work interactions. And any change in potential energy is negligible. That is:

$$q \approx 0, w_s = 0, g(z_2 - z_1) \approx 0$$

$$\frac{1}{2}(c_2^2 - c_1^2) = h_1 - h_2 \tag{3-21}$$

(5) Mixing chambers

In engineering applications, we often encounter two streams of fluids mixing. The section where the mixing process takes place is commonly referred to as a mixing chamber. It does not have to be a distinct "chamber". An ordinary T-elbow or a Y-elbow in a shower, for example, serves as the mixing chamber for the cold-and hot-water streams (Fig. 3-14).

The conservation of mass principle for a mixing chamber requires that the sum of the incoming mass flow rates equal to the mass flow rate of the outgoing mixture.

Mixing chambers are usually well insulated and usually do not involve any kind of work. Also, the kinetic and potential energies of the fluid streams are usually negligible. That is, $W_s = 0, m\Delta c^2/2 \approx 0, mg\Delta z \approx 0$. Then all that left in the energy equation is the total energies of the incoming streams and the outgoing mixture. Substituting above conditions into Eq.(3-14b), we can obtain the conservation of energy equation for this case. That is:

$$\Delta H = 0 \quad \text{or} \quad H_1 + H_2 = H_3 \quad \text{or} \quad m_1 h_1 + m_2 h_2 = m_3 h_3 \tag{3-22}$$

And the mass conservation equation is:

$$m_1 + m_2 = m_3 \tag{3-23}$$

Combining Eq.(3-22) and Eq.(3-23), the energy conservation equation gives:

$$h_3 = \frac{m_1 h_1 + m_2 h_2}{m_1 + m_2} \tag{3-24}$$

(6) Throttling devices

Throttling devices are any kinds of flow-restricting devices that cause a significant pressure drop in a flowing fluid without obtaining any shaft work. For example, the valve used in your kitchen faucet is a throttling device, which reduces the water pressure from the pressure of the water main to atmospheric pressure. Some other familiar examples include ordinary adjustable valves, capillary tubes, and porous plugs (Fig. 3-15).

Usually the fluid passing through a throttling device is moving very rapidly. As it does not remain within the device long enough to absorb, or give off, much heat. Therefore, it is reasonable to assume it to be adiabatic. In addition, throttling devices provide no work, to the surroundings, and the change in potential energy, if any, is very small ($g\Delta z \approx 0$). In many cases, the increase in kinetic energy is insignificant, $\Delta c^2/2 \approx 0$. Then the conservation of energy equation for this single-stream steady-flow device reduces to:

$$h_1 = h_2 \tag{3-25}$$

That is, enthalpy values at the inlet and outlet of a throttling valve are the same. The magnitudes of energy are identical at the inlet and outlet. However, the fluid experiences large changes in velocity near the throttling device and thus the enthalpy varies during throttling process. This process is not an isenthalpic process.

第4章　理想气体的热力性质与热力过程

本章导学

与自然界发生的一切事物一样，能量转换不能孤立地进行，总是在一定的内部条件和外部条件下发生。工质是能量转换的媒介，是实现能量转换的内部条件，工质的性质影响能量转换的效果。在热功转换中，主要以气相物质为工质，因为它们具有显著的膨胀能力。气相工质分理想气体和实际气体，对实际气体性质的研究将在后续章节介绍。仅有内部条件尚不足以实现能量转换，还必须让工质在一定的热力设备中通过状态的变化即热力过程，来实现预定的能量转换。对于热力过程的研究，其实质就是研究外部条件对能量转换的影响。本章研究理想气体的热力性质（内部条件）及热力过程（外部条件）。

学习目标

熟练掌握并正确应用理想气体状态方程。理解理想气体比热的概念，了解真实比热、平均比热及定值比热，熟练掌握应用比热计算热量的方法，以及理想气体内能、焓和熵的变化的计算方法。熟练掌握4种基本热力过程状态参数之间的关系，系统与外界交换的热量、功量的计算。了解多变过程的基本状态参数之间的关系。能将各过程表示在 p-v 图和 T-s 图上，并能正确应用 p-v 图和 T-s 图判断热力过程的特点，即 Δu、Δh、q 及 w 等的正负。

重点与难点

(1) 真实比热、平均比热与定值比热。

(2) 热力过程计算公式。

(3) 应用 p-v 图与 T-s 图分析热力过程。

4.1　理想气体状态方程

4.1.1　理想气体与实际气体

理想气体是一种经科学抽象、实际中不存在的假想气体。它被假设为：其分子是些弹性的、不具有体积的质点，分子体积与气体的总体积相比可忽略不计，分子间相互没有作用力，分子之间的互相碰撞以及分子与容器壁的碰撞都是弹性碰撞。

实验已经证明，当压力足够低、温度足够高时，气体的比容就足够大。此时，气体分子本身所占的体积相对气体总体积而言足够小，可忽略不计；气体分子之间的平均距离相当大，分子之间的相互作用力与分子间平均距离呈反比，因而分子之间的相互作用力也可忽略。可见，对气体提出这种理想化的模型是合理的。

对于工程上的各种热力过程来说，大多数情况下，气体的压力不太高，温度也不太低，可以作为理想气体处理，而不会带来较大误差。理想气体的假设，能够满足热力工程设计计算的要求，而计算的复杂性和工作量则大大减小。这是理想气体之所以在工程热力学中占有特殊地位的主要原因。

4.1.2　理想气体的状态方程

根据分子运动理论和对理想气体的假定，得出 1 kg 理想气体的状态方程为：

$$pv = RT \tag{4-1}$$

对质量为 m (kg)的理想气体，其状态方程形式为：

$$pmv = mRT$$

$$pV = mRT \tag{4-2}$$

式中　p——气体的绝对压力，Pa；

v——气体的比容，m^3/kg；

V——气体的体积，m^3；

T——气体的热力学温度，K；

R——气体常数，J/(kg・K)，只与气体种类有关，与气体状态无关。

在国际单位制中，常以 mol(摩尔)作为物质的量单位。1 kmol 物质的质量称为摩尔质量，以 M 表示，单位为 kg/kmol。1 kmol 物质的体积称为摩尔体积，以 V_m 表示，$V_m = Mv$。将式(4-1)两边乘以摩尔质量 M，即

$$pMv = MRT$$

令 $R_0 = MR$，得：

$$pV_m = R_0 T \tag{4-3}$$

R_0 称为摩尔气体常数，单位是 J/(kmol・K)，对所有的气体其值是一样的，可以在某一状态下求出。例如，在标准状态下(即 $p_0 = 1.013\ 25 \times 10^5$ Pa，$T_0 = 273.15$ K)，所有气体的摩尔体积 $V_m = 22.414\ m^3/kmol$，可求得摩尔气体常数 $R_0 = 8\ 314$ J/(kmol・K)。

这样一来，只要已知气体的摩尔质量，可以通过 $R = R_0/M$ 求得该气体的气体常数 R。例如，对于空气，有：

$$R = \frac{R_0}{M} = \frac{8\ 314}{28.97} = 287\ \text{J/(kg}\cdot\text{K)}$$

【例 4-1】　计算空气在标准状态下的比容和密度。

【解】　已知空气的气体常数 $R = 287$ J/(kg・K)。

在标准状态下，$p = 1.013\ 25 \times 10^5$ Pa，$T = 273.15$ K，根据理想气体状态方程，比容为：

$$v = \frac{RT}{p} = \frac{287 \times 273.15}{101\ 325} = 0.774\ (m^3/kg)$$

密度为：

$$\rho = \frac{1}{v} = 1.292\ (kg/m^3)$$

4.2 理想气体热力学性质

4.2.1 比热

4.2.1.1 比热的定义与单位

物体温度升高或降低 1 K 所需要吸收或放出的热量称为该物体的热容量，简称热容，单位为 kJ/K。

如果工质在一个微元热力过程中吸收或放出热量 δQ，温度升高或降低 $\mathrm{d}T$，则该工质的热容可表示为：

$$C = \frac{\delta Q}{\mathrm{d}T} \tag{4-4}$$

单位物量工质的热容称为该物体的比热容，简称比热。根据工质的物量单位不同，比热有以下几种：① 质量比热：1 kg 物质的比热，以 c 表示，单位为 kJ/(kg · K)；② 摩尔比热：1 kmol 物质的比热，以 c_m表示，单位为 kJ/(kmol · K)；③ 体积比热：标准状态下 1 m^3物质的比热，以 c'表示，单位为 kJ/(m^3 · K)。

三种比热之间的关系为：

$$22.4\,c' = c_m = Mc$$

因为热量是过程量，过程不同，相同温度变化情况下一定量的物质所吸收或放出的热量就不同，比热的数值也不同，即热力过程不同，比热也不同，在各种热力过程中，以定压比热和定容比热最为常用，也最重要。它们不仅是热工计算所需的物理量，而且还是计算内能变化、焓变和熵变的依据。

4.2.1.2 定压比热与定容比热

气体加热在容积不变的情况下进行，加入的热量全部用于增加气体的内能，使气体温度升高。气体加热在压力不变的情况下进行，加入的热量部分用于增加气体的内能，使其温度升高，部分用于推动活塞而对外做膨胀功。

由热力学第一定律知，对于可逆过程有：

$$\delta q = \mathrm{d}u + p\mathrm{d}v$$

$$\delta q = \mathrm{d}h - v\mathrm{d}p$$

因为定容过程 $\mathrm{d}v=0$，膨胀功 $\delta w = p\mathrm{d}v = 0$，定压过程 $\mathrm{d}p = 0$，技术功 $\delta w_t = -v\mathrm{d}p = 0$，根据比热定义，有：

$$c_v = \left(\frac{\delta q}{\mathrm{d}T}\right)_v = \left(\frac{\mathrm{d}u + p\mathrm{d}v}{\mathrm{d}T}\right)_v = \left(\frac{\mathrm{d}u}{\mathrm{d}T}\right)_v \tag{4-5}$$

$$c_p = \left(\frac{\delta q}{\mathrm{d}T}\right)_p = \left(\frac{\mathrm{d}h - v\mathrm{d}p}{\mathrm{d}T}\right)_p = \left(\frac{\mathrm{d}h}{\mathrm{d}T}\right)_p \tag{4-6}$$

由焓的定义知，对于理想气体有：

$$h = u + pv = u + RT$$

并且由上述定容比热及定压比热的计算式还可推得 c_p与 c_v之间的关系：

$$c_p = \frac{\mathrm{d}h}{\mathrm{d}T} = \frac{\mathrm{d}(u + pv)}{\mathrm{d}T} = \frac{\mathrm{d}u}{\mathrm{d}T} + \frac{\mathrm{d}(RT)}{\mathrm{d}T} = c_v + R$$

即

$$c_p - c_v = R \tag{4-7}$$

给式(4-7)两边乘以摩尔质量 M,可得：

$$Mc_p - Mc_v = R_0 \tag{4-8}$$

式(4-7)和式(4-8)称为迈耶公式,表示定压比热与定容比热之间的关系。可见,定压比热一般大于定容比热。这是因为气体在定压下受热膨胀,所吸收的热量在使内能增加的同时,还克服外力做功。定压比热与定容比热之间的差值就是使 1 kg 气体在定压下温度升高 1 K 时对外所做的功。

定压比热与定容比热的比值称为比热比,也被称为绝热指数。它是热力学理论研究和热工计算中常用的一个重要参数,用 κ 表示。

$$\kappa = \frac{c_p}{c_v} \tag{4-9}$$

将式(4-9)代入迈耶公式,得：

$$c_v = \frac{R}{\kappa - 1}, c_p = \frac{\kappa R}{\kappa - 1}$$

4.2.1.3　真实比热、平均比热与定值比热

(1) 真实比热

气体的比热与其压力和温度有关,但是理想气体分子间相互作用力很小,压力对比热的影响可以忽略,即理想气体的比热仅是温度的单值函数。一般来说,温度越高,比热越大,这是因为温度升高,双原子和多原子分子内部的原子振动动能增大。由实验可以测得它们之间的曲线关系,如图 4-1 所示。该曲线可以比较真实地反映比热与温度的关系,不同温度下的比热称为真实比热。

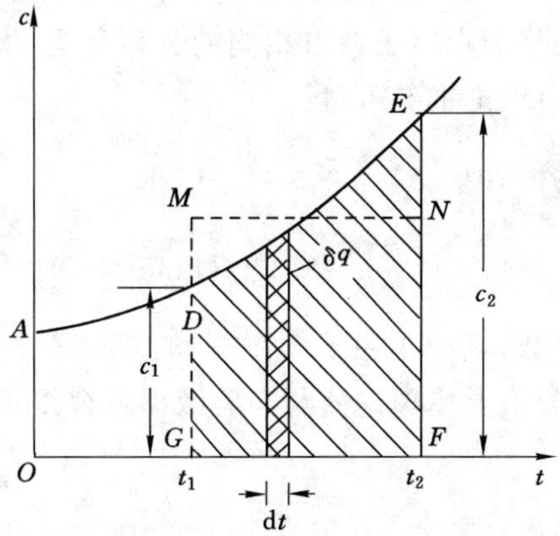

图 4-1　比热与温度的关系

工程上,这种函数关系通常近似表示成温度的多项式的形式,一般选三次多项式,即

$$c = a_0 + a_1 T + a_2 T^2 + a_3 T^3$$

式中,a_0,a_1,a_2,a_3 为常数,一般与工质的种类和温度范围有关,其值由实验确定,从物性手册上可查得。

根据真实比热的表达式,通过积分可计算每千克理想气体温度从 t_1 升高到 t_2 所吸收的热量。例如,对于定压过程,利用 $\delta q_p = c_p \mathrm{d}t$ 的定义式积分：

$$q=\int_{t_1}^{t_2}c_p\mathrm{d}t=\int_{T_1}^{T_2}(a_0+a_1T+a_2T^2+a_3T^3)\mathrm{d}T$$

可求得热力过程所需的热量。

(2) 平均比热

为了便于工程应用，引入平均比热的概念。每千克气体从温度 t_1 升高到 t_2 所需要的热量 q 除以温度变化量(t_2-t_1)所得的商，称为该气体在 $t_1\sim t_2$ 温度范围内的平均比热，表达式为：

$$c_m\Big|_{t_1}^{t_2}=\frac{q}{t_2-t_1}=\frac{\int_{t_1}^{t_2}c\mathrm{d}t}{t_2-t_1}$$

$$q=c_m\Big|_{t_1}^{t_2}\times(t_2-t_1)$$

由图 4-1，面积 $DEFGD$(表示气体由温度 t_1 升高到 t_2 吸收的热量)需要通过积分计算，比较复杂，为了简化计算，也可以用一块大小相等的矩形面积 $MNFGM$ 来代替，于是有：

$$q=\int_{t_1}^{t_2}c\mathrm{d}t=\overline{MG}(t_2-t_1)$$

矩形高度 $\overline{MG}$ 就是在 $t_1\sim t_2$ 温度范围内真实比热的平均值，即平均比热。

为了应用方便，可将各种常用气体的平均比热计算出来，并列成表格，用时可以直接查表。

(3) 定值比热

当气体温度较低且变化范围不大，或对计算精确度要求不高时，可以不考虑温度对比热的影响，将比热近似视为定值，不随温度而变化，这种情况下的比热称为定值比热。根据气体分子运动论及能量按运动自由度均分的原则，理想气体的比热只取决于气体的分子结构，而与气体所处状态无关。凡分子中原子数目相同的气体其运动自由度也相同，它们的摩尔比热相等，称为定值比热。由理论推导可得：

摩尔定容比热： $$Mc_v=\frac{i}{2}R_0,\ \mathrm{J/(kmol\cdot K)}$$

摩尔定压比热： $$Mc_p=\frac{2+i}{2}R_0,\ \mathrm{J/(kmol\cdot K)}$$

式中，i 为分子运动的自由度数，其中单原子气体 $i=3$，双原子气体 $i=5$，多原子气体 $i=7$。

实验表明，对于单原子分子上述理论值与实验数据符合得较好。

4.2.2 内能、焓和熵

4.2.2.1 内能和焓

理想气体分子间无作用力，不存在内势能。因此，理想气体的内能即等于内动能，是温度的单值函数，$u=f(T)$。根据焓的定义可知，对于理想气体，焓也是温度的单值函数，$h=f(T)$。

内能和焓是工质的状态参数，因此在热力过程中内能和焓的变化只与工质的初、终状态有关，而与过程的路径无关。对于理想气体来说，热力过程中工质的内能和焓的变化只取决于初、终状态的温度，而与所经历的过程和所处的状态均无关，初、终温度一定的过程，工质内能和焓的变化如图 4-2 所示。

在图 4-2 所示的 p-v 图上，有两条温度分别为 T_1 及 T_2 的等温线。根据上述分析可知，

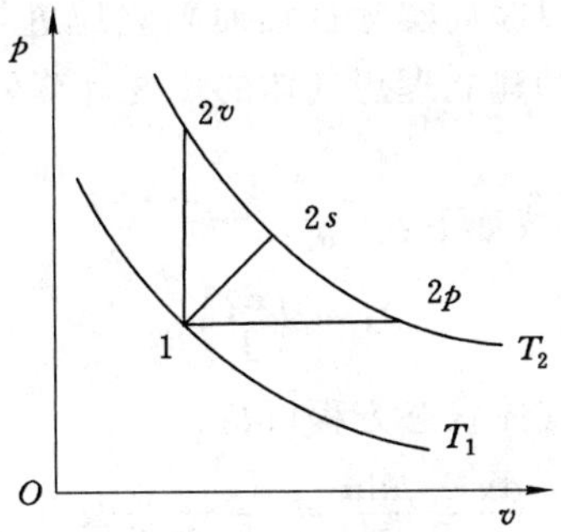

图 4-2　理想气体的内能变化与焓变化

在同一条等温线上，理想气体各点的内能值、焓值分别相等，而与所处的状态无关。即

$$u_{2s}=u_{2v}=u_{2p}\cdots,h_{2s}=h_{2v}=h_{2p}\cdots$$

显然，在各种参数坐标图上，理想气体的等温线也就是等内能线和等焓线。同时也可得出：在这两条等温线之间所经历的任何过程，理想气体的内能变化量都相等，焓变化量也均相等。进一步有：对于理想气体，任何一个过程的内能变化量等于相同温度变化条件下的定容过程的内能变化量，而任何一个过程的焓变化量则等于相同温度变化条件下的定压过程的焓变化量。

因此，根据理想气体定容比热和定压比热的定义式，可以得到微元热力过程内能和焓的增量：

$$du=c_v dT \tag{4-10}$$

$$dh=c_p dT \tag{4-11}$$

虽然式(4-10)和式(4-11)中内能和焓的增量计算用的分别是定容比热和定压比热，由于理想气体内能、焓的特性，故这两式不但适用于定容过程和定压过程，而且适用于理想气体的任何过程。

理想气体任意过程的内能和焓的变化量可分别由式(4-10)和式(4-11)积分求得，即

$$\Delta u=u_2-u_1=\int_{T_1}^{T_2}c_v dT \tag{4-12}$$

$$\Delta h=h_2-h_1=\int_{T_1}^{T_2}c_p dT \tag{4-13}$$

由此可见，理想气体的温度由 T_1 变化到 T_2，不论经过何种过程，也无须考虑压力和比容是否变化，其内能及焓的变化量即可按式(4-12)和式(4-13)确定。值得指出，对于非理想气体，式(4-12)只能用来计算定容过程的内能变化量，式(4-13)只能用来计算定压过程的焓的变化量，对于其他过程这些公式都是不适用的。

对于理想气体，若 c_p，c_v 为定值，则有：

$$\Delta u=c_v\Delta T=c_v(T_2-T_1) \tag{4-14}$$

$$\Delta h=c_p\Delta T=c_p(T_2-T_1) \tag{4-15}$$

4.2.2.2　熵

熵的热力性质与内能及焓的热力特性不同，理想气体的熵不仅仅是温度的函数，还与压力或比容有关。熵也是状态参数，当初、终状态确定时，系统熵变也就确定了，与过程性质及

路径无关。所以熵变的计算可以脱离实际过程而独立地进行。因此，可以选择最简单的热力学模型——可逆过程来推导和建立理想气体的熵变计算公式。由此得出的结论适用于具有相同初、终状态的任何过程。

在可逆的条件下，熵的定义式如下：

$$\mathrm{d}s=\left(\frac{\delta q}{T}\right)_{\mathrm{re}}$$

由热力学第一定律及理想气体状态方程可得：

$$\mathrm{d}s=\left(\frac{\delta q}{T}\right)_{\mathrm{re}}=\frac{\mathrm{d}u+p\mathrm{d}v}{T}=\frac{\mathrm{d}u}{T}+\frac{p}{T}\mathrm{d}v=c_{\mathrm{v}}\frac{\mathrm{d}T}{T}+R\frac{\mathrm{d}v}{v}\tag{4-16}$$

$$\mathrm{d}s=\left(\frac{\delta q}{T}\right)_{\mathrm{re}}=\frac{\mathrm{d}h-v\mathrm{d}p}{T}=\frac{\mathrm{d}h}{T}-\frac{v}{T}\mathrm{d}p=c_{\mathrm{p}}\frac{\mathrm{d}T}{T}-R\frac{\mathrm{d}p}{p}\tag{4-17}$$

将理想气体状态方程 $pv=RT$ 微分，可得：

$$\frac{\mathrm{d}p}{p}+\frac{\mathrm{d}v}{v}=\frac{\mathrm{d}T}{T}\tag{4-18}$$

将式(4-18)代入式(4-16)或式(4-17)，可得：

$$\begin{aligned}\mathrm{d}s&=c_{\mathrm{v}}\frac{\mathrm{d}T}{T}+R\frac{\mathrm{d}v}{v}\\&=c_{\mathrm{v}}\left(\frac{\mathrm{d}p}{p}+\frac{\mathrm{d}v}{v}\right)+R\frac{\mathrm{d}v}{v}\\&=c_{\mathrm{v}}\frac{\mathrm{dp}}{p}+(c_{\mathrm{v}}+R)\frac{\mathrm{d}v}{v}\\&=c_{\mathrm{v}}\frac{\mathrm{dp}}{p}+c_{\mathrm{p}}\frac{\mathrm{d}v}{v}\end{aligned}$$

或

$$\begin{aligned}\mathrm{d}s&=c_{\mathrm{p}}\frac{\mathrm{d}T}{T}-R\frac{\mathrm{d}p}{p}\\&=c_{\mathrm{p}}\left(\frac{\mathrm{d}p}{p}+\frac{\mathrm{d}v}{v}\right)-R\frac{\mathrm{d}p}{p}\\&=(c_{\mathrm{p}}-R)\frac{\mathrm{d}p}{p}+c_{\mathrm{p}}\frac{\mathrm{d}v}{v}\\&=c_{\mathrm{v}}\frac{\mathrm{d}p}{p}+c_{\mathrm{p}}\frac{\mathrm{d}v}{v}\end{aligned}\tag{4-19}$$

即

$$\Delta s=\int_{T_1}^{T_2}c_{\mathrm{v}}\frac{\mathrm{d}T}{T}+R\ln\frac{v_2}{v_1}\tag{4-20}$$

$$\Delta s=\int_{T_1}^{T_2}c_{\mathrm{p}}\frac{\mathrm{d}T}{T}-R\ln\frac{p_2}{p_1}\tag{4-21}$$

$$\Delta s=\int_{p_1}^{p_2}c_{\mathrm{v}}\frac{\mathrm{d}p}{p}+\int_{v_1}^{v_2}c_{\mathrm{p}}\frac{\mathrm{d}v}{v}\tag{4-22}$$

以上分别是以 T 和 v，T 和 p，p 和 v 表示的熵变的计算式。上述 3 个计算公式是等效的，已知初、终状态参数时，可以在上述 3 个公式中选择适当的公式精确计算 Δs。若考虑比热容与温度的关系，采用式(4-20)至式(4-22)积分计算熵变值会相对比较麻烦。工程上当温度变化范围不大时，可采用定值比热，即认为 c_{p}及 c_{v}为定值，这时有：

$$\Delta s = c_v \ln \frac{T_2}{T_1} + R \ln \frac{v_2}{v_1} \tag{4-23}$$

$$\Delta s = c_p \ln \frac{T_2}{T_1} - R \ln \frac{p_2}{p_1} \tag{4-24}$$

$$\Delta s = c_v \ln \frac{p_2}{p_1} + c_p \ln \frac{v_2}{v_1} \tag{4-25}$$

4.3 理想气体的典型热力过程

4.3.1 基本热力过程

理想气体的基本热力过程包括定容过程、定压过程、定温过程和定熵过程。

4.3.1.1 定容过程

气体比容保持不变的过程称为定容过程。例如，在刚性密闭容器(如压气机储气罐)中气体的加热或冷却过程。定容过程方程式为：

$$v = 常数$$

结合理想气体状态方程，可得：

$$\frac{p}{T} = \frac{R}{v} = 常数 \quad 或 \quad \frac{p_2}{p_1} = \frac{T_2}{T_1} \tag{4-26}$$

熵变：

$$\Delta s = \int_1^2 c_v \frac{dT}{T} + R \ln \frac{v_2}{v_1} = \int_1^2 c_v \frac{dT}{T} \tag{4-27}$$

若采用定值比热，则有：

$$\Delta s = c_v \ln \frac{T_2}{T_1} \tag{4-28}$$

因为 v=常数，$dv=0$，所以定容过程中气体的膨胀功为零，即

$$w = \int_1^2 p dv = 0 \tag{4-29}$$

技术功为：

$$w_t = -\int_1^2 v dp = v(p_1 - p_2) \tag{4-30}$$

即定容流动过程中技术功等于流体在进、出口处流动功之差。

吸、放热量为：

$$q = \Delta u = \int_1^2 c_v dT = c_v (T_2 - T_1) \tag{4-31}$$

由式(4-31)可知，定容过程中工质吸收的热量全部转换为工质的内能。这是定容能量转换的特点。

定容过程在 p-v 图上是一条垂直于 v 轴的直线，如图 4-3(a)所示，1—2 表示定容放热过程，1—2′表示定容吸热过程，定容线下面的面积为零，表明定容过程的膨胀功为零。

对于可逆过程，由熵的计算式可知：

$$T ds = c_v dT + p dv$$

因为 $dv = 0$，则有：

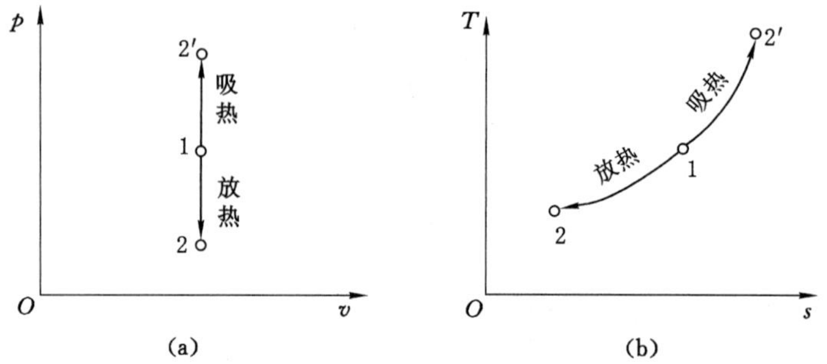

图 4-3 定容过程的 p-v 图与 T-s 图

$$\frac{dT}{ds}=\frac{T}{c_v} \tag{4-32}$$

可见，在 T-s 图上定容线是斜率为 T/c_v的指数函数曲线。斜率始终大于零，且斜率随温度的升高而增大。因此，在 T-s 图上的定容线随温度的升高越来越陡，如图 4-3(b)所示。

进一步推导定容过程 T 与 s 之间的关系为：

$$T=T_0 e^{(s-s_0)/c_v}$$

4.3.1.2 定压过程

气体压力保持不变的过程称为定压过程。

定压过程方程式为：

$$p=常数 \tag{4-33}$$

结合理想气体状态方程，可得：

$$\frac{v_2}{v_1}=\frac{T_2}{T_1} \tag{4-34}$$

熵变：

$$\Delta s=\int_1^2 c_p \frac{dT}{T} \tag{4-35}$$

若采用定值比热，则有：

$$\Delta s=c_p \ln\frac{T_2}{T_1} \tag{4-36}$$

定压过程中气体的膨胀功为：

$$w=\int_1^2 p dv=p(v_2-v_1)=R(T_2-T_1) \tag{4-37}$$

技术功为：

$$w_t=-\int_1^2 v dp=0 \tag{4-38}$$

吸、放热量为：

$$q=\Delta h=\int_1^2 c_p dT=c_p(t_2-t_1) \tag{4-39}$$

式(4-39)表明，定压过程中工质吸收或放出的热量等于工质的焓变。

定压过程在 p-v 图上是一条垂直于 p 轴的直线，如图 4-4(a)所示，1—2 表示定压膨胀过程，1—2′表示定压压缩过程。

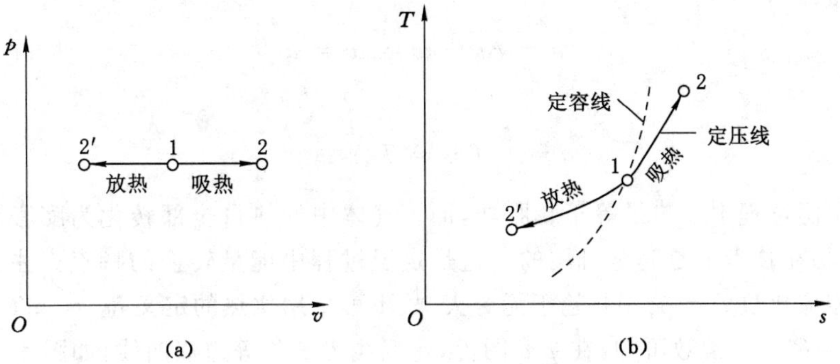

图 4-4　定压过程的 p-v 图与 T-s 图

由熵的计算式可知定压过程中 $T\mathrm{d}s = c_\mathrm{p}\mathrm{d}T - v\mathrm{d}p = c_\mathrm{p}\mathrm{d}T$，整理得：

$$\frac{\mathrm{d}T}{\mathrm{d}s} = \frac{T}{c_\mathrm{p}} \tag{4-40}$$

因此，在 T-s 图上定压线是斜率为 T/c_p 的指数函数曲线。斜率始终大于零，且斜率随温度的升高而增大。但由于 $\dfrac{T}{c_\mathrm{p}} < \dfrac{T}{c_\mathrm{v}}$，因此，在 T-s 图上，定压线 1—2 比定容线更平坦一些，如图 4-4(b)所示。

进一步推导定压过程 T 与 s 之间的关系为：

$$T = T_0 \mathrm{e}^{(s-s_0)/c_\mathrm{p}}$$

4.3.1.3　定温过程

气体温度保持不变的过程称为定温过程，过程方程式为：$T=$常数。对于理想气体，也可写成：$pv=$常数。

结合理想气体状态方程，可得：

$$\frac{p_1}{p_2} = \frac{v_2}{v_1} \tag{4-41}$$

$$\Delta u = \int_1^2 c_\mathrm{v}\mathrm{d}T = 0 \tag{4-42}$$

$$\Delta h = \int_1^2 c_\mathrm{p}\mathrm{d}T = 0 \tag{4-43}$$

可见，定温过程中，气体的内能和焓值不变。

熵变：

$$\Delta s = \int_1^2 R\frac{\mathrm{d}v}{v} = R\ln\frac{v_2}{v_1} = R\ln\frac{p_1}{p_2} \tag{4-44}$$

定温过程的膨胀功为：

$$w = \int_1^2 p\mathrm{d}v = RT\ln\frac{v_2}{v_1} = RT\ln\frac{p_1}{p_2} \tag{4-45}$$

技术功为：

$$w_\mathrm{t} = -\int_1^2 v\mathrm{d}p = RT\ln\frac{p_1}{p_2} = w \tag{4-46}$$

可见，定温过程中膨胀功等于技术功。

吸、放热量为：

$$q = \Delta u + w = w = w_t \tag{4-47}$$

或

$$q = \int_1^2 T\mathrm{d}s = T(s_2 - s_1) \tag{4-48}$$

可见，定温过程中的热量等于膨胀功，加入气体中的热量全部转化为膨胀功（技术功），并且该膨胀功在技术上是完全可用的。这是定温过程中能量转换的特点。注意：定温过程温升为零，相应比热 $c = \delta q/\mathrm{d}T$ 趋于无穷大，故不可采用比热的定义 $\delta q = c\mathrm{d}T$ 计算热量。

由 $pv = RT =$ 常数可知，在 p-v 图上，定温线为一条等边双曲线，如图 4-5(a)所示。气体膨胀，压力降低。1—2 表示定温吸热膨胀过程，1—2′表示定温压缩放热过程。在 T-s 图上定温过程是一条垂直于 T 轴的直线，如图 4-5(b)所示。

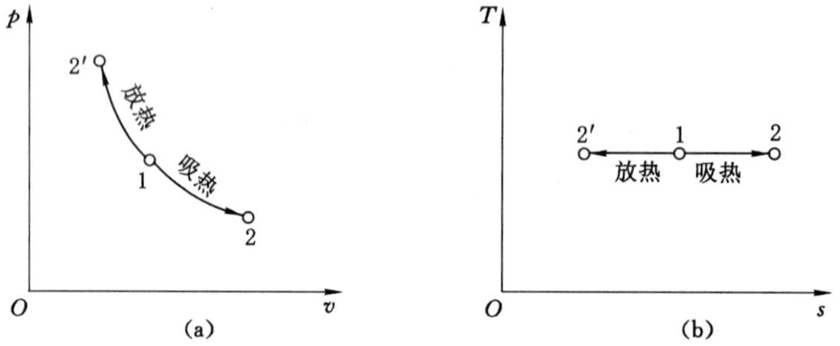

图 4-5　定温过程的 p-v 图和 T-s 图

【例 4-2】　2 m³空气初始时压力 $p_1=0.2$ MPa，温度 $t_1=150$ ℃，在定温下膨胀到原体积的两倍，设比热为定值。求终压 p_2，过程中的熵变、膨胀功及热量。

【解】　由定温过程的状态参数关系式可得：

$$p_2 = p_1 \frac{v_1}{v_2} = p_1 \frac{V_1}{V_2} = 0.2 \times \frac{1}{2} = 0.1 \text{ (MPa)}$$

过程中的熵变为：

$$\Delta s = R\ln\frac{v_2}{v_1} = 287 \times \ln 2 = 198.93 \text{ [J/(kg·K)]}$$

空气质量为：

$$m = \frac{p_1 V_1}{RT_1} = \frac{0.2 \times 10^6 \times 2}{287 \times (273 + 150)} = 3.29 \text{ (kg)}$$

$$\Delta S = m\Delta s = 3.29 \times 198.93 = 654.48 \text{ (J/K)} = 0.654 \text{ kJ/K}$$

过程中的膨胀功为：

$$W = mw = T\Delta S = (273 + 150) \times 0.654 = 276.64 \text{ (kJ)}$$

定温过程中吸收的热量等于过程的膨胀功，为：

$$Q = W = 276.64 \text{ kJ}$$

4.3.1.4　绝热过程

气体与外界没有热量交换的过程称为绝热过程。

对于可逆绝热过程，$ds = \frac{\delta q}{T} = 0$，也就是说可逆绝热过程是定熵过程。

依据能量守恒关系，对于定熵过程，$\delta q = c_v dT + p dv = 0$ 及 $\delta q = c_p dT - v dp = 0$，移项后两式相除可得：

$$\frac{dp}{p} + \frac{c_p}{c_v}\frac{dv}{v} = \frac{dp}{p} + \kappa\frac{dv}{v} = 0$$

或

$$\frac{dp}{p} = -\frac{c_p}{c_v}\frac{dv}{v}$$

若 c_p，c_v取定值或平均值，则 κ 为定值或平均值。上式积分得：

$$\ln p + \kappa \ln v = 常数$$

或

$$\ln pv^{\kappa} = 常数$$

即：

$$pv^{\kappa} = 常数 \tag{4-49}$$

式(4-49)即定熵过程方程式。其中，κ 为比热比，在这里也称为绝热指数。当比热取为定值比热时，对于单原子气体，$\kappa=1.667$；对于双原子气体，$\kappa=1.4$；对于多原子气体，$\kappa=1.286$。

进而可得定熵过程中初、终状态参数之间的关系如下：

$$\frac{p_2}{p_1} = \left(\frac{v_1}{v_2}\right)^{\kappa} \tag{4-50}$$

$$\frac{T_2}{T_1} = \left(\frac{v_1}{v_2}\right)^{\kappa-1} \tag{4-51}$$

$$\frac{T_2}{T_1} = \left(\frac{p_2}{p_1}\right)^{\frac{\kappa-1}{\kappa}} \tag{4-52}$$

可见，当气体绝热膨胀($v_2 > v_1$)时，p 与 T 均降低；当气体被绝热压缩($v_2 < v_1$)时，p 与 T 均增高。定熵过程中压力增大，温度升高，压力与温度的变化趋势是一致的。

可逆绝热过程为定熵过程，因而有：

$$q = \int_1^2 T ds = T(s_2 - s_1) = 0 \tag{4-53}$$

膨胀功为：

$$w = q - \Delta u = -\Delta u \tag{4-54}$$

技术功为：

$$w_t = q - \Delta h = -\Delta h \tag{4-55}$$

可见，定熵过程中，系统对外所做的膨胀功等于工质的内能的减少量，输出的技术功等于工质的焓降。

若取定值比热，则有：

$$\begin{aligned} w &= -\Delta u = c_v(T_1 - T_2) = \frac{R}{\kappa-1}(T_1 - T_2) \\ &= \frac{RT_1}{\kappa-1}\left[1 - \left(\frac{p_2}{p_1}\right)^{\frac{\kappa-1}{\kappa}}\right] = \frac{RT_1}{\kappa-1}\left[1 - \left(\frac{v_1}{v_2}\right)^{\kappa-1}\right] \end{aligned} \tag{4-56}$$

$$w_t = -\Delta h = c_p(T_1 - T_2) = \frac{\kappa R}{\kappa - 1}(T_1 - T_2)$$

$$= \frac{\kappa R T_1}{\kappa - 1}\left[1 - \left(\frac{p_2}{p_1}\right)^{\frac{\kappa-1}{\kappa}}\right] = \frac{\kappa R T_1}{\kappa - 1}\left[1 - \left(\frac{v_1}{v_2}\right)^{\kappa-1}\right] \tag{4-57}$$

对比可知：

$$w_t = \kappa w$$

这表明定熵过程中的技术功为膨胀功的 κ 倍。

在 p-v 图上定熵线为高次双曲线，根据过程方程式可导出定熵线在 p-v 图上的斜率：

$$\frac{dp}{dv} = -\kappa\frac{p}{v}$$

可知，定熵线斜率为负值。而定温线在 p-v 图上的斜率为：

$$\frac{dp}{dv} = -\frac{p}{v}$$

由于 κ 值总是大于 1，定熵线斜率的绝对值大于定温线斜率的绝对值，即定熵线比定温线更陡一些，如图 4-6(a)所示。1—2 为定熵膨胀过程，压力和温度均降低；1—2′ 为定熵压缩过程，压力增加，温度升高。在 T-s 图上，定熵线为一条垂直于 s 轴的直线。

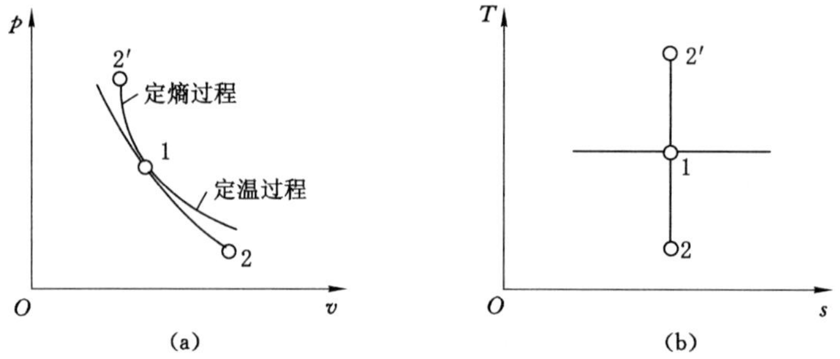

图 4-6　定熵过程的 p-v 图和 T-s 图

4.3.2　多变过程

上述四种典型热力过程的共同特点是过程进行中有一个状态参数保持不变或者与外界无热量交换。但实际热力过程中，各种参数都会变化，而且变化规律与上述四种典型的热力过程不符。因此，有必要进一步研究既能近似于大多数实际过程，而状态参数又仍按一定规律变化的理想过程。通过研究发现，许多过程可以近似地用下面的关系式描述：

$$pv^n = 常数 \tag{4-58}$$

式中，n 为多变指数。满足这一规律的过程就称为多变过程，式(4-58)即多变过程方程式。

不同的多变过程，具有不同的 n 值。理论上，n 可以是 $-\infty \sim +\infty$ 之间的任何一个实数，相应的多变过程也可以有无穷多种。当多变指数为某些特定的值时，多变过程便表现为某些典型的热力过程。如：

$n = 0$：$p =$ 常数，为定压过程；

$n = 1$：$pv =$ 常数，为定温过程；

$n = \kappa$：$pv^\kappa =$ 常数，为定熵过程；

$n = \pm\infty$：$v =$ 常数，为定容过程。

工程中，一个复杂的实际过程往往无法用一个确定的 n 来表示。对这类情况，过程可近似看作由几个多变过程组成，每个过程的多变指数 n 各不相同。

由多变过程方程式 pv^n = 常数可得 p-v 图中多变过程曲线的斜率：

$$\frac{\mathrm{d}p}{\mathrm{d}v}=-n\frac{p}{v} \tag{4-59}$$

如果从同一初态出发，其 p，v 值相同，过程线的斜率取决于 n 值，如图 4-7(a)所示。即

$n=0$：$\frac{\mathrm{d}p}{\mathrm{d}v}=0$，定压线为一水平线；

$n=1$：$\frac{\mathrm{d}p}{\mathrm{d}v}=-\frac{p}{v}<0$，定温线为一斜率为负的等边双曲线；

$n=\kappa$：$\frac{\mathrm{d}p}{\mathrm{d}v}=-\kappa\frac{p}{v}$，定熵线为一高次双曲线，其斜率的绝对值大于定温线；

$n=\pm\infty$：$\frac{\mathrm{d}p}{\mathrm{d}v}=\infty$，定容线为一垂直线。

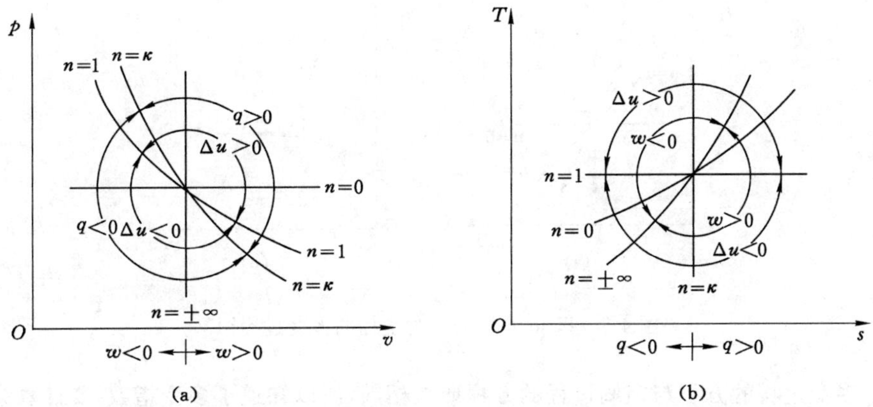

图 4-7　多变过程的 p-v 图和 T-s 图

由图 4-7(a)可以看出，在 p-v 图上，多变过程线的分布规律为：从定容线出发，n 由 $-\infty\to0\to1\to\kappa\to+\infty$，按顺时针方向递增。

在 T-s 图上，多变过程线的斜率可根据 $\delta q=T\mathrm{d}s=c_\mathrm{n}\mathrm{d}T$ 得出，即

$$\frac{\mathrm{d}T}{\mathrm{d}s}=\frac{T}{c_\mathrm{n}}$$

式中，c_n 为多变比热，$c_\mathrm{n}=\frac{n-\kappa}{n-1}c_\mathrm{v}$。多变过程线的斜率也随 n 而变化，即

$n=0$：$c_\mathrm{n}=c_\mathrm{p}$，$\frac{\mathrm{d}T}{\mathrm{d}s}=\frac{T}{c_\mathrm{p}}>0$，定压线为一斜率为正的指数曲线；

$n=1$：$c_\mathrm{n}\to\infty$，$\frac{\mathrm{d}T}{\mathrm{d}s}=0$，定温线为一水平线；

$n=\kappa$：$c_\mathrm{n}\to0$，$\frac{\mathrm{d}T}{\mathrm{d}s}\to\infty$，定熵线为一垂直线；

$n=\pm\infty$：$c_\mathrm{n}=c_\mathrm{v}$，$\frac{\mathrm{d}T}{\mathrm{d}s}=\frac{T}{c_\mathrm{v}}>0$，定容线也是一斜率为正的指数曲线，从同一状态出发，

定容线的斜率大于定压线的斜率。

在 T-s 图上，多变过程线的分布规律也是从定容线开始，多变指数 n 按顺时针方向递增。

对式(4-59)积分得：

$$\ln p + n\ln v = 常数$$

即

$$\ln p = -n\ln v + 常数$$

上式表明，如果将多变过程画在以 $\ln p$ 为纵坐标，以 $\ln v$ 为横坐标的平面坐标系中，那么所有的多变过程线都是直线(图 4-8)，而每条直线的斜率正好等于多变指数的负值，即

$$\frac{\mathrm{d}(\ln p)}{\mathrm{d}(\ln v)} = -n$$

这就提供了分析任意过程的一种方法：将任意过程画到 $\ln p$-$\ln v$ 对数坐标系中，不管过程线如何不规则，它总可以用几条相互衔接的曲线来代替。这就是说，不管热力过程中压力与比容的变化如何复杂，总可以用几个多变过程来近似地描述这些过程。

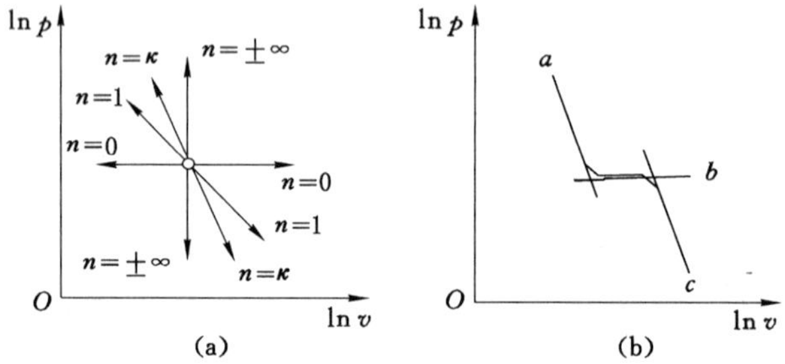

图 4-8　以 $\ln p$ 与 $\ln v$ 为坐标的多变过程图

由于多变过程的方程与定熵过程的方程形式相同，所以知道了多变指数，要计算多变过程初、终状态参数、膨胀功、技术功和热量时，只需要将定熵过程有关公式中的 κ 改为 n 即可。

$$\frac{p_2}{p_1} = \left(\frac{v_1}{v_2}\right)^n, \frac{T_2}{T_1} = \left(\frac{v_1}{v_2}\right)^{n-1}, \frac{T_2}{T_1} = \left(\frac{p_2}{p_1}\right)^{\frac{n-1}{n}} \tag{4-60}$$

将多变过程方程 $p = \dfrac{p_1 v_1^n}{v^n}$ 代入膨胀功的定义式 $w = \int_1^2 p\mathrm{d}v$，当 $n \neq 1$ 时，整理得：

$$\begin{aligned} w = \int_1^2 p\mathrm{d}v &= \frac{1}{n-1}(p_1 v_1 - p_2 v_2) = \frac{R}{n-1}(T_1 - T_2) \\ &= \frac{RT_1}{n-1}\left[1 - \left(\frac{p_2}{p_1}\right)^{\frac{n-1}{n}}\right] = \frac{RT_1}{n-1}\left[1 - \left(\frac{v_1}{v_2}\right)^{n-1}\right] \end{aligned} \tag{4-61}$$

技术功为：

$$w_t = -\int_1^2 v\mathrm{d}p = \int_1^2 np\,\mathrm{d}v = nw \tag{4-62}$$

多变过程以上几个量的公式推导结果与用 n 取代定熵过程相应量计算公式中的 κ 得到的结果相同。

多变过程中交换的热量可由热力学第一定律 $q = \Delta u + w$ 得出，对于理想气体：

$$q = c_v(T_2 - T_1) + \frac{R}{n-1}(T_1 - T_2) \tag{4-63}$$

又　　$$R=(\kappa-1)c_{\mathrm{v}}$$

于是：

$$q=c_{\mathrm{v}}(T_2-T_1)-\frac{\kappa-1}{n-1}c_{\mathrm{v}}(T_2-T_1)=\frac{n-\kappa}{n-1}c_{\mathrm{v}}(T_2-T_1)=c_{\mathrm{n}}(T_2-T_1) \tag{4-64}$$

式中，$c_{\mathrm{n}}=\frac{n-\kappa}{n-1}c_{\mathrm{v}}$ 为多变比热。当 $1<n<\kappa$ 时，c_{n}为负值，过程吸热，温度反而会降低。

将从同一初态出发的 4 个基本过程和多变过程描绘在同一坐标图上，如图 4-7 所示，就可确定 n 为任意值的多变过程在坐标图上的相对位置，从而判断过程中状态参数、热量、功量以及内能的变化规律，判断原则如下：

始于某初始状态的过程，判断其内能 Δu 的正负以过初始点的定温线（$n=1$）为基准。若终点位于该定温线上，该过程温度不变，$\Delta u=0$；若终点位于 T-s 图中该线的上方或 p-v 图中该线的右上方，该过程温度升高，$\Delta u>0$，内能增加；若终点位于 T-s 图中该线的下方或 p-v 图中该线的左下方，该过程温度降低，$\Delta u<0$，内能减小。焓变 Δh 与 Δu 具有相同的规律。

膨胀功 w 的正负判断以过初始点的定容线（$n=\pm\infty$）为基准。若终点位于该定容线上，该过程容积不变，膨胀功 $w=0$；若终点位于 p-v 图中该线的右方或 T-s 图中该线的右下方，该过程容积增大，为膨胀过程，$w>0$；若终点位于 p-v 图中该线的左方或 T-s 图中该线的左上方，该过程容积减小，为压缩过程，$w<0$。

传热量 q 的正负判断以过初始点的定熵线（$n=\kappa$）为基准。若终点位于该定熵线上，该过程 $\Delta s=0$，$q=0$；若终点位于 T-s 图中该线的右方或 p-v 图中该线的右上方，该过程 $\Delta s>0$，$q>0$，为吸热过程；若终点位于 T-s 图中该线的左方或 p-v 图中该线的左下方，该过程 $\Delta s<0$，$q<0$，为放热过程。

由图 4-7 还可以看出，当 $1<n<\kappa$ 时，气体的膨胀过程线 ΔT 为负值，而 q 为正值，即过程吸热，但温度降低；气体的压缩过程线 ΔT 为正值，而 q 为负值。这两种情况下，过程的多变比热为负值。c 为负值的物理意义可理解为：当气体膨胀做功时，对外做的功大于加入的热量，故气体的内能减小而温度降低；压缩时，外界对气体做的功大于气体向外界放出的热量，故其内能增加而温度升高。

为应用方便，将常用的基本热力过程的主要计算公式（定值比热）汇总在表 4-1 中。

表 4-1　气体基本热力过程的主要计算公式

过程	定容过程	定压过程	定温过程	定熵过程	多变过程
过程指数 n	∞	0	1	κ	n
过程方程	$v=$常数	$p=$常数	$pv=$常数	$pv^{\kappa}=$常数	$pv^{n}=$常数
p,v,T 关系	$\frac{T_2}{T_1}=\frac{p_2}{p_1}$	$\frac{T_2}{T_1}=\frac{v_2}{v_1}$	$p_1v_1=p_2v_2$	$p_1v_1^{\kappa}=p_2v_2^{\kappa}$ $\frac{T_2}{T_1}=\left(\frac{v_1}{v_2}\right)^{\kappa-1}$ $=\left(\frac{p_2}{p_1}\right)^{\frac{\kappa-1}{\kappa}}$	$p_1v_1^{n}=p_2v_2^{n}$ $\frac{T_2}{T_1}=\left(\frac{v_1}{v_2}\right)^{n-1}$ $=\left(\frac{p_2}{p_1}\right)^{\frac{n-1}{n}}$
Δu	$\Delta u=c_{\mathrm{v}}(T_2-T_1)$	$\Delta u=c_{\mathrm{v}}(T_2-T_1)$	0	$\Delta u=c_{\mathrm{v}}(T_2-T_1)$	$\Delta u=c_{\mathrm{v}}(T_2-T_1)$

表 4-1(续)

过程	定容过程	定压过程	定温过程	定熵过程	多变过程
Δh	$\Delta h=c_p(T_2-T_1)$	$\Delta h=c_p(T_2-T_1)$	$\Delta h=0$	$\Delta h=c_p(T_2-T_1)$	$\Delta h=c_p(T_2-T_1)$
Δs	$\Delta s=c_v\ln\frac{T_2}{T_1}$	$\Delta s=c_p\ln\frac{T_2}{T_1}$	$\Delta s=R\ln\frac{v_2}{v_1}$ $=-R\ln\frac{p_2}{p_1}$	0	$\Delta s=c_v\ln\frac{T_2}{T_1}+R\ln\frac{v_2}{v_1}$ $=c_p\ln\frac{T_2}{T_1}-R\ln\frac{p_2}{p_1}$ $=c_p\ln\frac{v_2}{v_1}+c_v\ln\frac{p_2}{p_1}$
膨胀功 w	$w=0$	$w=p(v_2-v_1)$ $=R(T_2-T_1)$	$w=RT\ln\frac{v_2}{v_1}$ $=RT\ln\frac{p_1}{p_2}$	$w=\frac{1}{\kappa-1}(p_1v_1-p_2v_2)$ $=\frac{R}{\kappa-1}(T_1-T_2)$ $=\frac{RT_1}{k-1}\left[1-\left(\frac{p_2}{p_1}\right)^{\frac{\kappa-1}{\kappa}}\right]$	$w=\frac{1}{n-1}(p_1v_1-p_2v_2)$ $=\frac{R}{n-1}(T_1-T_2)$ $=\frac{RT_1}{n-1}\left[1-\left(\frac{p_2}{p_1}\right)^{\frac{n-1}{n}}\right]$ $(n\neq1)$
技术功 w_t	$w_t=v(p_1-p_2)$	0	$w_t=w$	$w_t=\kappa w$	$w_t=nw$
热量 q	$q=\Delta u$ $=c_v(T_2-T_1)$	$q=\Delta h$ $=c_p(T_2-T_1)$	$q=T\Delta s$ $=w$	$q=0$	$q=\frac{n-\kappa}{n-1}c_v(T_2-T_1)$ $(n\neq1)$

备注:c_v,c_p为定值比热,需精确计算,可用平均比热代替定值比热。

4.3.3 热力过程的一般分析方法

(1) 根据热力过程的特性,确定过程中状态参数的变化规律,建立表达这种状态参数变化规律的方程,即过程方程式。

(2) 根据已知参数及过程方程式,确定未知的参数以及过程中内能、焓和熵的变化。理想气体的内能与焓都是温度的单值函数。对于任何过程,若已知过程初、终状态的温度,采用定值比热 c_v 和 c_p,可由 $\Delta u=c_v(T_2-T_1)$ 和 $\Delta h=c_p(T_2-T_1)$ 计算得到其内能和焓的变化量。理想气体的熵变仅取决于过程的初、终状态,熵变可由式(4-20)至式(4-22)计算得到。

(3) 将过程中状态参数的变化规律表示在 p-v 图和 T-s 图上。

(4) 根据可逆过程的特征,应用热力学第一定律计算过程中的热量、膨胀功和技术功。

【例 4-3】 1 kg 空气初态为 $p_1=0.8$ MPa,$t_1=500$ ℃,分别经历① 定熵过程;② 定温过程;③ 多变指数为 1.2 的多变过程达到同一终态压力 $p_2=0.1$ MPa。试求终态的温度、过程中熵的变化量以及对外所做的技术功,并将过程表示在 p-v 图和 T-s 图上。设过程可逆,比热为定值比热。

【解】 已知空气的物性参数:$R=0.287$ kJ/(kg·K),$c_p=1.004$ kJ/(kg·K),$\kappa=1.4$。

(1) 定熵过程

$$T_2=T_1\left(\frac{p_2}{p_1}\right)^{\frac{\kappa-1}{\kappa}}=(500+273)\left(\frac{0.1}{0.8}\right)^{\frac{1.4-1}{1.4}}=426.7\ (\mathrm{K})$$

定熵过程 $\Delta s=0$。

对外所做的技术功为：

$$w_t = \frac{\kappa R}{\kappa - 1}(T_1 - T_2) = \frac{1.4 \times 0.287}{1.4 - 1}(773 - 426.7) = 347.9\ (\text{kJ/kg})$$

(2) 定温过程

$$T_2 = T_1 = 773\ \text{K}$$

$$\Delta s = c_p \ln\frac{T_2}{T_1} - R\ln\frac{p_2}{p_1} = -R\ln\frac{p_2}{p_1} = -0.287\ln\frac{0.1}{0.8} = 0.597\ [\text{kJ/(kg·K)}]$$

$$w_t = -\int_1^2 v\mathrm{d}p = RT_1\ln\frac{p_1}{p_2} = 0.287 \times 773 \times \ln\frac{0.8}{0.1} = 461.3\ (\text{kJ/kg})$$

(3) 多变过程

$$T_2 = T_1\left(\frac{p_2}{p_1}\right)^{\frac{n-1}{n}} = (500 + 273)\left(\frac{0.1}{0.8}\right)^{\frac{1.2-1}{1.2}} = 546.6\ (\text{K})$$

$$\Delta s = c_p\ln\frac{T_2}{T_1} - R\ln\frac{p_2}{p_1} = 1.004\ln\frac{546.6}{773} - 0.287\ln\frac{0.1}{0.8} = 0.249\ [\text{kJ/(kg·K)}]$$

$$w_t = \frac{n}{n-1}R(T_1 - T_2) = \frac{1.2}{1.2 - 1} \times 0.287 \times (773 - 546.6) = 389.9\ (\text{kJ/kg})$$

三个过程在 p-v 图和 T-s 图上分别如图 4-9(a)和图 4-9(b)所示，图中 2_s，2_n 和 2_T 分别对应定熵过程、多变过程和定温过程的终态。

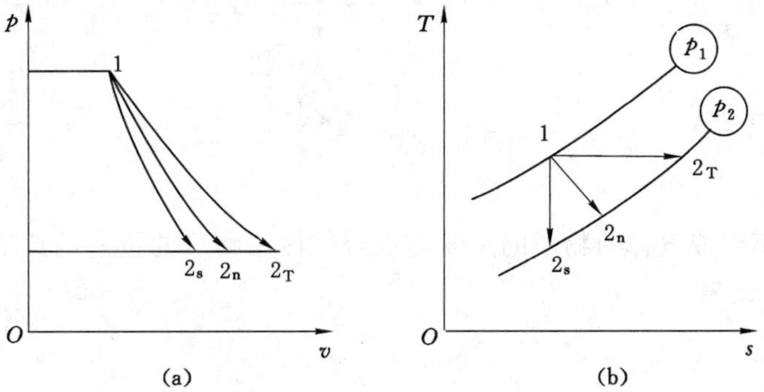

图 4-9　例 4-3 图

4.4　理想混合气体

热力工程上使用的气体工质往往不是单一的，而是气体混合物。例如，内燃机、燃气轮机装置中的燃气，这些混合气体中各组成气体之间不发生化学反应，而是一种均匀混合物。由理想气体组成的混合物称为理想混合气体，组成混合气体的各种气体称为组分或组元。混合气体的性质取决于混合气体中各组分的成分及热力性质。由理想气体组成的混合气体，仍然具有理想气体特性，理想混合气体服从理想气体的各种定律。

4.4.1　混合气体成分

混合气体中各组成气体的含量与混合气体总量之比值，称为混合气体的成分。按物理单位的不同，有质量成分、容积成分和摩尔成分。

(1) 质量成分

混合气体中某组成气体的质量 m_i 与混合气体总质量 m 的比值，称为该组成气体的质量成分，用 ω_i 表示，即

$$\omega_i = \frac{m_i}{m} \tag{4-65}$$

因为混合气体的总质量等于各组成气体的质量之和，即

$$m_1 + m_2 + \cdots + m_n = \sum_{i=1}^{n} m_i = m$$

所以：

$$\sum_{i=1}^{n} \omega_i = \sum_{i=1}^{n} \frac{m_i}{m} = 1$$

(2) 容积成分

混合气体中某组成气体的分容积 V_i 与混合气体总容积 V 的比值，称为该组成气体的容积成分，用 r_i 表示，即

$$r_i = \frac{V_i}{V} \tag{4-66}$$

根据分容积定律：

$$V = V_1 + V_2 + \cdots + V_n = \sum_{i=1}^{n} V_i$$

则有：

$$r_1 + r_2 + \cdots + r_n = \sum_{i=1}^{n} r_i = 1$$

(3) 摩尔成分

混合气体中某组成气体的物质的量 n_i 与混合气体总物质的量 n 的比值，称为该组成气体的摩尔成分，用 x_i 表示：

$$x_i = \frac{n_i}{n} \tag{4-67}$$

因为：

$$n = n_1 + n_2 + \cdots + n_n = \sum_{i=1}^{n} n_i$$

则有：

$$x_1 + x_2 + \cdots + x_n = \sum_{i=1}^{n} x_i = 1$$

上述三式说明，混合气体中各组成气体成分之和都等于 1。

(4) 各组成成分之间的换算关系

① 容积成分与摩尔成分数值相等

由：

$$r_i = \frac{V_i}{V} = \frac{n_i V_{mi}}{n V_m}$$

式中，V_{mi}，V_m 分别为某组成气体与混合气体的摩尔体积。

根据阿伏伽德罗定律，同温同压下，各种气体的摩尔容积相等，即 $V_{mi} = V_m$。

于是得：

$$r_i = \frac{n_i}{n} = x_i \tag{4-68}$$

② 质量成分与容积成分(摩尔成分)的换算

由：

$$\omega_i = \frac{m_i}{m} = \frac{n_i M_i}{nM} = x_i \frac{M_i}{M} = r_i \frac{M_i}{M}$$

式中，M_i，M 分别为某组成气体与混合气体的摩尔质量(即相对分子质量)。

根据摩尔气体常数 $MR = M_i R_i = R_0$ 可得：

$$\omega_i = r_i \frac{M_i}{M} = r_i \frac{R}{R_i} \tag{4-69}$$

4.4.2　混合气体的温度、压力和容积

(1) 温度

处于平衡态的均匀的理想混合气体的温度 T 与混合物中各组元的温度相等。

(2) 压力与分压力

如图 4-10 所示，温度为 T、容积为 V 的混合物的压力 p 是混合物的总压力。各组元气体单独存在并具有与混合物相同的温度 T 和容积 V 时所具有的压力称为该组元的分压力，用 p_i 表示。理想混合气体的总压力等于混合物中各组元的分压力之和，即

$$p = p_1 + p_2 + p_3 + \cdots + p_n = \sum_{i=1}^{n} p_i \tag{4-70}$$

式(4-70)在 1801 年被道尔顿实验所证实，被称为道尔顿分压定律。可表述为，彼此不起化学反应的不同气体，若共置于同一容器中，混合物的总压力等于各组元的分压力之和。道尔顿分压定律只适用于理想混合气体。

(3) 容积与分容积

如图 4-11 所示，温度为 T、压力为 p 的混合物占有的容积 V 是混合物的总容积。各组元气体单独存在并具有与混合物相同的温度 T 和压力 p 时所占有的容积称为该组元的分容积，用 V_i 表示，如图 4-11 中的 V_1，V_2 和 V_3。

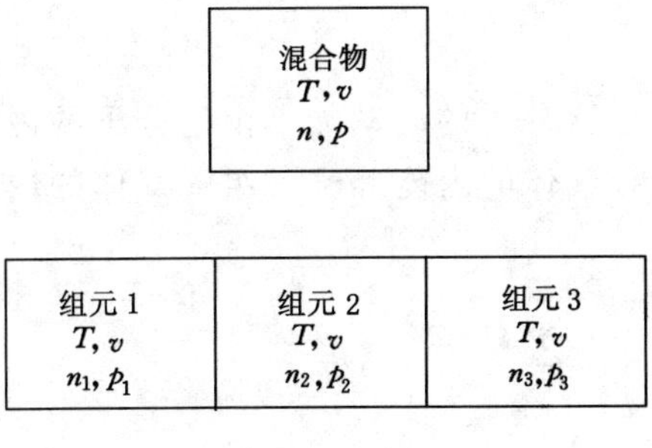

图 4-10　混合气体分压力

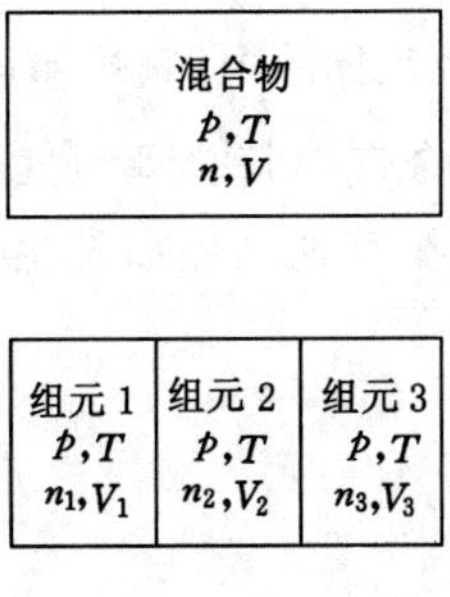

图 4-11　混合气体分容积

利用理想气体状态方程得：

$$pV_i = p_i V$$

$$p\sum_i V_i = V\sum_i p_i$$

由道尔顿分压定律得：

$$V = \sum_i V_i \tag{4-71}$$

式(4-71)称为分容积定律，即理想混合气体的总容积等于各组元气体的分容积之和。分容积定律也只适用于理想混合气体。

4.4.3 混合气体的折合相对分子质量与折合气体常数

任何物质的相对分子质量在数值上都等于质量(g)除以物质的量(mol)。据此，可求得混合气体的折合相对分子质量(平均相对分子质量)：

$$M = \frac{m}{n} = \frac{\sum_{i=1}^{n} n_i M_i}{n} = \sum_{i=1}^{n} x_i M_i = \sum_{i=1}^{n} r_i M_i \tag{4-72}$$

得到折合相对分子质量后，根据摩尔气体常数可得混合气体的折合气体常数：

$$R = \frac{R_0}{M} = \frac{8\ 314}{M}, \text{J/(kg} \cdot \text{K)}$$

4.4.4 混合气体的比热

混合气体温度升高所需热量，等于各组成气体所需热量的总和，由此可得混合气体比热的计算式。

若各组成气体的质量比热和质量成分已知，则混合气体的质量比热为：

$$c = \omega_1 c_1 + \omega_2 c_2 + \cdots + \omega_n c_n = \sum_{i=1}^{n} \omega_i c_i \tag{4-73}$$

同理，混合气体的容积比热为：

$$c' = r_1 c_1' + r_2 c_2' + \cdots + r_n c_n' = \sum_{i=1}^{n} r_i c_i' \tag{4-74}$$

混合气体的摩尔比热：

$$c_{\text{m}} = M\sum_{i=1}^{n} g_i c_i = \sum_{i=1}^{n} x_i M_i c_i \tag{4-75}$$

4.4.5 混合气体的内能、焓和熵

理想混合气体符合理想气体的假设，内能只包含分子的内动能。由于分子的内动能并不因为存在其他气体而发生变化，所以理想混合气体的内能等于各组元气体的内能之和，即

$$U_1 + U_2 + \cdots + U_n = \sum_{i=1}^{n} U_i$$

$$u = \frac{U}{m} = \frac{U_1 + U_2 + \cdots + U_n}{m} = \frac{\sum_{i=1}^{n} m_i u_i}{m} = \sum_{i=1}^{n} \omega_i u_i \tag{4-76}$$

同样，理想混合气体的焓等于各组元气体的焓之和，即

$$h = \sum_{i=1}^{n} \omega_i h_i \tag{4-77}$$

理想混合气体的熵等于各组元气体的熵之和，即

$$s = \sum_{i=1}^{n} \omega_i s_i \tag{4-78}$$

本 章 小 结

本章讨论了理想气体的热力性质和热力过程。主要包括理想气体状态方程及其应用；理想气体的比热；理想气体的内能、焓和熵；理想气体的混合物及理想气体的热力过程。

理想气体状态方程的基本形式为 $pv=RT$，气体常数 R 是随工质而异的常数，工质一定，其值是一个确定的常数。

理想气体的比热有真实比热、平均比热及定值比热，可根据精度要求选用。

理想气体的内能和焓只是温度的单值函数，计算关系式为：

$$\Delta u = \int_1^2 c_v \mathrm{d}T \quad 及 \quad \Delta h = \int_1^2 c_p \mathrm{d}T$$

理想气体的熵不仅与温度有关，而且与压力或比容有关。

$$\Delta s = \int_{T_1}^{T_2} c_v \frac{\mathrm{d}T}{T} + R\ln\frac{v_2}{v_1} \text{ 或 } \Delta s = \int_{T_1}^{T_2} c_p \frac{\mathrm{d}T}{T} - R\ln\frac{p_2}{p_1} \text{ 或 } \Delta s = \int_{p_1}^{p_2} c_v \frac{\mathrm{d}p}{p} + \int_{v_1}^{v_2} c_p \frac{\mathrm{d}v}{v}$$

在理想气体的热力过程部分主要讨论了 4 种典型基本过程，即定容过程、定压过程、定温过程、定熵过程，以及具有一般意义的多变过程；得出了过程方程式、状态参数之间的关系及过程中所交换的热量和功量的计算关系；分析了各过程在 p-v 图和 T-s 图上的表示方法。

理想混合气体具有理想气体的一切特性，利用理想混合气体的成分可以求解折合气体常数和折合相对分子质量。

思考与练习题

(1) 什么是理想气体？将工质按理想气体处理在计算上有什么方便之处？

(2) 摩尔气体常数与气体常数有何不同？

(3) 有一充满气体的容器，容积 $V=4.5\ \mathrm{m^3}$，气体压力根据压力表的读数为 $p_g=245.2$ Pa，温度计读数为 $t=40$ ℃。问在标准状态下气体容积为多少？

(4) 质量为 3 kg 的空气，经过一系列的状态变化过程，其压力增加了一倍，体积减小了一半。试求过程中熵的变化（按定值比热计算）。

(5) 定温过程是等内能和等焓过程，这一结论对任意工质都成立吗？

(6) 理想气体从同一初态膨胀到同一比容的终态，定温膨胀与绝热膨胀相比，哪个过程做功多？若为压缩过程，耗功又如何？

(7) 质量为 4 kg 的空气由初态 $p_1=0.4$ MPa，$t_1=25$ ℃，经过下列不同过程膨胀到同一终压力 $p_2=0.1$ MPa：① 定温过程；② 定熵过程。试计算这两个过程中空气对外做的膨胀功、所进行的热量交换、终态参数和空气熵的变化。

(8) 空气的容积 $V_1=2\ \mathrm{m^3}$，由 $p_1=0.2$ MPa，$t_1=40$ ℃，压缩到 $p_2=1$ MPa，$V_2=0.5\ \mathrm{m^3}$。求过程的多变指数、压缩功、气体在过程中所放出的热量，以及气体熵的变化。设空气的比热为定值比值 $c_v=0.717$ kJ/(kg·K)，空气的气体常数 $R=0.287$ kJ/(kg·K)。

(9) 试将满足以下要求的理想气体多变过程在 p-v 图和 T-s 图上表示出来(先画出 4 个基本热力过程):① 气体受压缩、升温和放热;② 气体的多变指数 $n=0.8$,膨胀;③ 气体受压缩、降温又降压;④ 气体的多变指数 $n=1.2$,受压缩;⑤ 气体膨胀、降压且放热。

知识拓展:比热容

知识拓展:绝热过程压力-温度关系

知识拓展:真实比热、平均比热与定值比热

Chapter 4 Thermodynamic Properties and Processes of Ideal-Gas

Heat energy into mechanical energy depends on the expansion of the working medium can be realized. And gas has the best thermal expansion, the most suitable as the working medium, So the gas properties of working medium is one of the important part of the study of thermodynamics.

Volume change of a working medium is indispensible for the transformation from thermal energy to other forms of energy. Any thermodynamic process can not be accomplished without working medium. Thus, we will study properties of working medium before we step further to study processes which may be concerned. In this chapter, we will discuss thermodynamic properties of ideal gases and related processes. These are of practical significance and also form the base to study real-gas processes.

4.1 The Ideal Gas State Equation

4.1.1 Ideal Gas and Real Gas

Ideal gas is an imaginary substance. The ideal-gas model can be denoted as follows: the molecules of an ideal gas are elastic particles that have no volume; there are no any force interactions among the molecules of an ideal gas.

It has been experimentally observed that at low pressure and high temperature, the specific volume is large enough for the gas to behave as an ideal gas. In the range of practical engineering, many familiar gases such as oxygen, nitrogen, argon, carbon dioxide and even gas mixture such as air, exhaust fume can be treated as ideal gases with acceptable error. Therefore, in air-conditioning applications, water vapor in the air can be treated as ideal gas essentially with no error since the pressure of the water vapor is very low. In steam power plant, however, the pressures involved are usually very high, ideal-gas relations can not be used.

4.1.2 Ideal-Gas Equation of State

The ideal-gas equation of state relates the properties in the simplest form. This relation was obtained from experimental data at the beginning and was confirmed theoretically later. For per unit ideal gas, equation of state can be demoted as:

$$pv = RT \tag{4-1}$$

For m kg mass, it can be written as:

$$pV = mRT \tag{4-2}$$

Where p is the absolute pressure, Pa; v is the specific volume, m^3/kg; V is the volume, m^3; T is the absolute temperature, K; R is the gas constant, J/(kg · K) and its value just depends on the variety of the gas but not related to the state of the gas. It is a very important parameter.

In SI unit, mol is often used as the unit of a gas. M is the molar mass that can be defined as the mass of one kmol of a substance in kilograms, kg/kmol. V_m is the molar specific volume, that is the volume per unit mole, $m^3/kmol$. Thus, it can also be written as:

$$pMv = MRT$$

Let $R_0 = MR$, then:

$$pV_m = R_0 T \tag{4-3}$$

For n kmol gas, it can be written as:

$$pV = nR_0 T$$

Where R_0 is the universal gas constant, J/(kmol · K). It is the same for all kinds of ideal gases. Its value can be obtain at a certain state, say, the standard condition ($p_0 = 1.013\ 25 \times 10^5$ Pa, $T_0 = 273.15$ K) by using the equation of state. At this state, the molar specific volume of any gas is 22.4 $m^3/kmol$. Thus:

$$R_0 = 8\ 314 \text{ J/(kmol} \cdot \text{K)}$$

$$R = R_0 / M$$

And if the molar mass M is known, then the gas constant R can be obtained. For air:

$$R = R_0/M = 8\ 314/28.97 = 287 \text{ J/(kg} \cdot \text{K)} = 0.287 \text{ kJ/(kg} \cdot \text{K)}$$

4.2 The Thermodynamic Properties of Ideal-Gas

4.2.1 Specific Heat

4.2.1.1 The Definition and Unit of the Specific Heat

The specific heat is defined as the energy required to raise the temperature of per unit substance by one degree. In terms of the unit of the substance used, there are different kinds of specific heats as following. If the specific heat is based on a unit mass, then it has the unit kJ/(kg · K) and is denoted as:

$$C = \frac{\delta Q}{\mathrm{d}T} \tag{4-4}$$

Volume is often used to express the amount of gases. And the specific heat based on volume at the standard state is denoted by c' and has the unit kJ/(m^3 · K).

The relations among these specific heats can be expressed as:

$$22.4c' = c_m = Mc$$

4.2.1.2 Specific Heat at Constant Volume and Specific Heat at Constant Pressure

In general, specific heat is a path function and depends on how the process is executed. In thermodynamics, we are interested in two kinds of process specific heats: specific heat at constant volume c_v and specific heat at constant pressure c_p.

Physically, the specific heat at constant pressure c_p can be viewed as the energy required to raise the temperature of a unit mass of a substance by one degree as the pressure is maintained constant. It is:

$$c_p = \left(\frac{\delta q}{dT}\right)_p$$

The energy required to do the same as the volume heat is maintained constant is the specific c_v at constant volume. It is:

$$c_v = \left(\frac{\delta q}{dT}\right)_v$$

Energy balance equation can be expressed in the differential form as:

$$\delta q = du + p dv$$

Considering a reversible process of an ideal-gas, say, a constant-volume process, the boundary work is $\delta w = p dv = 0$. The specific heat at constant volume c_v of ideal-gas can be defined as:

$$c_v = \left(\frac{\delta q}{dT}\right)_v = \left(\frac{du + p dv}{dT}\right)_v = \left(\frac{du}{dT}\right)_v \tag{4-5}$$

Similarly, an expression for the specific heat at constant pressure can be obtained by considering a constant-pressure process. It yields:

$$c_p = \left(\frac{\delta q}{dT}\right)_p = \left(\frac{dh - v dp}{dT}\right)_p = \left(\frac{dh}{dT}\right)_p \tag{4-6}$$

Using the definition of enthalpy and the ideal-gas equation of state, we know:

$$h = u + pv = u + RT$$

Then, the relationship between the specific heat c_v and c_p of ideal gases is:

$$c_p = \frac{dh}{dT} = \frac{d(u + pv)}{dT} = \frac{du}{dT} + \frac{d(RT)}{dT} = c_v + R \tag{4-7}$$

On a molar basis, the relationship between the specific heats is

$$Mc_p - Mc_v = R_0 \tag{4-8}$$

This is an important relationship for ideal gases. The specific heat at constant pressure c_p is always greater than c_v, because the energy for the expansion work must be supplied to the system by heat when the system is allowed to expand at constant pressure.

At this point, we will introduce another ideal gas property called the specific heat ratio κ, defined as:

$$\kappa = \frac{c_p}{c_v} \tag{4-9}$$

And c_v and c_p can be calculated by:

$$c_v = \frac{R}{\kappa - 1}, c_p = \frac{\kappa R}{\kappa - 1}$$

4.2.1.3 Actual Specific Heat, Average Specific Heat and Constant Specific Heat

(1) Actual specific heat

The ideal-gas specific heats depend on two independent properties, which are T and p. But the influence of pressure on ideal-gas specific heat is small and can be ignored. It can be deemed that specific heats depend on temperature only (Fig. 4-1). And the value of specific heat at any temperature can be defined as the actual specific heats. The accurate analytical expressions of the actual specific heat for ideal gases are based on measurements or on calculations from statistical behavior of molecules. In general, they are given as a third-degree polynomials as follow:

$$c = a_0 + a_1 T + a_2 T^2 + a_3 T^3$$

Where, a_0, a_1, a_2, a_3 are constant obtained experimentally.

(2) Average specific heat

To avoid the time consuming integration, we define $c_m |_{t_1}^{t_2}$ as the average specific heat between t_1 to t_2, expressed by:

$$c_m \Big|_{t_1}^{t_2} = \frac{q}{t_2 - t_1} = \frac{\int_{t_1}^{t_2} c\mathrm{d}t}{t_2 - t_1}$$

$$q = c_m \Big|_{t_1}^{t_2} \times (t_2 - t_1)$$

The specific heats of a number of gases have been calculated and tabulated over small temperature intervals. These tables are obtained by choosing a reference point, and performing the integrations on the above equations by treating state 1 as the reference state.

(3) Constant specific heat

If the temperature is not high and the temperature interval is not very large, the influence of temperature on specific heat value can be ignored. The specific heat can be treated as constant. Constant specific heat on a molar basis and corresponding specific heat ratio are as follow:

$$Mc_v = \frac{i}{2} R_0, \mathrm{J/(kmol \cdot K)}$$

$$Mc_p = \frac{2+i}{2} R_0, \mathrm{J/(kmol \cdot K)}$$

Where, $i=3$ for monatomic gases, $i=5$ for diatomic gases, and $i=7$ for polyatomic gases.

4.2.2 Internal Energy, Enthalpy and Entropy of Ideal-Gas

4.2.2.1 Internal Energy and Enthalpy of Ideal-Gas

It has been demonstrated that ideal-gas internal energy is a function of temperature only. That is:

$$\mathrm{d}u = c_v \mathrm{d}T \tag{4-10}$$

In terms of the definition of enthalpy and the ideal-gas equation of state, we have:

$$dh = c_p dT \tag{4-11}$$

During a process from state 1 to state 2:

$$\Delta u = u_2 - u_1 = \int_{T_1}^{T_2} c_v dT \tag{4-12}$$

$$\Delta h = h_2 - h_1 = \int_{T_1}^{T_2} c_p dT \tag{4-13}$$

As long as the relations for specific heat are established, integrations can be conducted easily to determine the change in internal energy and enthalpy of ideal-gas. If c_p and c_v are constant, they yields:

$$\Delta u = c_v \Delta T = c_v (T_2 - T_1) \tag{4-14}$$

$$\Delta h = c_p \Delta T = c_p (T_2 - T_1) \tag{4-15}$$

4.2.2.2 Entropy of Ideal-Gas

In terms of the definition of enthalpy, the ideal-gas equation of state and the First Law of Thermodynamics, we obtain the differential entropy change of an ideal gas as:

$$ds = \left(\frac{\delta q}{T}\right)_{re} = \frac{du + p dv}{T} = \frac{du}{T} + \frac{p}{T} dv = c_v \frac{dT}{T} + R \frac{dv}{v} \tag{4-16}$$

$$ds = \left(\frac{\delta q}{T}\right)_{re} = \frac{dh - v dp}{T} = \frac{dh}{T} - \frac{v}{T} dp = c_p \frac{dT}{T} - R \frac{dp}{p} \tag{4-17}$$

$$\frac{dp}{p} + \frac{dv}{v} = \frac{dT}{T} \tag{4-18}$$

$$\begin{aligned} ds &= c_v \frac{dT}{T} + R \frac{dv}{v} \\ &= c_v \left(\frac{dp}{p} + \frac{dv}{v}\right) + R \frac{dv}{v} \\ &= c_v \frac{dp}{p} + (c_v + R) \frac{dv}{v} \\ &= c_v \frac{dp}{p} + c_p \frac{dv}{v} \end{aligned}$$

$$\begin{aligned} ds &= c_p \frac{dT}{T} - R \frac{dp}{p} \\ &= c_p \left(\frac{dp}{p} + \frac{dv}{v}\right) - R \frac{dp}{p} \\ &= (c_p - R) \frac{dp}{p} + c_p \frac{dv}{v} \\ &= c_v \frac{dp}{p} + c_p \frac{dv}{v} \end{aligned} \tag{4-19}$$

The entropy change between states 1 and 2 can be obtained by integration:

$$\Delta s = \int_{T_1}^{T_2} c_v \frac{dT}{T} + R \ln \frac{v_2}{v_1} \tag{4-20}$$

$$\Delta s = \int_{T_1}^{T_2} c_p \frac{dT}{T} - R \ln \frac{p_2}{p_1} \tag{4-21}$$

$$\Delta s = \int_{p_1}^{p_2} c_v \frac{dp}{p} + \int_{v_1}^{v_2} c_p \frac{dv}{v} \tag{4-22}$$

If c_p and c_v are constant, they become:

$$\Delta s = c_v \ln \frac{T_2}{T_1} + R \ln \frac{v_2}{v_1} \tag{4-23}$$

$$\Delta s = c_p \ln \frac{T_2}{T_1} - R \ln \frac{p_2}{p_1} \tag{4-24}$$

$$\Delta s = c_v \ln \frac{p_2}{p_1} + c_p \ln \frac{v_2}{v_1} \tag{4-25}$$

4.3 Typical Thermodynamic Processes of Ideal-Gas

4.3.1 Basic Thermodynamic Processes

4.3.1.1 Isochoric Process

The specific volume remains constant during the process, which is called isochoric process or constant volume process. The process equation can be expressed as:

$$v = \text{const}$$

In terms of the ideal-gas equation of state, we obtain the properties relation between the initial and final state:

$$\frac{p}{T} = \frac{R}{v} = \text{const} \quad \text{or} \quad \frac{p_2}{p_1} = \frac{T_2}{T_1} \tag{4-26}$$

$$\Delta s = \int_1^2 c_v \frac{dT}{T} + R \ln \frac{v_2}{v_1} = \int_1^2 c_v \frac{dT}{T} \tag{4-27}$$

$$\Delta s = c_v \ln \frac{T_2}{T_1} \tag{4-28}$$

Since $dv = 0$, the expansion work done during a constant volume process is zero.

$$w = 0 \tag{4-29}$$

The technical work can be calculated by:

$$w_t = -\int_1^2 v dp = v(p_1 - p_2) \tag{4-30}$$

The amount of heat absorbed or rejected during the process is equal to:

$$q = \Delta u = \int_1^2 c_v dT = c_v (t_2 - t_1) \tag{4-31}$$

This formula reveals that the heat absorbed/rejected by the working fluid in an isochoric process is converted entirely into the change in internal energy.

For isochoric process, the First Law of Thermodynamics can be expressed as:

$$T ds = c_v dT + p dv$$

As $dv=0$, it reduces to:

$$\frac{dT}{ds} = \frac{T}{c_v} \tag{4-32}$$

On T-s diagram isochoric process is the curve whose slope is T/c_v. The slope is always positive and increases along with the in temperature as shown in Fig. 4-3.

4.3.1.2 Isobaric Process

The pressure remains constant during a process, which is called isobaric process or constant pressure process.

The process equation can be expressed as:

$$p = \text{const} \tag{4-33}$$

The properties relation between the initial and final states is:

$$\frac{v_2}{v_1} = \frac{T_2}{T_1} \tag{4-34}$$

$$\Delta s = \int_1^2 c_p \frac{dT}{T} \tag{4-35}$$

$$\Delta s = c_p \ln \frac{T_2}{T_1} \tag{4-36}$$

Since p=const, the expansion work in an isobaric process is

$$w = \int_1^2 p dv = p(v_2 - v_1) = R(T_2 - T_1) \tag{4-37}$$

And the technical work is:

$$w_t = -\int_1^2 v dp = 0 \tag{4-38}$$

The heat absorbed or rejected can be calculated by

$$q = \Delta h = \int_1^2 c_p dT = c_p (t_2 - t_1) \tag{4-39}$$

These formulas denote that the heat absorbed/rejected by the working fluid in an isobaric process is converted entirely into the change in enthalpy.

For isobaric process, the energy balance equation can be expressed as:

$$T ds = c_p dT - v dp = c_p dT$$

Thus:

$$\frac{dT}{ds} = \frac{T}{c_p} \tag{4-40}$$

On T-s diagram, the slope of an isobaric curve is T/c_p and $c_p > c_v$. The slope of an isobaric process curve is smaller than that of an isochoric process curve. The curve of an isobaric process is flatter than an isochoric process curve, as shown in Fig. 4-4.

4.3.1.3 Isothermal Process

Isothermal process isa process in which the temperature is maintained constant. The process equation can be expressed as T=const.

Then, the relation between the initial and final states is:

$$\frac{p_1}{p_2} = \frac{v_2}{v_1} \tag{4-41}$$

$$\Delta u = \int_1^2 c_v dT = 0 \tag{4-42}$$

$$\Delta h = \int_1^2 c_p dT = 0 \tag{4-43}$$

$$\Delta s = \int_1^2 R\frac{dv}{v} = R\ln\frac{v_2}{v_1} = R\ln\frac{p_1}{p_2} \tag{4-44}$$

The expansion work output in an isothermal process is:

$$w = \int_1^2 p dv = RT\ln\frac{v_2}{v_1} = RT\ln\frac{p_1}{p_2} \tag{4-45}$$

And the technical work is:

$$w_t = -\int_1^2 v dp = RT\ln\frac{p_1}{p_2} = w \tag{4-46}$$

The expansion work is equal to the technical work. And the heat absorption and rejection during an isothermal process is:

$$q = \Delta u + w = w = w_t \tag{4-47}$$

or

$$q = \int_1^2 T ds = T(s_2 - s_1) \tag{4-48}$$

It indicates the heat transferred during an isothermal process is entirely converted to expansion/compression work.

According to $pv=RT=\text{const}$, the curve of an isothermal process on p-v diagram is an equilateral hyperbola, as shown in Fig. 4-5.

4.3.1.4 Isentropic Processes

Isentropic process is an internally reversible and adiabatic process. In such a process the entropy remains constant.

For an isentropic process:

$$\delta q = 0$$

and

$$ds = \frac{\delta q}{T} = 0$$

The property relation can be obtained from Eq. (4-25) by employing the definition of differential entropy change of ideal gas:

$$\frac{dp}{p} + \kappa\frac{dv}{v} = 0$$

It can also be expressed in a compact form as:

$$pv^{\kappa} = \text{const} \tag{4-49}$$

κ is the specific heat ratio.

The properties relation between the initial and final state is:

$$\frac{p_2}{p_1} = \left(\frac{v_1}{v_2}\right)^{\kappa} \tag{4-50}$$

It can be rearranged as:

$$\frac{T_2}{T_1} = \left(\frac{v_1}{v_2}\right)^{\kappa-1} \tag{4-51}$$

and

$$\frac{T_2}{T_1} = \left(\frac{p_2}{p_1}\right)^{\frac{\kappa-1}{\kappa}} \tag{4-52}$$

The heat absorbed or rejected is:

$$q = \int_1^2 T\mathrm{d}s = T(s_2 - s_1) = 0 \tag{4-53}$$

The expansion work during an isentropic process is:

$$w = q - \Delta u = -\Delta u \tag{4-54}$$

And

$$w_t = q - \Delta h = -\Delta h \tag{4-55}$$

It indicates that the amount of expansion work is equal to the decrease in internal energy. And the technical work output is equal to the decrease in the enthalpy. Thus:

$$w = \int_1^2 p\mathrm{d}v = \frac{RT_1}{\kappa - 1}\left[1 - \left(\frac{p_2}{p_1}\right)^{\frac{\kappa-1}{\kappa}}\right] = \frac{R}{\kappa - 1}(T_1 - T_2) \tag{4-56}$$

And the technical work is:

$$w_t = c_p(T_1 - T_2) = \frac{\kappa RT_1}{\kappa - 1}\left[1 - \left(\frac{p_2}{p_1}\right)^{\frac{\kappa-1}{\kappa}}\right] = \frac{\kappa R}{\kappa - 1}(T_1 - T_2) \tag{4-57}$$

The relationship between the expansion work and the technical work is:

$$w_t = \kappa w$$

According to $pv^{\kappa} = \text{const}$, an isentropic process curve on p-v diagram is a high hyperbola, and $\kappa > 1$, as shown in Fig. 4-6. The slope of an isentropic process curve is greater than that of an isothermal process curve.

4.3.2 Poly-Tropic Process

The above mentioned four basic thermodynamic processes have one characteristic in common. That is, a particular property remains constant. But in actual thermodynamic processes of gases, properties will change according to various different rules. It is necessary to establish an idealized process model which can approximate most of actual processes. Four basic processes can be fitted in this model as well.

A process is called a poly-tropic process during which the relationship between pressure and volume can be expressed as:

$$pv^n = \text{const} \tag{4-58}$$

Where n is the poly-tropic exponent. Every of the four basic thermodynamic processes is a special case of the poly-tropic process model:

When $n=0$, $p=\text{const}$ and it turns to be the isobaric process equation;

When $n=1$, $pv=\text{const}$ and it turns to be the isothermal process equation;

When $n=\kappa$, $pv^{\kappa}=\text{const}$ and it is be the isentropic process equation;

When $n=\pm\infty$, $v=\text{const}$ and it turns to be the isochoric process equation.

n may be a value with in $0 \sim \pm\infty$ at discretion. An actual thermodynamic process cannot be denoted by one poly-tropic process with a certain n. However, we can break up the process into several poly-tropic processes so that every sub-process corresponds to a certain n.

We can position n on p-v and T-s diagrams by virtue of four basic processes, as shown in Fig. 4-7. The slope of a poly-tropic process curve on p-v diagram can be expressed as:

$$\frac{dp}{dv} = -n\frac{p}{v} \tag{4-59}$$

From the base line of an isochoric process, the value of n increases clock-wisely from $-\infty \to 0 \to 1 \to \kappa \to +\infty$. as shown in Fig. 4-7(a).

$n=0$, $\frac{dp}{dv}=0$, The line pressing for a horizontal line;

$n=1$, $\frac{dp}{dv}=-\frac{p}{v}<0$, The fixed temperature line is an equilateral hyperbola with negative slope;

$n=\kappa$, $\frac{dp}{dv}=-\kappa\frac{p}{v}<0$, The entropy line is a higher order hyperbolic with negative slope.

The greater the polytropic index, therefore, the value of n, process line is the absolute value of the slope.

Similarly, the slope of a poly-tropic process curve on T-s diagram can be expressed as:

$$\frac{dT}{ds} = \frac{T}{c_n}$$

The value of n also increases clock-wisely, as shown in Fig. 4-7(b).

The equation of poly-tropic process has the similar form with that of an isentropic process, simply poly-tropic exponent n replacing the specific heat ratio κ. Substituting n into Eq. (4-50)~ Eq. (4-52), we get the properties relations for a poly-tropic process:

$$\frac{p_2}{p_1} = \left(\frac{v_1}{v_2}\right)^n, \frac{T_2}{T_1} = \left(\frac{v_1}{v_2}\right)^{n-1}, \frac{T_2}{T_1} = \left(\frac{p_2}{p_1}\right)^{\frac{n-1}{n}} \tag{4-60}$$

And the expansion work output during a poly-tropic process is:

$$w = \int_1^2 p dv = \frac{1}{n-1}(p_1v_1 - p_2v_2) = \frac{R}{n-1}(T_1 - T_2)$$

$$= \frac{RT_1}{n-1}\left[1-\left(\frac{p_2}{p_1}\right)^{\frac{n-1}{n}}\right] = \frac{RT_1}{n-1}\left[1-\left(\frac{v_1}{v_2}\right)^{n-1}\right] \tag{4-61}$$

$$w_t = -\int_1^2 v dp = \int_1^2 np dv = nw \tag{4-62}$$

The amount of heat can be determined in terms of the First Law of Thermodynamics:

$$q = c_v(T_2 - T_1) + \frac{R}{n-1}(T_1 - T_2) \tag{4-63}$$

$$q = c_v(T_2 - T_1) - \frac{\kappa-1}{n-1}c_v(T_2 - T_1) = \frac{n-\kappa}{n-1}c_v(T_2 - T_1) = c_n(T_2 - T_1) \tag{4-64}$$

Where, $c_n = \frac{n-\kappa}{n-1}c_v$ is the specific heat in a poly-tropic process.

From the denotation on p-v and T-s diagrams, we can intuitively judge whether q, w and Δu are positive or negative during a poly-tropic process, as shown in Fig. 4-7.

The isotherms ($n=1$) is the separatrix between the positive and negative change in internal energy, Δu. The changes in internal energy during isothermal processes are zero.

From the same initial point, if the process line is above the isotherm, the change in internal energy is positives and if the process line is below the isotherm, the change in internal energy is negative.

The isochore ($n=\pm\infty$) is the separatrix between the positive and negative value of the work output, w. From the same initial point, the expansion work during an isochore process is zero. We can judge that: if the process line is on the right of the isochore, then the expansion work is positive; and if the process line is on the left of the isochor, then the expansion work is negative.

The isentropic ($n=\kappa$) is the separatrix between the positive and negative value of the heat transferred, q. From the same initial point, the amount of heat transfer during an isentropic process is zero. If the process line is on the right of the isentropic, the amount of heat transferred is positive; and if the process line is on the left of the isentropic, the amount of heat transferred is negative.

It can also be seen from Fig. 4-7, when $1<n<\kappa$, during an expansion process the change in temperature, ΔT, is negative while q is positive. It means that the gas absorbs heat in an expansion process but the temperature drops. Similarly in a compression process, the change in temperature, ΔT, is positive while q is negative. It means that the specific heats of the poly-tropic processes are negative in these cases. It is because the expansion work is greater than the heat absorption in this expansion process. As a result, the internal energy is reduced, which leads to the drop in temperature. It is similar for compressed process.

4.3.3 The General Steps of Thermal Process Analysis

Energy conversion is fulfilled through thermal processes by working fluid in some thermal equipment. Therefore, it is important to determine the relation between properties of different states during thermal processes and to obtain the energy conversion relations, including the amount of heat addition/rejection and of the work output/input. The fundamental tools for ideal-gas processes analysis include the First Law of Thermodynamics, the ideal-gas equation of state and the characteristics of processes. The steps to take are as following.

(1) According to the characteristics of the processes, determine the properties relationship between the initial and final states, that is, the processes equations.

(2) Based on the given properties and process equation, determine other unknown properties and the change in internal energy, enthalpy and entropy during the process.

(3) Illustrate the process on p-v and T-s diagrams.

(4) Determine the amount of work output/input and heat addition/rejection during the process.

4.4 Ideal-Gas Mixture

Many important thermodynamic appliances involve mixtures of several pure substances. For example, air is conventionally treated as an ideal gas in the range where nitrogen and oxygen behave as ideal gases. If the individual gases (called components or constituents) are ideal gases, the mixture behaves as ideal gas and it can be treated as an ideal-gas mixture. The properties of a gas mixture obviously depend on the properties of the constituents as well as on the amount of each gas in the mixture. Therefore, it is necessary to develop some rules to determine the mixture properties from the knowledge of the mixture compositions and the properties of the individual gases.

4.4.1 The Compositions of Ideal-Gas Mixture

The properties of a mixture depend on the composition of the mixture as well as the properties of the individual components. There are three ways to describe the composition of a mixture: mass, volume and molar fractions.

(1) Mass fraction

the ratio of the mass of one component m_i and the total mass of the ideal-gas mixture m is called the mass fraction, denoted as ω_i, namely:

$$\omega_i = \frac{m_i}{m} \tag{4-65}$$

Since:

$$m_1 + m_2 + \cdots + m_n = \sum_{i=1}^{n} m_i = m$$

So:

$$\sum_{i=1}^{n} \omega_i = \sum_{i=1}^{n} \frac{m_i}{m} = 1$$

(2) Volume fraction

the ratio of one forming gas volume V_i and the total volume of the ideal-gas mixture V is called the volume fraction, denoted as r_i, namely:

$$r_i = \frac{V_i}{V} \tag{4-66}$$

According to the law of partial volume: $V = V_1 + V_2 + \cdots + V_n = \sum_{i=1}^{n} V_i$, it can be derived that:

$$r_1 + r_2 + \cdots + r_n = \sum_{i=1}^{n} r_i = 1$$

(3) Mole fraction

the ratio of one forming gas moles n_i and the total moles of the ideal-gas mixture n is called the mole fraction, denoted as x_i, namely:

$$x_i = \frac{n_i}{n} \tag{4-67}$$

Since $n = n_1 + n_2 + \cdots + n_n = \sum_{i=1}^{n} n_i$, it can be found that:

$$x_1 + x_2 + \cdots + x_n = \sum_{i=1}^{n} x_i = 1$$

(4) The conversion relation among the three compositions

① the volume fraction is equal to mole fraction

From:

$$r_i = \frac{V_i}{V} = \frac{n_i V_{mi}}{n V_m}$$

V_{mi} is the molar volume of one forming gas, V_m is the molar volume of ideal-gas mixture.

According to the Avogadero's law, under the same temperature and pressure, all kinds of gas molar volume is equal, that is $V_{mi} = V_m$.

It can be found that:

$$r_i = \frac{n_i}{n} = x_i \tag{4-68}$$

② The conversion between mass fraction and mole fraction

From:

$$\omega_i = \frac{m_i}{m} = \frac{n_i M_i}{nM} = x_i \frac{M_i}{M} = r_i \frac{M_i}{M}$$

M_i, M is the molar mass of forming gas and mixed gas.

According to the universal gas constant $MR = M_i R_i = R_0$ and Avogadero's law, under the same temperature and pressure, gas density is proportional to the molecular weight.

It can be found that:

$$\omega_i = r_i \frac{M_i}{M} = r_i \frac{R}{R_i} \tag{4-69}$$

4.4.2 Temperature, Pressure and Volume of Ideal-Gas Mixture

(1) The temperature of ideal-gas mixture

Temperature of ideal-gas mixture at an equilibrium state is equal to the temperature of each gas composition.

(2) The pressure of ideal-gas mixture and the component pressures

The pressure that each gas exerts is called the component pressure if it exists alone at the same temperature and volume as those of the mixture. It is also called the partial pressure. The pressure of a gas mixture is equal to the sum of the component pressures, as shown in Fig. 4-10.

$$p = p_1 + p_2 + p_3 + \cdots + p_n = \sum_{i=1}^{n} p_i \tag{4-70}$$

Eq. (4-70) is known as Dalton's Law of Additive Pressures. It is described that the pressure of a gas mixture is equal to the sum of the component pressures of each gases. Note that Dalton's law holds exactly for ideal-gas mixtures.

(3) The volume of ideal-gas mixture and the component volume

The volume each gas occupies is called the component volume (also called partial volume) of a component in a gas mixture if it exists alone at the same temperature and pressure as those of the mixture. It is denoted by V_i, as shown in Fig. 4-11.

In terms of the ideal-gas equation of state, we get:

$$pV_i = p_i V$$

$$p\sum_i V_i = V\sum_i p_i$$

Applying Dalton's law of additive pressures, we get:

$$V = \sum_i V_i \tag{4-71}$$

It is called the Amagat's Law of Additive Volumes. The volume of a gas mixture is equal to the sum of the volumes each gas would occupy if it exists alone at the mixture temperature and pressure.

4.4.3 The Apparent Molar Mass and the Gas Constant of Ideal-Gas Mixture

The mass of a substance can be expressed in terms of the mole number n and molar mass M of the substance. Then the apparent (or average) molar mass of a mixture can be expressed as:

$$M = \frac{m}{n} = \frac{\sum_{i=1}^{n} n_i M_i}{n} = \sum_{i=1}^{n} x_i M_i = \sum_{i=1}^{n} r_i M_i \tag{4-72}$$

The gas constant of a mixture can also be calculated by:

$$R = \frac{R_0}{M} = \frac{8\ 314}{M}, \mathrm{J/(kg \cdot K)}$$

4.4.4 Specific Heat of Ideal-Gas Mixture

The mass specific heat of ideal-gas mixture:

$$c = \omega_1 c_1 + \omega_2 c_2 + \cdots + \omega_n c_n = \sum_{i=1}^{n} \omega_i c_i \tag{4-73}$$

The volume specific heat of ideal-gas mixture:

$$c' = r_1 c_1' + r_2 c_2' + \cdots + r_n c_n' = \sum_{i=1}^{n} r_i c_i' \tag{4-74}$$

The molar specific heat of ideal-gas mixture:

$$c_{\mathrm{m}} = M\sum_{i=1}^{n} g_i c_i = \sum_{i=1}^{n} x_i M_i c_i \tag{4-75}$$

4.4.5 Total Internal Energy, Enthalpy, and Entropy of Ideal-Gas Mixture

The gas mixture and its components can be treated as ideal gases, then the total internal anergy, enthalpy, and entropy of a gas mixture can be expressed as:

$$u = \frac{U}{m} = \frac{U_1 + U_2 + \cdots + U_n}{m} = \frac{\sum_{i=1}^{n} m_i u_i}{m} = \sum_{i=1}^{n} \omega_i u_i \tag{4-76}$$

$$h = \sum_{i=1}^{n} \omega_i h_i \tag{4-77}$$

$$s = \sum_{i=1}^{n} \omega_i s_i \tag{4-78}$$

第 5 章　热力学第二定律

本章导学

热力学第一定律揭示了这样一个自然规律，即在热力过程中，参与转换与传递的各种能量在数量上是守恒的。但它并没有说明，满足能量守恒原则的过程是否都能实现。经验告诉我们，自然过程是有方向性的。揭示热力过程方向、条件与限度的定律是热力学第二定律。热力学第二定律揭示了能量传递和转换中品质高低的问题，其表现形式是热力过程的方向性和不可逆性。只有同时满足热力学第一定律和热力学第二定律的过程才是可以实现的过程。热力学第二定律与热力学第一定律共同组成了热力学的理论基础。

学习目标

深刻理解热力学第二定律的实质，认识能量不仅有“量”的多少，还有“质”的高低；掌握卡诺循环和卡诺定理及其意义。掌握熵的意义、计算和应用。能利用孤立系的熵增原理进行不可逆过程和循环的分析计算。了解做功能力损失的计算方法。

重点与难点

(1) 热力学第二定律的实质及表述。

(2) 热力学第二定律表述的等效性。

(3) 热力学第二定律与熵。

(4) 热力学第二定律的数学表达式。

热力学第一定律揭示了能量在转换与传递过程中数量守恒的客观规律，但是是否任何不违反热力学第一定律的过程都可以实现呢？事实并非如此。例如，一块烧红的铁板，在空气中自然冷却，显然热能从铁板散发到周围空气中，周围空气获得的热量等于铁板放出的热量，这完全遵守热力学第一定律。但设想这个已经冷却了的铁板从周围空气中收回那部分散失的热量，重新炽热起来，这并不违反热力学第一定律，但很明显这样的过程是不能实现的。

再有，一个转动的飞轮，如不继续用外力推动它旋转，那它的转速就会逐渐减小，最终停止，飞轮原先具有的动能由于飞轮轴和轴承之间的摩擦以及飞轮表面和空气的摩擦，变成热能散发到周围空气中，飞轮失去的动能等于周围空气获得的热能，这完全符合热力学第一定律。但反过来，周围空气是否可以将原先获得的热能变为动能，还给飞轮，使飞轮重新转起来，尽管这不违反热力学第一定律，但显然也是不可能的。

可见，热力学第一定律有两方面的问题没有涉及：其一，该定律不能判断热力过程的方向性，即任何热力过程都具有方向性——可以自发进行的热力过程，而其反向过程则不能自发进行。其二，该定律强调的是能量在数量上的守恒，没有考虑不同类型能量在做功能力上的差别，例如，同样数量的机械能与热能其价值并不相等，机械能具有直接可用性，可以无条件地转换为热能(属优质能)，而热能必须在一定的补充条件下才可能部分地转换为机械能(属低质能)。不同质的能量直接相加，严格地说并不合理。

人们从无数实践中总结出了热力学第二定律，该定律揭示了能量在转换与传递过程中的方向性和能质不守恒的客观规律。所有热力过程只有同时遵守热力学第一定律和热力学第二定律才能得以实现。

5.1 热力学第二定律的实质与表述

5.1.1 自发过程的方向性与不可逆性

自然界中的热力过程归纳起来可分为两大类。其中一类是不需要任何附加条件就可以自动地进行的过程，称为自发过程，例如，温差传热、摩擦生热、自由膨胀、扩散混合等都是典型的自发过程。针对不同的物理问题，可以有各种自发过程，那么，这些过程的反向过程能否自动进行呢？

例如，两个温度不同的物体，$T_1>T_2$，热量在温差的作用下由高温物体传至低温物体；若要使其反向进行，即将热量由低温物体返还给高温物体，该过程却不能自发进行，但是，若采用制冷机就可以实现这一过程，但必须有功的耗散，如图 5-1 所示。又如，高处的水在势差的作用下自动地流向低处，欲使水由低处自动地流向高处，这是不可能的，但是若采用水泵即可实现这一过程。

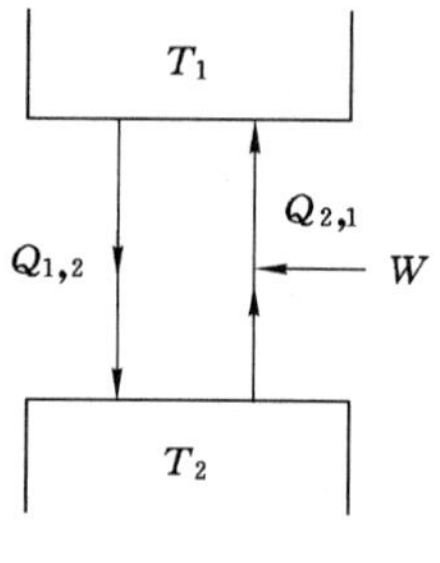

图 5-1　温差传热的方向性

自然界中不同的物理范畴具有许多自发过程，忽略其物理内容，形形色色的自发过程均可以抽象为这样的过程模型：热力系统在不平衡势差的作用下，会自动地趋于其外界对应的平衡状态变化，不平衡会自发地趋于平衡。显然，自发过程均具有方向性，只能自发朝着使系统与外界趋于平衡的方向进行。例如，热量只能由高温物体自发地传至低温物体，直至达到热平衡。

自发过程的反方向过程称为非自发过程。自发过程不需要任何补偿条件就可以自动地进行，直至热力系统的不平衡势差消失。非自发过程无法自动地进行，必须具有一定的补偿条件才能进行。例如，制冷和热泵循环虽实现了热量由低温物体向高温物体的传递，却消耗了功，这部分功连同从低温物体提取的热量一起传递给高温物体。功转换为热是这一非自发过程得以实现的补偿条件。由可逆过程的概念"如果系统完成了某一过程之后，再沿着原路逆行而回到原来的状态，外界也随之回到原来的状态而不留下任何变化，则这一过程称为可逆过程。"可知，自发过程是不可逆过程。

导致过程不可逆的因素称为不可逆因素，大体可分为两大类：一类是系统与外界存在不平衡势差时，过程中存在不平衡损失。热现象涉及的不平衡势差包括温差、压差和化学势差，凡是有不平衡势差必然引起不平衡损失。例如，有限温差下的传热过程，工质吸热、放热时，$T_{热源} \neq T_{工质}$，在温差的作用下，高品位能转化为低品位能，导致能质的贬值。温差传热、混合、自由膨胀、溶解过程都是不平衡势差引起的不可逆过程。这种出现在热力系统与环境间的不平衡势差被称为外部不可逆因素。热力学分析中引入的准平衡过程的概念，该过程中 $\Delta T \to 0, \Delta p \to 0$，因此，准平衡过程没有不平衡损失。另一类是过程中存在耗散效应而引起的不可逆损失。耗散损失是热力过程中由于机械摩擦、流体的黏性、电阻、磁阻等作用而产生的损失。耗散损失是一种与物质性质有关的不可逆损失，这种出现在热力系统内部的不可逆因素称为内部不可逆因素。若热力过程既无不平衡损失，又无耗散损失，则该过程就是可逆过程。可逆过程只是一种理想化概念，是为了便于实际过程的分析计算而引入的概念。任何实际过程都不可避免地包含上述一种或几种不可逆因素，因此任何涉及热现象的实际宏观过程都是不可逆的。

5.1.2　热力学第二定律的实质

热力过程具有方向性这一客观规律，归根结底是由于不同类型或不同状态下的能量具有质的差别，而过程的方向性正缘于较高品位能向较低品位能的转化。例如，热量由高温物体传至低温物体，机械能转化为热能，按热力学第一定律能量的数量保持不变，但是，以做功能力为标志的能质却降低了，称之为能质的退化或贬值。因此，热力学第二定律的实质便是论述热力过程的方向性及能质退化或贬值的客观规律。所谓过程的方向性，除指明自发过程进行的方向外，还包括对实现非自发过程所需要的条件，以及过程进行的最大限度等内容。

热力学第二定律告诉我们，自然界的物质和能量只能沿着一个方向转换，即从可利用到不可利用，从有效到无效，这说明了节能的必要性。只有热力学第二定律才能充分解释事物变化的性质和方向，以及变化过程中所有事物的相互关系。热力学第二定律除广泛应用于分析热力过程和能源工程外，还被应用于分析社会、经济发展及生物进化等许多领域，可以预料该定律今后还将得到更广泛的应用。

5.1.3　热力学第二定律的表述

热力学第二定律有各种不同的表述。最经典的表述是 1850—1851 年间，从工程应用角度归纳总结出来的两种表述：

(1) 克劳修斯(Clausius)说法(从热量传递的角度)：不可能把热量从低温物体传到高温物体而不引起其他变化。

(2) 开尔文-普朗克(Kelvin-Planck)说法(从热功转换的角度)：不可能从单一热源取热，并使之完全转化为机械能而不引起其他变化。

只从单一热源取热而连续不断地做功的机器称为第二类永动机，实践证明这种发动机是造不出来的。热力学第二定律也可表述为第二类永动机是制造不出来的。

上述两种经典说法分别针对不同的现象进行表述，但其实质是一致的，可以用反证法证明。假设克劳修斯说法不成立(违反克劳修斯说法)，即热量可以自发地、不付代价地从低温物体传到高温物体，如图 5-2(a)所示，热量 Q_2 由冷源自发地流向热源，另外有一热机从热源吸收 Q_1 热量，并使其传给冷源的热量正好等于 Q_2，热机将做净功 $W_0 = Q_1 - Q_2$。取热源、冷源及热机为一热力系统，则整个热力系统在完成一个循环时，冷源没有变化，产生的唯一结果是热机从单一热源取得热量 $(Q_1 - Q_2)$ 全部变为循环功 W_0，整个热力系变成了第二类永动机，违背了开尔文-普朗克说法。这也证明克劳修斯说法不成立，则开尔文-普朗克说法也不成立。反之，若违反开尔文-普朗克说法，如图 5-2(b)所示，热机从热源取热量 Q_1，全部变成净功 W_0，不向冷源放热，热效率为 100%。若用这部分功 W_0 带动一台制冷机工作，从冷源提取热量 Q_2，同时向热源放出热量 Q。根据热力学第一定律可写出 $Q = W_0 + Q_2 = Q_1 + Q_2$。整体而言，就相当于有 Q_2 的热量从低温冷源流向了高温热源，而未消耗外功，这显然违反了克劳修斯说法。

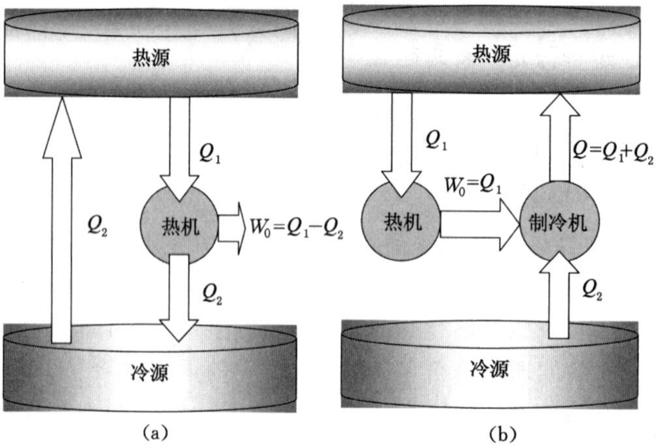

图 5-2　热力学第二定律两种经典表述的一致性

显然，开尔文-普朗克说法和克劳修斯说法是完全等效的，当一种说法不成立时，另一种说法也不成立。

综上所述，各种自发过程之间是有联系的，不可逆过程是物质的同一客观属性在不同情况下的体现，从一种自发过程的不可逆性可以推断另一种自发过程的不可逆性，即热力学第二定律的各种表述彼此是等效的。热力学第二定律的实质表明了自发过程是不可逆的，非自发过程不能够自动进行，但并非无法实现，要使其得以实现必须有一个适当的自发过程作为补偿条件相伴随。

热力学第二定律和热力学第一定律一样，是建立在长期积累的无数经验的基础上的，不能从任何更普遍的定律中推导出来，但整个热力学的发展过程令人信服地表明，它的一切推论都符合客观实际，反映了客观规律的真理。

以上各种热力学第二定律的表述都是通过对客观实践的总结而得出的，定性地阐明了自发过程具有方向性的实质，但尚不能定量地给出非自发过程得以实现所需要的补偿条件。

5.2　卡诺循环与卡诺定理

热功转换是热力学的主要研究内容，根据热力学第二定律的开尔文-普朗克说法，热不能连续地全部转换为功，即热机的效率一定小于 100%。那么，在一定温度范围内的高温热源和低温热源之间，热转换为功的最高效率是多少呢？1824 年法国工程师卡诺(Carnot)回答了这个问题。

5.2.1　卡诺循环

卡诺循环是在实践的基础上经过科学抽象提出的理想循环。如图 5-3 所示，卡诺热机工作于温度为 T_1及 T_2的两个恒温热源之间，卡诺循环是可逆循环。

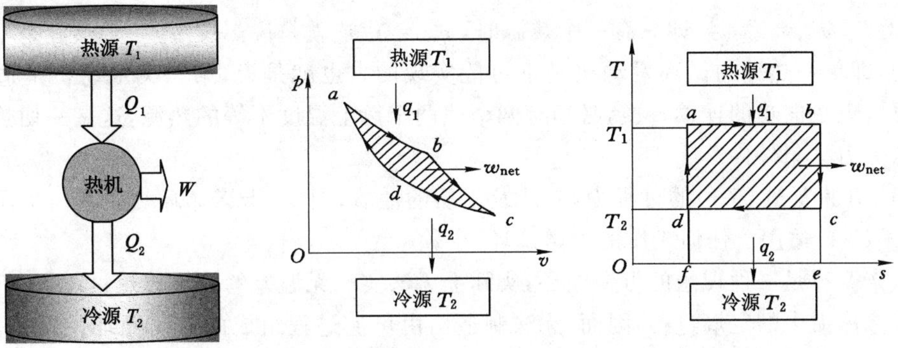

(a) 工作于热源与冷源之间的热机　　(b) 卡诺循环的 p-v 图　　(c) 卡诺循环的 T-s 图

图 5-3　卡诺热机和卡诺循环

温差传热是不可逆过程，只有定温吸热、放热过程才是可逆的。为使热机从一个热源(T_1)向一个冷源(T_2)的吸、放热过程可逆，工质与热源或冷源间的传热温差分别均为无限小。但仅两个定温过程不可能组成封闭循环，需要加入其他过程。所加入的过程介于两条等温线之间。该过程将经历从 T_1到 T_2(或从 T_2到 T_1)的温度变化，若该过程中工质与热源(T_1)或冷源(T_2)有热交换，则就是有限温差传热，过程不可逆。因此，这两个过程应是绝热的。如图 5-3 所示，卡诺循环包括如下过程：

a—b：工质从热源 T_1的可逆定温吸热过程；

b—c：工质的可逆绝热(定熵)膨胀过程；

c—d：工质向冷源 T_2的可逆定温放热过程；

d—a：工质的可逆绝热(定熵)压缩过程回到初始状态。

工质在整个循环中从热源 T_1 吸热 q_1，向冷源 T_2 放热 q_2，对外界做功 w_1，外界对系统做功 w_2。按热力学第一定律 $\oint \delta q = \oint \delta w$ 即 $q_1 - q_2 = w_1 - w_2 = w_{net}$ (循环净功)。

循环热效率：

$$\eta_t = \frac{w_{net}}{q_1} = 1 - \frac{q_2}{q_1}$$

$$q_1 = T_1(s_b - s_a) = \text{面积 } abefa$$

$$q_2 = T_2(s_c - s_d) = \text{面积 } cdfec$$

因为：

$$s_b - s_a = s_c - s_d$$

则卡诺循环热效率：

$$\eta_{tc} = 1 - \frac{T_2}{T_1} \tag{5-1}$$

由卡诺循环热效率公式(5-1)可得到以下结论：

(1) 卡诺循环热效率只决定于热源温度 T_1 及冷源温度 T_2，要提高卡诺循环热效率可通过提高 T_1 及降低 T_2 的办法来实现。

(2) 卡诺循环热效率总是小于 1。只有当 $T_1=\infty$ 或 $T_2=0$ 时，热效率才能等于 1，但这都是不可能的。

(3) 当 $T_1=T_2$ 时，即只有一个热源时，$\eta_{tc}=0$。这就是说，只冷却一个热源是不能进行循环的，即单一热源的循环发动机是不可能实现的。也就是说，第二类永动机是造不出来的。要实现热向功的连续转换，必须有两个或两个以上温度不等的热源，这是一切热机必不可少的热力学条件。

(4) 在推导式(5-1)的过程中，未涉及工质的性质，因此，卡诺循环热效率与工质的性质无关，式(5-1)适用于任何工质的卡诺循环。

卡诺循环是一种理想的可逆循环，实际上无法实现无温差的等温传热过程，也不可能实现没有摩擦损失的定熵过程，因而，无法制造出由可逆过程组成的卡诺循环发动机。然而，卡诺循环在热力学中具有重要的热力学意义，它首先奠定了热力学第二定律的基本概念，其次为如何提高各种热力发动机的热效率指明了方向，具有重要的理论价值。

5.2.2 卡诺逆循环

卡诺循环是可逆循环，故可使循环沿相反方向进行，逆向进行的卡诺循环称为卡诺逆循环，如图 5-4 所示。如同在正循环中从高温热源吸收热量 q_1，向冷源放出热量 q_2，对外做净功 w_{net} 一样，在逆循环中只需要消耗同量的净功 w_{net}，就能从冷源取回同量的 q_2，而向热源送回同量的 q_1。

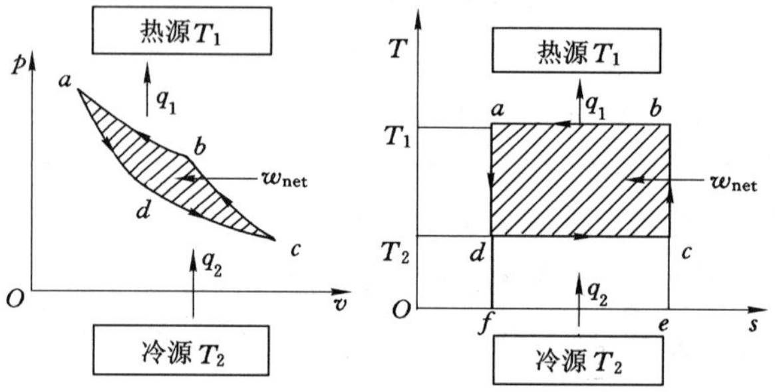

(a) p-v 图　　(b) T-s 图

图 5-4　卡诺逆循环的 p-v 图及 T-s 图

卡诺逆循环由下列四个可逆过程所组成：

c—b：工质被定熵压缩；

b—a：工质向热源(T_1)可逆定温放热；

a—d：工质定熵膨胀；

d—c：工质从冷源(T_2)可逆定温吸热。

在整个逆循环中，工质向热源放热 q_1，从冷源吸热 q_2(即冷量)，外界消耗功 w_1，对外界做功 w_2，外界消耗净功 $w_{net}=w_1-w_2$。从低温冷藏室吸取热量排向大气环境所用的机械称制冷机，如图 5-5 中的 12341 所示；热量由大气传送到高温室所用的机械称为热泵，如图 5-5 中的 $1'2'3'4'1'$。制冷机和热泵都将热量由低温物体传向高温物体，并消耗外功，它们都是按逆循环工作的，只是工作温度范围不同而已。

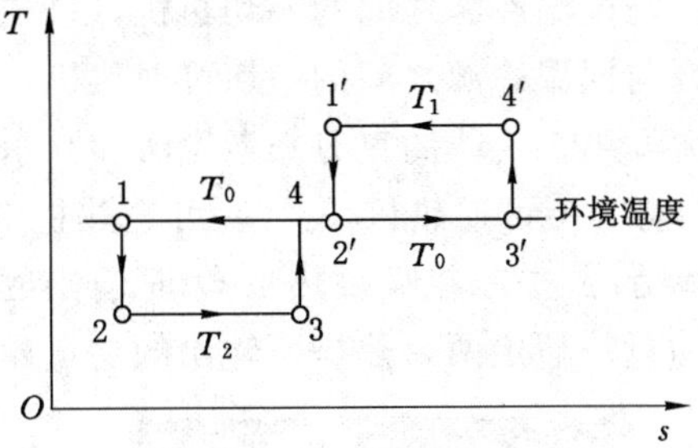

图 5-5　制冷循环与热泵循环

卡诺逆循环用作制冷循环时，由低温热源吸取的热量 q_2 与消耗净功 w_{net} 之比为制冷系数：

$$\varepsilon_c = \frac{q_2}{w_{net}} = \frac{q_2}{q_1 - q_2} = \frac{T_2(s_c - s_d)}{T_1(s_b - s_a) - T_2(s_c - s_d)}$$

因为：

$$s_b - s_a = s_c - s_d$$

则：

$$\varepsilon_c = \frac{T_2}{T_1 - T_2} \tag{5-2}$$

卡诺逆循环用于供热（热泵）循环时，传送到高温热源的热量 q_1 与消耗净功 w_{net} 之比为供热系数：

$$\varepsilon_h = \frac{q_1}{w_{net}} = \frac{q_1}{q_1 - q_2} = \frac{T_1}{T_1 - T_2} \tag{5-3}$$

由式(5-2)及式(5-3)可得下列结论：

(1) 卡诺逆循环的性能系数只决定于热源温度 T_1 及冷源温度 T_2，它随 T_1 的降低及 T_2 的升高而增大。

(2) 卡诺逆循环的制冷系数 ε_c 可以大于 1、等于 1 或小于 1，而供热系数 ε_h 总是大于 1，二者之间的关系为 $\varepsilon_h = 1 + \varepsilon_c$。

(3) 因实际情况下 $T_1 - T_2$ 总是小于 T_2，因此，卡诺逆循环的制冷系数 ε_c 也总是大于 1。

(4) 卡诺逆循环可以用来制冷，也可以用来供热，这两个目的可以单独实现，也可以在同一设备中交替实现，即冬季用来作为热泵采暖，夏季作为制冷机用于空调制冷。以这种方式工作的机械可整年服务于建筑物的空气调节。

所有实际运行在相同温差之间的制冷机或热泵的性能系数均小于卡诺逆循环的性能系数。卡诺逆循环也是一种理想循环，实际上是不能实现的，但却为提高制冷机和热泵的完善程度指明了方向，具有重要的理论意义。

5.2.3　卡诺定理

前面我们得出了以理想气体为工质的卡诺循环的热效率，那么采用水蒸气等其他工质时，卡诺循环的热效率为多少呢？两个热源间工作的一切可逆循环热效率是否相同？不可逆循环的热效率又如何呢？卡诺定理阐明了这些问题。

卡诺定理分为定理一和定量二，分别表达为：

定理一：工作于同温热源与同温冷源之间的一切热机，以可逆热机的热效率最高。

定理二：工作于同温热源与同温冷源之间的一切可逆热机，其热效率均相等。

下面采用反证法，以热力学第二定律的克劳修斯说法和开尔文-普朗克说法为依据来证明卡诺定理。如图 5-6 所示，有一不可逆热机 A 及一可逆热机 B 工作在温度为 T_1 及 T_2 的两个恒温热源之间。用 W_A 表示热机 A 在吸收热量 Q_1 时输出的功量，用 W_B 表示可逆热机 B 在吸收热量 Q_1' 时输出的功量。假定两台热机所输出的功量相等，即 $W_A = W_B$。

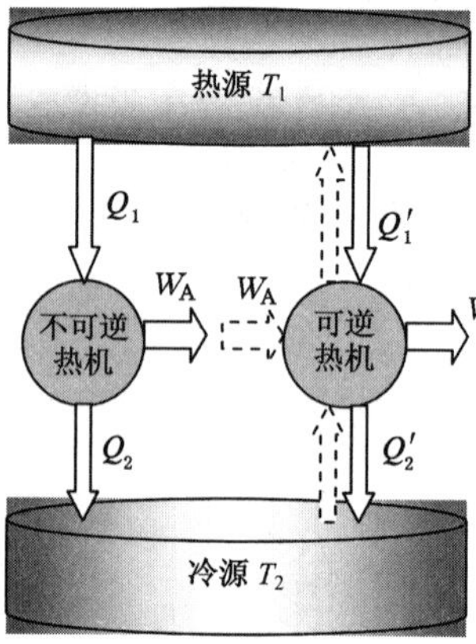

图 5-6　卡诺定理证明

【定理一证明】　假定定理一不正确，也就是说不可逆热机的热效率大于或等于可逆热机的热效率，即 $\eta_{t,A} \geqslant \eta_{t,B}$。

依据热力学第一定律，有：

$$W_A = W_B = Q_1 - Q_2 = Q_1' - Q_2'$$

将两台热机联合工作，利用不可逆热机 A 带动可逆热机 B 工作，则进行正循环的不可逆热机 A 输出的功可以供可逆热机 B(图 5-6 中虚线所示)。因为 B 是可逆热机，使其按逆循环(制冷机)反向工作。

假定 $\eta_{t,A} > \eta_{t,B}$，则根据循环热效率的公式有：

$$\frac{W_A}{Q_1} > \frac{W_B}{Q_2}$$

进而可得 $Q_1' > Q_1$，$Q_2' > Q_2$ 且 $Q_1' - Q_1 = Q_2' - Q_2 > 0$。

可见，不可逆热机 A 与可逆热机 B 联合运行的结果，使热量 $Q_2' - Q_2$ 自动地从低温热源传向高温热源，这是违反热力学第二定律的克劳修斯说法的。因此，$\eta_{t,A} > \eta_{t,B}$ 的假设不成立。

若 $\eta_{t,A} = \eta_{t,B}$，有 $Q_1' - Q_1 = Q_2' - Q_2 = 0$，也就是说，不可逆热机 A 带动可逆热机 B 联合运行的结果，使工质、热源、冷源都回到初态而不留下任何变化。这一结果与热机 A 不可逆的假设相矛盾，因此，$\eta_{t,A} = \eta_{t,B}$ 也不能成立。唯一的可能是 $\eta_{t,A} < \eta_{t,B}$，即在相同的热源与相同的冷源之间，可逆热机的热效率总是大于不可逆热机的热效率。

用同样的方法也可以证明相同热源与相同冷源之间的一切可逆热机其热效率均应相等。

卡诺定理在热力学的研究中具有重要的理论意义和实用价值：

(1) 解决了热机热效率的极限值问题，定量地指明了热机的最高极限热效率。在两热源间工作的热机的最高热效率为卡诺循环的热效率，其值仅与冷、热源的热力学温度有关，热效率恒小于 1；一切实际循环的热效率均低于卡诺循环的热效率。因此，要想设计制造高于卡诺循环热效率的热机是不可能的，因为一切实际热机进行的都是不可逆循环。

(2) 指明了提高循环热效率的方向和途径。热机的热效率虽不可能高于卡诺循环的热效率，但可以通过尽量提高热源温度 T_1 或尽量降低冷源温度 T_2 以及尽可能减少不可逆因素等方式来提高。然而，提高热源温度 T_1 受到材料耐高温性能的制约，降低低温冷源的温度 T_2 也受到大气温度 T_0 的限制，且自然界中不可逆因素总是不可避免的。因此，提高循环热效率时只能尽可能使其接近卡诺循环的热效率。

(3) 指出了热量的最大可用能的大小。大气环境温度 T_0 是不花费任何代价的低温热源的温度，因此即使在最理想的情况下，从热源吸取的热量 Q 也不可能全部转变为功，其中最多只有 $Q\left(1-\dfrac{T_0}{T_1}\right)$ 的能量可转化为技术上可以利用的功，即热量的最大可用能，剩余的热量 $Q\dfrac{T_0}{T_1}$ 将排至大气环境。

(4) 反映了能量有质的差别。同样数量的热和功并不等价，功可以无条件地完全转化为热，但热却只能部分地转化为功；而且即使同样数量的热量，由不同温度的热源释放出来，其做功能力也是有差别的，高温热源放出的热量品质更高。

(5) 为证明熵是状态参数奠定了基础。

5.3 熵

前面已提到熵是系统的状态参数以及理想气体熵变计算，那么，熵参数是如何导出来的？有何物理意义？它与热力学第二定律有何关系？下面就进行讨论。

5.3.1 熵的导出与克劳修斯积分等式

熵参数可由卡诺循环和卡诺定理导出，该方法是 1865 年克劳修斯提出的。图 5-7 表示一任意可逆循环 $a—b—c—d—a$。假设用许多定熵线分割该循环，相邻的定熵线间距无线小，在两定熵线间配合上定温线，从而构成一系列微元卡诺循环。取其中一个微元卡诺循环(如图中斜阴影线所示)，则有：

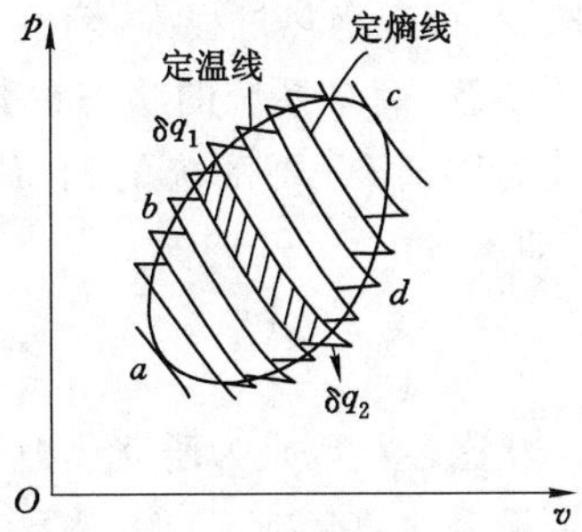

图 5-7　熵参数导出图

$$\eta_{t,c}=1-\frac{\delta q_2}{\delta q_1}=1-\frac{T_2}{T_1}$$

可得：

$$\frac{\delta q_1}{T_1}=\frac{\delta q_2}{T_2}\quad 或\quad \frac{\delta q_1}{T_1}-\frac{\delta q_2}{T_2}=0$$

式中，吸热量 δq_1 与放热量 δq_2 都是绝对值，按其意义，放热量 δq_2 应为负值，将其绝对值改为代数值，则：

$$\frac{\delta q_1}{T_1}+\frac{\delta q_2}{T_2}=0$$

对上式积分，对于整个可逆循环：

$$\int_{abc}\frac{\delta q_1}{T_1}+\int_{cda}\frac{\delta q_2}{T_2}=\oint\left(\frac{\delta q}{T}\right)_{re}=0 \tag{5-4}$$

式(5-4)称为克劳修斯积分等式。克劳修斯提出 $\frac{\delta q}{T}$ 为热温商，其中 δq 是指工质从热源获得的或向冷源排出的热量，其符号的"+""−"分别表示循环吸热、放热，热温商中的 T 是指工质交换热量时的热源温度，过程可逆时，$T_{热源}=T_{工质}$。式(5-4)表明，任何可逆循环，以传热时的绝对温度除各无限小传热量所得到的商的代数和等于零，即可逆循环的热温商可以互相抵消。

式(5-4)中被积函数 $\oint\left(\frac{\delta q}{T}\right)_{re}$ 的循环积分为零，表明该函数与积分路径无关，是一个状态函数。令：

$$ds=\left(\frac{\delta q}{T}\right)_{re} \tag{5-5}$$

1865 年，克劳修斯将这个新的状态参数定名为熵，式(5-5)中 s 是对单位质量工质而言的，称为比熵。对系统总质量而言的总熵则为：

$$S=ms,\text{J/K}$$

于是式(5-4)可写成：

$$\oint ds=0 \tag{5-6}$$

这样，对于可逆的吸热或放热过程，工质的熵变等于传热量与热源温度的比值，工质温度与热源温度相等，即 $dS=\left(\frac{\delta Q}{T}\right)_{re}$。可见，熵的变化反映了可逆过程中热交换的方向和大小。系统从外界可逆吸热，则 $\delta Q>0$，系统熵增加；系统可逆地向外界放热，则 $\delta Q<0$，系统熵减少；可逆绝热过程中，熵不变，即为定熵过程。

注意，熵是系统的状态参数，熵变仅取决于系统的初、终状态，与过程的性质及途径无关。因而，只要系统的初、终状态一定，无论过程可逆与否，系统的熵变都有确定的值，即

$$\int_1^2 ds=s_2-s_1 \tag{5-7}$$

状态参数熵是为研究热力学第二定律引入的表征自然规律的特征量，是热力学第二定律定量表达的基础。

5.3.2 克劳修斯不等式与不可逆过程的熵变

对于任意不可逆循环，如图 5-8 所示，图中虚线 1—a—2 表示不可逆过程，根据卡诺定理，对于每一个微元不可逆循环，均有：

$$\eta_t=1-\frac{\delta q_2}{\delta q_1}<1-\frac{T_2}{T_1}$$

用代数值表示可得：

$$\frac{\delta q_1}{T_1}+\frac{\delta q_2}{T_2}<0$$

对于整个不可逆循环：

$$\int_{1a2}\frac{\delta q_1}{T_1}+\int_{2b1}\frac{\delta q_2}{T_2}=\oint\left(\frac{\delta q}{T}\right)_{\text{irr}}<0 \tag{5-8}$$

即工质经过任意不可逆循环，以传热时的绝对温度除各无限小传热量所得到的商的代数和小于零，即不可逆循环的克劳修斯积分小于零。

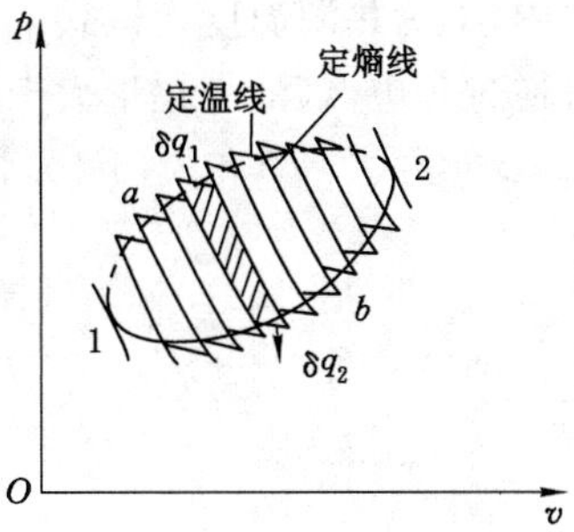

图 5-8　不可逆循环

必须注意，对于不可逆循环，在吸热和放热过程中热源与工质之间都存在着温差，即 $T_{热源}\neq T_{工质}$，式(5-8)中 T 均采用高、低温热源的温度，而不是工质的温度(可逆循环中热源温度与工质温度相等)。

综合式(5-4)及式(5-8)可得克劳修斯导出的关于循环特性的数学表达式：

$$\oint\frac{\delta q}{T}\leqslant 0 \tag{5-9}$$

也就是说：

$$\oint\frac{\delta q}{T}\leqslant\oint \mathrm{d}s=0$$

式(5-9)就是著名的克劳修斯不等式，式中，T 为热源温度(K)。其中，“$=$”适用于可逆循环，“$<$”适用于不可逆循环。式中被积函数表明任何循环的 $\oint\frac{\delta q}{T}$ 的循环积分永远小于零，它与积分路径无关；在极限状况下，即可逆时积分为零。克劳修斯不等式可以判断循环过程是否可逆、是否能够实现。如，$\oint\frac{\delta q}{T}>0$ 的循环不能自发实现。

应该注意：运用克劳修斯积分不等式分析循环时，应从循环的角度确定热量及功的方向，循环中工质吸热，热温商取正值；工质放热，热温商取负值。这与后面介绍的孤立系熵增原理的分析角度恰好相反。

由前面分析可知，可逆过程的熵变等于克劳修斯积分，那么不可逆过程的熵变与克劳修斯积分是什么关系呢？下面进行推导。

对有限过程，如图 5-8 所示的不可逆循环由 1—a—2 不可逆过程和 2—b—1 可逆过程构成，按克劳修斯不等式(5-9)有：

$$\oint\left(\frac{\delta q}{T}\right)_{\text{irr}}=\int_1^2\left(\frac{\delta q}{T}\right)_{\text{irr}}+\int_2^1\left(\frac{\delta q}{T}\right)_{\text{re}}=\int_1^2\left(\frac{\delta q}{T}\right)_{\text{irr}}-\int_1^2\left(\frac{\delta q}{T}\right)_{\text{re}}<0$$

因为 1—b—2 为可逆过程，有：

$$\int_1^2\left(\frac{\delta q}{T}\right)_{\text{re}}=s_2-s_1$$

因而：

$$\int_1^2\left(\frac{\delta q}{T}\right)_{\text{irr}}-(s_2-s_1)<0$$

即

$$s_2-s_1>\int_1^2\left(\frac{\delta q}{T}\right)_{\text{irr}} \tag{5-10}$$

式中，T 为热源温度。该式说明当过程不可逆时，系统熵变大于克劳修斯积分。即在任意不可逆过程中，熵的变化量大于该过程中加入系统的热量除以热源温度。

结合式(5-6)和式(5-10)，得：

$$ds \geqslant \frac{\delta q}{T} \tag{5-11a}$$

即

$$s_2 - s_1 \geqslant \int_1^2 \frac{\delta q}{T} \tag{5-11b}$$

式(5-11b)是针对有限过程的克劳修斯不等式，表明任何热力过程熵的变化总是大于或等于热温商 $\frac{\delta q}{T}$，$ds < \frac{\delta q}{T}$ 的过程不能进行。该式也可用作判断过程是否可逆、是否可以进行的依据。

引起过程熵变化的原因有两类：如果热力系统与外界交换热量时产生 $\int_1^2 \frac{\delta q}{T}$，这部分的熵变化称为熵流 s_f，其微分形式为 δs_f；熵流的方向与过程中所交换热量的方向相同，即吸热时 δs_f 为正，放热时 δs_f 为负，绝热时 δs_f 为零。如果过程中出现不可逆因素，导致功的损失，如摩擦，是由于功的耗散而变成热量，此热量仍为系统本身所吸收，必然引起系统熵的增加。这种原因产生的熵变化称作熵产 s_g，其微分形式为 δs_g。熵产是不可逆过程的标志，当出现不可逆因素时 $\delta s_g > 0$，可逆过程时 $\delta s_g = 0$，使得熵产为负值的过程是违背热力学第二定律的。对于单位质量工质，任意过程熵的变化可以用熵流与熵产的代数和表示：

微分形式：

$$ds = \delta s_f + \delta s_g \tag{5-12a}$$

积分形式：

$$\Delta s_{12} = s_f + s_g \tag{5-12b}$$

热力过程经历可逆变化时 $s_g = 0$，即 $\Delta s = s_f$，过程中熵是守恒的。在不可逆过程中熵产永远是正值，$s_g > 0$，即 $\Delta s > s_f$，熵是不守恒的，不可逆性越大，熵产也越大。这一结论适用于任何不可逆因素引起的系统熵产。因此，熵产可作为过程不可逆程度的度量。

5.4 孤立系的熵增原理与做功能力损失

5.4.1 孤立系的熵增原理

热力学第二定律最著名的数学表达式是孤立系的熵增原理。

孤立系与外界没有任何能量和物质的交换，由系统熵变计算式与克劳修斯不等式 $\Delta s \geqslant \int_1^2 \frac{\delta q}{T}$ 可以看出：

$$\Delta s_{iso} \geqslant 0 \quad 或 \quad ds_{iso} \geqslant 0 \tag{5-13}$$

式中，“＝”适用于可逆过程，“＞”适用于不可逆过程。它表明孤立系中进行的任何实际过程，熵只会增加，不会减少，理想情况(可逆过程)下，则保持不变。也就是说，孤立系的熵只能增加(不可逆过程)或保持不变(可逆过程)，而绝不能减少。这就是著名的孤立系熵增原理。

孤立系是熵增原理的应用前提，通常可将所研究的物系以及外部环境一并取在热力系统内，这样热力系统与外界既没有能量交换也没有质量交换，不存在任何的相互作用。所谓的系统熵增 ΔS_{iso}，是指组成热力系统各物系熵变的代数和。热力系统内的物体，可以因其吸热而熵增大，放热而熵减小，但是，整个系统的熵一定朝着规定的方向变化。

孤立系与外界没有任何能量和质量的传递，显然，孤立系的熵流 S_f 为零，因此：

$$\Delta S_{iso} = S_f + S_g = S_g \tag{5-14}$$

式(5-14)说明，孤立系的熵增等于孤立系的熵产，也就是说孤立系的熵产可以通过该系统各组成部分的熵变进行计算。

孤立系熵增原理的意义：

(1) 熵增原理可用以判断孤立系中过程进行的方向。孤立系中的过程总是沿着使其熵增大的方向进行。如果某过程进行的结果是使孤立系熵增加，即 $\Delta S_{iso}>0$，则过程在孤立系中可自发进行而不需要补充条件；如果使孤立系熵保持不变，即 $\Delta S_{iso}=0$，则理论上为可逆过程；如果使孤立系熵减少，即 $\Delta S_{iso}<0$，则过程在孤立系中不能自发进行。

(2) 熵增原理指出了过程进行的条件。根据熵增原理分析可知，热力过程中诸如高温物体向低温物体传热、功转变为热等这类自发过程，都是使系统熵增大的过程，故自发过程可以单独进行。反之，如低温物体向高温物体传热、热转变为功等这类非自发过程，是使系统熵减小的过程，故非自发过程不能单独进行。要使 $\Delta S_{iso}<0$ 的过程得以实现，就需要一个熵增的自发过程与原孤立系中的过程相伴进行，而且使原孤立系与该过程组成的新的孤立系的熵变大于零，或至少维持不变。该伴随过程即非自发过程得以实现的补偿条件。

(3) 熵增原理可作为孤立系平衡的判据。孤立系内由于存在不平衡势差，所以会有自发地趋于平衡态的变化发生，自发过程的进行将促使孤立系的熵增加。当达到平衡态时，孤立系的熵达到最大，即热力系统达到平衡态，过程即告终止。孤立系在变化时，达到平衡的充要条件是系统熵达最大值。熵函数可以作为判定平衡态的判据，可以用于分析热力过程进行的程度。熵函数达最大值时，其微分应等于零。因此，热力系统达到平衡的条件为：

$$dS_{iso}=0$$

(4) 熵增原理反映孤立系中过程不可逆的程度。孤立系与外界没有任何物质和能量的交换，其熵的增加完全是由其中的不可逆因素引起的，不可逆程度越大，熵增也越大，由此可以了解其偏离可逆过程的程度。可逆过程中热功转换的效率最高，热力学完善度为 1。因此，过程的不可逆程度越大，热功转换效率就越低，热力学完善度越小。也就是说，用熵增可以定量地评价过程的热力学完善度。

在应用熵增原理时还应注意，计算孤立系的熵增是站在热力系统内各物体熵变化的角度，以组成热力系统的各物体为对象。例如，热源放热，$\Delta S_{热源}$ 为负；工质吸热，$\Delta S_{工质}$ 为正。孤立系的熵增符号的规定与克劳修斯积分不等式中热温商符号的规定恰好相反，这一点在运用孤立系的熵增原理分析问题时需要特别注意。

孤立系的熵增原理全面透彻地揭示了热力过程的方向、限度和条件，这正是热力学第二定律的实质。因此，常把式(5-13)称为热力学第二定律的数学表达式。

【例 5-1】 试讨论使热从低温热源传向高温热源的过程能否自发实现？并说说在何种条件下才能实现？

【解】 取高温热源和低温热源共同组成的孤立系，如图 5-9(a)所示，其总熵变是两个热源熵变的代数和。设低温热源放出热量 δQ 被高温热源所吸收。

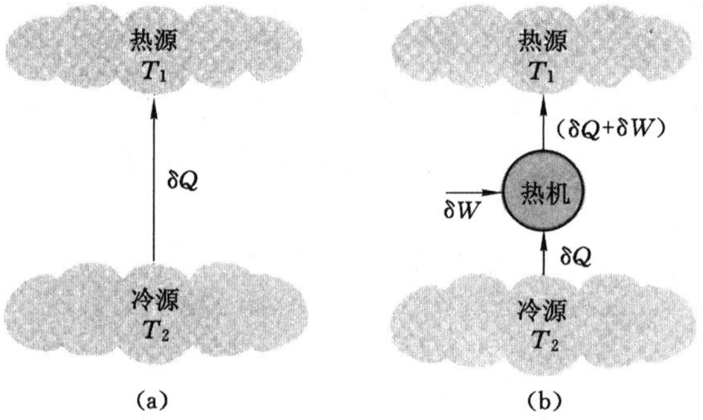

图 5-9 热从低温热源向高温热源传递的过程

低温热源放热后，其熵变：

$$dS_2 = -\frac{\delta Q}{T_2}$$

高温热源吸热后，其熵变：

$$dS_1 = \frac{\delta Q}{T_1}$$

那么，孤立系的熵变为：

$$dS_{iso} = dS_1 + dS_2 = \frac{\delta Q}{T_1} - \frac{\delta Q}{T_2} = \delta Q\left(\frac{1}{T_1} - \frac{1}{T_2}\right) < 0$$

根据熵增原理，该过程不能自发进行。要使热量从低温热源传向高温热源，则需要补偿条件。现增加一台耗功量为 δW 的制冷机，取低温热源、高温热源和制冷机为新的孤立系，如图 5-9(b)所示，则 $dS_{iso} = dS_1 + dS_2 + dS_r$。

其中，低温热源放热 δQ 后，其熵变：

$$dS_2 = -\frac{\delta Q}{T_2}$$

高温热源吸热 $\delta Q + \delta W$ 后，其熵变：

$$dS_1 = \frac{\delta Q + \delta W}{T_1}$$

制冷机中工质经历循环后，熵变 $dS_r = 0$。

可知，当 $dS_{iso} = dS_1 + dS_2 + dS_r = \frac{\delta Q}{T_1} + \frac{\delta W}{T_1} - \frac{\delta Q}{T_2} = \delta Q\left(\frac{1}{T_1} - \frac{1}{T_2}\right) + \frac{\delta W}{T_1} \geqslant 0$ 时，该过程方可实现，即

$$\delta W \geqslant \delta Q\left(\frac{T_1 - T_2}{T_2}\right)$$

也就是说，制冷机消耗的功必须满足上式，方能使热量从低温热源传向高温热源。

5.4.2 做功能力损失

当工质在给定的高温热源与低温热源(环境)之间进行可逆循环时，工质从高温热源所吸收的热量 Q_1 中，最大限度地转换为可用功的那一部分热能，称为在给定热源条件下的做

功能力，或称为热量 Q_1 的可用能。

根据热力学第二定律的论述，任何过程只要有不可逆因素存在，就将造成系统做功能力的损失，而不可逆过程进行的结果又将使包含该系统在内的孤立系的熵增加。一切实际过程都是不可逆过程，都伴随着熵的产生和做功能力的损失，这二者之间必然存在内在联系。可以从熵增原理进一步导出不可逆过程与孤立系做功能力损失之间的关系。

通常取环境状态作为衡量系统做功能力大小的参考状态，即认为系统与环境状态相平衡时，系统不再有做功能力。

假设在温度为 T 的热源与温度为 T_0 的环境之间同时工作着可逆热机 R 和不可逆热机 IR，各自与热源交换的热量、对外输出的功如图 5-10 所示。由卡诺定理可知，$\eta_{t,R} > \eta_{t,IR}$，即

$$\frac{W_R}{Q_1} > \frac{W_{IR}}{Q_1'} \tag{a}$$

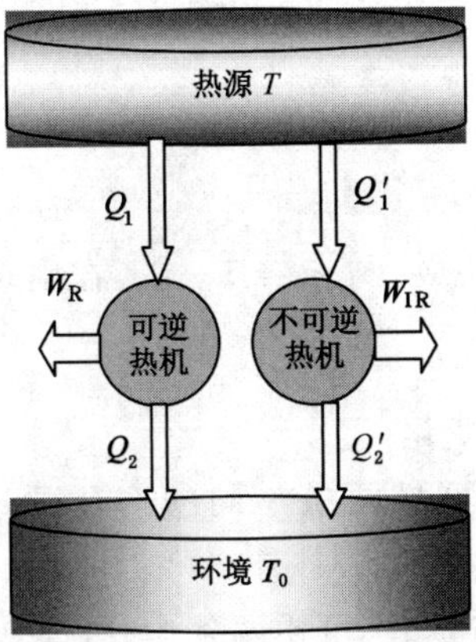

图 5-10　做功能力损失导出示意图

令两热机从热源吸收的热量相同，即 $Q_1 = Q_1'$，由式(a)可得 $W_R > W_{IR}$。于是，由不可逆因素引起的功的损失 L 为：

$$L = W_R - W_{IR} = (Q_1 - Q_2) - (Q_1' - Q_2') = Q_2' - Q_2 \tag{b}$$

将温度为 T 的热源、温度为 T_0 的环境、可逆热机 R、不可逆热机 IR 及蓄功器一起看作一个孤立系，则经过一个工作循环后此孤立系的熵增为：

$$\Delta S_{iso} = \Delta S_T + \Delta S_{T_0} = -\frac{Q_1}{T} - \frac{Q_1'}{T} + \frac{Q_2}{T_0} + \frac{Q_2'}{T_0} \tag{c}$$

根据卡诺定理，对可逆热机，有 $Q_1/T = Q_2/T_0$，同时考虑 $Q_1' = Q_1$，则式(c)可改写为：

$$\Delta S_{iso} = \frac{Q_2'}{T_0} - \frac{Q_2}{T_0} = \frac{Q_2' - Q_2}{T_0} \tag{d}$$

将式(d)代入式(b)，得：

$$L = T_0 \Delta S_{iso} \tag{5-15}$$

由此可见，当环境的热力学温度 T_0 确定后，做功能力损失与孤立系的熵增呈正比，比例常数就是环境温度 T_0。所以，孤立系的熵增是衡量系统做功能力损失的尺度。式(5-15)适用于计算任何不可逆因素引起的系统做功能力的损失。

本章小结

本章讨论了热能转换所涉及的基本定律，即热力学第二定律。热力学第二定律经典的表述是克劳修斯说法和开尔文-普朗克说法。虽然这两种说法表述上不同，但实质是相同的，具有等效性。

卡诺循环和卡诺定理是热力学第二定律的重要内容之一，它不仅指出了热机的最高效率，而且奠定了热力学第二定律的基础。

利用卡诺循环和卡诺定理可以导出状态参数熵。

热力学第二定律的数学表达式有：用于判断循环是否可行、可逆的克劳修斯不等式 $\oint \frac{\delta Q}{T} \leqslant 0$；用于判断过程是否可行、可逆的 $dS \geqslant \frac{\delta Q}{T}$；以及孤立系熵增原理 $dS_{iso} \geqslant 0$。

孤立系熵增与做功能力损失之间存在着一定的关系，即 $L = T_0 \Delta S_{iso}$。

思考与练习题

(1) 循环热效率公式 $\eta_t = 1 - \frac{q_2}{q_1}$ 与 $\eta_t = 1 - \frac{T_2}{T_1}$ 在用于分析循环时，有何区别？各适用于什么场合？

(2) 试分析下列情况下熵的变化：

① 从某一初态经可逆与不可逆两条不同的途径到达相同的终态，则不可逆途径的 $\Delta S_{不可逆}$ 是否大于可逆途径的 $\Delta S_{可逆}$？

② 若从某初态出发，经历可逆和不可逆两条途径，这是一个吸热过程，如果热源条件相同、途径中吸热量相同，工质终态熵是否相同？

③ 从初态分别经可逆、不可逆绝热过程到达相同的终态压力，两过程的终态熵、终态温度及比容关系如何？

④ 在相同的初、终态之间，进行可逆过程与不可逆过程，则不可逆过程中工质熵的变化量是否大于可逆过程中工质熵的变化量？

(3) 下列说法是否正确？

① 系统熵增大的过程必须是不可逆过程。

② 系统熵减小的过程无法进行。

③ 系统熵不变的过程必须是绝热过程。

④ 对不可逆循环，工质熵的变化 $\oint ds \geqslant 0$。

⑤ 工质经任何一种循环，其熵变等于零。

(4) 设计某新型热机希望能达到 75% 的热效率，设计时工质所能达到的最高温度被材料的耐热温度限制在 1 050 ℃，冷源温度为 100 ℃。试根据卡诺定理判断是否能达到预期的热效率，分析之。

(5) 有 A、B 两个可逆循环，如图 5-11 所示。已知 T_1、T_2，求 A、B 两循环的热效率比，并求当 $T_1 \to \infty$ 时两循环热效率比的极限值；若 $T_1 = 1\ 000$ K，$T_2 = 300$ K，求两循环的热效率比。

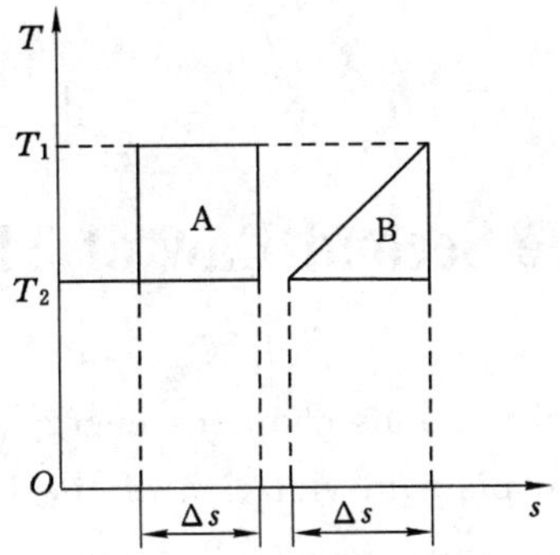

图 5-11　习题(5)图

(6) 有一循环发动机，工作于热源(T_1＝1 000 K)及冷源(T_2＝400 K)之间，从热源取热 1 000 kJ 而做功 700 kJ。问该循环发动机能否实现？

(7) 某气缸活塞热力装置内有 10 kg 空气，初始温度为 20 ℃，温度为 373 K 的热源向热力系供热 100 kJ，若气缸活塞上压有恒定质量的重物，求热力系在这一变化过程中熵的变化、熵产及做功能力损失。(比热为定值比热，环境温度为 300 K)

知识拓展：热力学温标

知识拓展：热力学第三定律

知识拓展：补充例题

Chapter 5 The Second Law of Thermodynamics

As mentioned repeatedly in previous chapters, energy is a conserved property and no process is known to have taken place in violation of the First Law of Thermodynamics. Therefore, it is reasonable to conclude that a process must satisfy the first law to occur. However, satisfying the first law alone does not ensure that the process actually take place. For example, the heat is always transferred naturally from warmer to colder bodies in the absence of other influence. Though the reversed heat flow certainly is not in violation of the first law, we all know that this process never takes place naturally.

It is clear from these arguments that processes proceed in a certain direction and not in the reverse direction. The first law places no restriction on the direction of a process, but satisfying the first law does not ensure that the process can actually occur. The inadequacy of the first law to identify whether a process can takes place is remedied by introducing another general principle, the Second Law of Thermodynamics. A process cannot occur unless it satisfies both the first and the second laws of thermodynamics. However, the use of the Second Law of Thermodynamics is not limited to identifying the direction of processes. It also asserts that energy has quality as well as quantity. The first law is concerned with the quantity of energy and the transformations of energy from one form to another with no regard to its quality. Preserving the quality of energy is a major concern to engineers, and the second law provides the necessary means to determine the quality as well as the degree of degradation of energy during a process. As discussed later in this chapter, more of high-temperature energy can be converted to work, and thus it has a higher quality than the same amount of energy at a lower temperature. The Second Law of Thermodynamics is also used in determining the theoretical limits for the performance of commonly used systems in engineering, such as heat engines and refrigerators, as well as predicting the degree of completion of chemical reactions.

5.1 Essence and Statements of the Second Law of Thermodynamics

5.1.1 Direction of Natural Processes and Irreversibility

All the processes occurring in nature only proceed in a certain direction naturally. Once having taken place, those processes cannot reverse themselves spontaneously and restore the system to its initial state. For this reason, they are classified as irreversible process. The factors that cause a process to be irreversible are called irreversibility. They include friction, unrestrained expansion, mixing of matters of different composition or state, heat transfer across a finite temperature difference, electric resistance, inelastic deformation of solids, and chemical reactions.

Heat transfer can occur only when there is a temperature difference between a system and its surroundings (Fig. 5-1). Therefore it is physically impossible to have a reversible transfer process. But a heat transfer process become less and less irreversible as the temperature difference between the two bodies approaches zero.

From the above discussions, we can find that all actual processes only proceed in the direction which makes the system tend to be in equilibrium with the surroundings. For example, heat is always transferred from high temperature body to low temperature one until two bodies achieve equilibrium. Moreover, all processes in nature are irreversible. A system can not restore to the initial state after it experiences any spontaneous process. Some condition is needed to ensure the reversed process to occur and it may leave influence on the surroundings. For instance, refrigerator and heat pump do accomplish the heat transfer from low temperature body to high temperature body, however, energy is consumed. And the work consumed along with the heat absorbed from the low temperature body is transferred to compensation which makes the non-spontaneous process become true.

In addition, there are various different irreversibility which cause actual process to be irreversible. They can be categorized into two kinds. One is the unbalanced potential, which may bring about energy losses. When heat is transferred from a high temperature body to a lower temperature one, it is degraded since less of it now can be converted to work. Heat transfer through a finite temperature difference, unrestrained expansion and mixing of two fluids are all irreversible process caused by unbalanced potential. The presence of these effects renders a process irreversible. This kind of effect is called internal irreversibility. Actual processes involve one or several kings of irreversibility unavoidably, thus any process concerning thermal phenomena are all irreversible.

5.1.2 Essence and Statements of the Second Law of Thermodynamics

There are numerous valid statements of the Second Law of Thermodynamics. Although each may not appear to have much connecting with the others, they are equivalent. Listed below are two classical statements of the second law often encountered—the Clausius statement, which is related to refrigerators or heat pumps and the Kelvin-Planck statement, which is related to heat engines.

(1) Clausius statement

The Clausius statement is expressed as follows: it is impossible to construct a device that operates in a cycle and produces no effect other than the transfer of heat from a lower-temperature body to a higher-temperature body.

(2) Kelvin-Planck statement

The Kelvin-Planck statement is expressed as follow: it is impossible for any device that operates on a cycle to receive heat from a single reservoir and produce a net amount of work.

Note that the impossible of having a 100 percent efficient heat engine is not due to friction or other dissipative effects. It is a limitation that applies to both the idealized and the actual heat engines. Later in this chapter, we will develop a relation for the maximum thermal efficiency of a heat engine. It is also demonstrated that the maximum value depends on the reservoir temperatures only.

A process cannot take place unless it satisfies both the first and second laws of thermodynamics. Any device that violates either law is called a perpetual-motion machine. And despite numerous attempts, no perpetual-motion machine is known to have worked. But this has no stopped inventors from trying to create new ones. A device that violates the First Law of Thermodynamics is called a perpetual-motion machine of the first kind (PMM1), and a device that violates the Second Law of Thermodynamics is called a perpetual-motion machine of the second kind (PMM2). The Second Law of Thermodynamics can also be expressed as PMM2 can not be created.

The Kelvin-Planck and the Clausius statements are equivalent in their consequences, and either statement can be used as the expression of the Second Law of Thermodynamics. Any devices that violates the Kelvin-Planck statement also violates the Clausius statement, and vice versa. This can be demonstrated as follows.

Consider the heat-engine-refrigerator combination shown in Fig. 5-2, operating between the same two reservoirs. The heat engine is assumed to have, in violation of the Kelvin-Planck statement. It absorbs the heat Q_1 from the high temperature reservoir. If its thermal efficiency is of 100 percent, and therefore it converts all the Q_1 it receives to work W_0. This work is now supplied to a refrigerator that removes heat in the amount of Q_2 from the low-temperature reservoir and rejects heat in the amount of Q_2+W_0 to the high-temperature reservoir. During this process, the high temperature reservoir receives a net amount of heat Q_2 (the difference between Q_1 and W_0). Thus, the combination of these two devices can be viewed as a refrigerator, as shown in Fig. 5-2, that transfers heat in an amount of Q_2 from a cooler body to a warmer one without requiring any input from outside. This is clearly a violation of the Clausius statement. Therefore, a violation of the Kelvin-Planck statement results in the violation of the Clausius statement. A reversible heat engine converts the Q_2 it receives to work W_0 and rejects Q_2+W to a low temperature sink. The work is supplied to a refrigerator.

It can also be shown in a similar manner that a violation of the Clausius statement leads to the violation of the Kelvin-Planck statement. Therefore, the Clausius and the Kelvin-Planck statements are two equivalent expressions of the Second Law of Thermodynamics.

In conclusion, the Second Law of Thermodynamics reveals that all spontaneous processes are irreversible. Non-spontaneous processes can not take place naturally. It can occur but need some proper spontaneous process accompanying as compensation.

5.2 The Carnot Cycle and Carnot Theorems

The Second Law of Thermodynamics states that no heat engine can achieve an efficiency of 100 percent. Then one may ask, what is the highest efficiency that a heat engine can possibly have?

5.2.1 The Carnot Cycle

The Carnot's cycle was proposed by Sadi Carnot (1796-1832), who is considered to be the founder of thermodynamics. The cycle describes a sequence of steps that can be performed only in an ideal engine, which transforms heat into work. The theoretical heat engine that operates on the Carnot cycle is called the Carnot heat engine. The Carnot cycle is composed of four reversible processes-two isothermal and two isentropic. The *p-v* diagram of this cycle is shown in Fig. 5-3. The four reversible processes that make up the Carnot cycle are as follows: reversible isothermal expansion process *a-b* ($T_1 = \text{const}$), reversible adiabatic expansion process *b-c* (temperature drops from T_1 to T_2), reversible isothermal compression process *c-d* ($T_2 = \text{const}$), reversible adiabatic compression process *d-a* (temperature rises from T_2 to T_1).

From *T-s* diagram of Carnot cycle, the efficiency of a Carnot engine is:

$$\eta = 1 - \frac{T_2}{T_1} \tag{5-1}$$

Then, you may ask what the thermal efficiency is if the Carnot engine does not work with ideal gases? The Carnot Principle answers the question.

5.2.2 The Reversed Carnot Cycle

The Carnot heat-engine cycle just described is a totally reversible cycle. Therefore, all the processes that comprise it can be reversed, in which case it becomes the Carnot refrigeration cycle. In this case, this cycle remains exactly the same as the Carnot cycle, except that the directions of any heat work interactions are reversed: heat in the amount of q_2 is absorbed from a low-temperature reservoir, heat in the amount of q_1 is rejected to a high-temperature reservoir, and a work input of w_{net} is required to accomplish all this. The *p-v* diagram of the reversed Carnot cycle is the same as the one given for the Carnot cycle, except that the directions of the process are reversed, as shown in Fig. 5-4.

The function of a refrigerator is to remove heat q_2 from a refrigerated space at a low temperature. The efficiency of a refrigerator is expressed by the coefficient of performance (COP), denoted by ε_c. The COP of a refrigerator can be expressed as:

$$\varepsilon_c = \frac{\text{Desired output}}{\text{Required input}} = \frac{q_2}{w_{net}} = \frac{q_2}{q_1 - q_2} = \frac{T_2(s_c - s_d)}{T_1(s_b - s_a) - T_2(s_c - s_d)} = \frac{T_2}{T_1 - T_2} \tag{5-2}$$

The function of a heat pump is to maintain a heated space at a high temperature. Its performance is also expressed in terms of the coefficient of performance, denoted by ε_h and it is defined as:

$$\varepsilon_h = \frac{q_1}{w_{net}} = \frac{q_1}{q_1 - q_2} = \frac{T_1}{T_1 - T_2} \tag{5-3}$$

COP_R is a positive quantity. Note that the amount of heat removed from the refrigerated space can be greater, less than or equal to the amount of work input. That is, the value of COP_R can be greater, less than or equal to unity. The definition Eq. (5-3) implies that the coefficient of performance of a heat pump is always greater than unity since COP_R is positive.

The same air-conditioning unit can be used as a heat pump in winter by installing it backwards. In this mode, the unit absorbs heat from the cold outside and delivers it into the room. Air-conditioning systems that are equipped with proper controls and a reversing valve can operate as air conditioners in summer and as heat pumps in winter.

5.2.3 The Carnot Theorems

The Carnot theorems include two conclusions pertaining to thermal efficiency of reversible and irreversible (i. e., actual) heat engines, expressed as follows:

① The thermal efficiency of an irreversible heat engine is always less than that of a reversible one operating between the reservoirs.

② The thermal efficiencies of all reversible heat engines operating between the same two reservoirs are the same.

These two statements can be proved by demonstrating that the violation of either statement results in the violation of the Second Law of Thermodynamics.

From the Carnot cycle and the Carnot theorems, we can make the following conclusions:

① Thermal efficiency of the Carnot cycle is just determined by the absolute temperatures of the heat source and sink, while it has nothing to do with the properties fluid and the type heat engine. The thermal efficiency of a Carnot engine with ideal gas as working medium is $\eta_c = 1 - \frac{T_2}{T_1}$. And thermal efficiencies of all reversible heat engine working between the two thermal reservoirs should be the same, that is $\eta_c = 1 - \frac{T_2}{T_1}$.

② When there is no temperature difference between the heat source and sink, that is $T_1 = T_2$, meaning only one heat source at constant temperature, the thermal efficiency becomes zero. This indicates that a heat engine absorbing heat from a single thermal reservoir can not fulfill the energy transformation to work. At lest two thermal reservoirs are indispensible for the conversion of heat to work.

The Carnot cycle is important since it describes an optimal system, where thermal energy is used at its best. Being a reversible cycle, the Carnot cycle is the most efficient one operating between two specified temperature limits. In thermodynamics, Carnot cycle and the Carnot Theorems have very important practical significances, such as:

① The Carnot efficiency is the highest efficiency a heat engine operating between two thermal energy reservoirs at temperatures T_2 and T_1 can achieve. It is:

$$\eta = 1 - \frac{T_2}{T_1}$$

Note that T_2 and T_1 are absolute temperatures. The Carnot heat engine is known to be the best reversible engine. All irreversible (i. e., actual) heat engines operating between the same temperature limits (T_2 and T_1) process lower efficiencies. An actual heat engine cannot reach the highest theoretical efficiency because it is impossible to entirely eliminate all the irreversibilities associated with the actual cycle.

② The Carnot efficiency can also be use to describe theoretical limits of the transformation of heat into mechanical work. Even though the Carnot cycle cannot be achieved in reality, the efficiency of actual cycles can be improved by attempting to approximate the Carnot cycle more closely. In order to improve the thermal efficiency of a heat engine, we should try to increase the heat source temperature T_1, to lower the temperature of the heat sink T_2, or to eliminate the irreversibilities at the most. However, the increase in T_1 is limited by thermal resistance performance of the equipment materials used. And T_2 should not be lowered to below the environmental temperature, T_0.

③ It indicates that only part of the heat absorbed by a system can be converted to work. And the largest amount of heat can be transformed into useful work is $W = Q(1 - \frac{T_0}{T_1})$, the rest pare $\frac{T_0}{T_1}Q$ has to be discharged into the atmosphere.

④ It also indicates that energy has quality as well as quantity. Different types of energies have different qualities even though their quantities are the same. As we know, the same amount of heat and work are not equivalent. Work can transform into heat totally without any conditions. However, heat can only be converted partly into work under certain specified conditions. Furthermore, the same amount of heat absorbed from heat sources at different temperatures can be transformed into different amount of useful work. The heat emitted by a heat source at higher temperature possesses better quality.

5.3 Entropy

5.3.1 Derivation of Entropy

Entropy can be derived and verified to be a sate property based on Carnot cycle and Carnot theorems. Let us consider a reversible cycle *a-b-c-d-a* shown in Fig. 5-7. Assume there are innumerable isentropic lines intersecting this cycle path and the distances between every two intersection lines are infinitesimal. Then, the processes line between them can be deemed as isothermal processes. Thus a series of differential Carnot cycle are formed. As for each of them, the thermal efficiency is:

$$\eta_{t,c} = 1 - \frac{\delta q_2}{\delta q_1} = 1 - \frac{T_2}{T_1}$$

Thus:

$$\frac{\delta q_1}{T_1} + \frac{\delta q_2}{T_2} = 0$$

By integrating along the whole reversible cycle, it yields:

$$\int_{abc} \frac{\delta q_1}{T_1} + \int_{cda} \frac{\delta q_2}{T_2} = \oint \left(\frac{\delta q}{T}\right)_{re} = 0 \tag{5-4}$$

Eq. (5-4) is called the Clausius equality, in which the cyclic integral of $\frac{\delta q}{T}$ can be viewed as the sum of all these differential amounts of heat transfer divided by the temperature at the boundary. And the cyclic integral of $\left(\frac{\delta q}{T}\right)_{re}$ is equal to zero. It independent that the $\left(\frac{\delta q}{T}\right)_{re}$ is independent of the path along which the integration is performed and it possesses the performance of a property. It just depends on the state and the $\left(\frac{\delta q}{T}\right)_{re}$ must represent a property in differential form. Clausius named this property as entropy, a new thermodynamic property, in 1865. The specific entropy is designated by s and it has the unit J/(kg · K). It is defined as:

$$ds = \left(\frac{\delta q}{T}\right)_{re} \tag{5-5}$$

Thus, Eq. (5-4) can also be written as:

$$\oint ds = 0 \tag{5-6}$$

As the total entropy s is an extensive property, $S = m \cdot s$, J/K. Note that entropy is a property, and like all other properties, it has fixed values at fixed states. Therefore, the entropy change ΔS between two specified states is the same no matter what path, reversible or irreversible, is followed during a process. The entropy change of a system during a process can be determined by integrating Eq. (5-5) between the initial and final state:

$$\int_1^2 ds = s_2 - s_1 \tag{5-7}$$

From Eq. (5-4) we know that for any reversible heat absorption or rejection processes, the change in entropy of the working fluid is equal to the amount of heat transfer divided by the temperature of the heat sources or heat sinks, that is $\left(\frac{\delta q}{T}\right)_{re}$. The change of the entropy reveals the direction and the amount of heat transferred. The entropy increase indicates that a system absorbs heat from its surroundings reversibly as $\delta Q > 0$; And the entropy decrease indicates that a system rejects heat to its surroundings reversibly as $\delta Q < 0$.

Note that the integral of $\frac{\delta q}{T}$ gives us the value of entropy change only if the integration is carried out along a reversible path between the two states. The integral of $\frac{\delta q}{T}$ along an irreversible path is not a property.

5.3.2 Clausius Inequality and Entropy Change during an Irreversible Process

Consider a cycle that is made up of two processes: an irreversible process 1-*a*-2 and a reversible process 2-*b*-1, as shown in Fig. 5-8. Assume there are innumerable isentropic lines intersecting the cycle path and the distances between every two intersection lines are infinitesimal. Thus, the processes line between them can be deemed as isothermal processes. Then, a series of differential irreversible cycles are formed, whose integration effect is the original cycle. In terms of Carnot theorem, an actual (i. e. irreversible) heat engine, is less efficient than a reversible one operating between the same two thermal reservoirs. Then, we have:

$$\eta_t = 1 - \frac{\delta q_2}{\delta q_1} < 1 - \frac{T_2}{T_1}$$

Thus:

$$\frac{\delta q_1}{T_1} + \frac{\delta q_2}{T_2} < 0$$

Integrating along the whole irreversible cycle, we obtain that:

$$\int_{1a2} \frac{\delta q_1}{T_1} + \int_{2b1} \frac{\delta q_2}{T_2} = \oint \left(\frac{\delta q}{T}\right)_{irr} < 0 \tag{5-8}$$

That is, the cyclic integral of $\delta Q/T$ is always less than zero for irreversible cycles.

Combining Eq. (5-4) and Eq. (5-8) yields:

$$\oint \frac{\delta q}{T} \leqslant 0 \tag{5-9}$$

It mean that the cyclic integral of $\delta Q/T$ is never greater than zero. This inequality is valid of all cycles, reversible or irreversible. It is the Clausius Inequality. The first integral in Eq. (5-8) is recognized as the entropy change $s_2 - s_1$. Thus:

$$\int_1^2 \left(\frac{\delta q}{T}\right)_{irr} - (s_2 - s_1) < 0$$

That is:

$$s_2 - s_1 > \int_1^2 \left(\frac{\delta q}{T}\right)_{irr} \tag{5-10}$$

Note that T is the absolute temperature of the heat source. Eq. (5-10) states that the entropy changes of a system during irreversible processes are always greater than the integral of $\left(\frac{\delta Q}{T}\right)_{irr}$ from the initial to the final state. It means that the entropy change in an irreversible process is greater than the heat added to the system divided by the heat source temperature.

It can also be expressed in a differential form:

$$ds \geqslant \frac{\delta q}{T} \tag{5-11a}$$

and

$$s_2 - s_1 \geqslant \int_1^2 \frac{\delta q}{T} \tag{5-11b}$$

This is also called the Clausius Inequality, first stated by the German physicist R. J. E. Clausius (1822-1888), one of the founders of thermodynamics. In Eq. (5-11), the equality holds for an reversible process and the inequality for an irreversible process. We may conclude from these equations that the entropy change of a system during an irreversible process is greater than the integral of $\delta q/T$ evaluated for that process. In the limiting case of a reversible process, these two quantities become equal. We again emphasize that T in these relations is the thermodynamic temperature at the boundary where the differential heat δq is transferred between the system and the surroundings.

The above equation can serve as a criterion in determining whether a process is irreversible or impossible. A process never takes place in nature if $ds > \frac{\delta q}{T}$. It means that Eq. (5-11c) can be used to judge the characteristics of a process. That is:

$$\text{if} \begin{cases} ds > \dfrac{\delta q}{T} \\ ds = \dfrac{\delta q}{T} \\ ds < \dfrac{\delta q}{T} \end{cases} \text{ in a process, then the process is} \begin{cases} \text{irreversible} \\ \text{reversible} \\ \text{impossible to occur in nature} \end{cases} \tag{5-11c}$$

Define $\delta s_f = \frac{\delta q}{T}$ as the entropy transfer by heat (T is the temperature of the heat source). It is the entropy change caused by the heat transfer between a system and its surroundings. The direction of the entropy transfer by heat is the same as that of heat transfer. Heat transfer to a system increases the entropy of that system, and heat transfer from a system decreases it. If a system is adiabatic, then the entropy transfer by heat is zero. The increase of entropy principle states that an irreversible process occurs in the direction in which the entropy increases. The entropy increase entirely due to the presence of irreversibilities during the process is called entropy generation, denoted by s_g or in differential form as δs_g. Irreversibilities, such as friction, mixing, chemical reactions, heat transfer through a finite temperature difference, unrestrained expansion, non-quasi equilibrium compression and expansion, etc., always cause the entropy of a system to increase, and entropy generation is a measure of the entropy created by such effects during a process.

Therefore, the entropy change of a system during a process is equal to the net entropy transfer through the system boundary and the entropy generated within the system.

$$ds = \delta s_f + \delta s_g \tag{5-12a}$$

$$\Delta s_{12} = s_f + s_g \tag{5-12b}$$

5.4 The Increase of Entropy Principle and Loss of Conversion Ability of Heat to Work

5.4.1 The Increase of Entropy Principle

There is no mass and energy transfer between an isolated system and its surroundings. From the Clausius inequality $\Delta s \geqslant \int_1^2 \frac{\delta q}{T}$, it can be found that isolated systems it reduces to:

$$\Delta s_{iso} \geqslant 0 \quad \text{or} \quad ds_{iso} \geqslant 0 \tag{5-13}$$

This equation can be expressed as the entropy of an isolated system during a process alway increases or, in the limiting case of a reversible process, remains constant. In other words, it never decreases. This is known as the increase of entropy principle. Note that entropy change in an isolated system is due to irreversibilities only, and their effects are always to increase entropy. Thus, the entropy change of an isolated system is also called entropy generation. It can be seen from Eq. (5-13) that the entropy generation is always positive.

The increase of entropy principle has very important theoretical significances. While using it, we should note that:

(1) The increase of entropy principle can be used to judge the direction along which a natural process proceeds. Processes can only occur in a certain direction, not in any direction. In nature, a process must proceed in the direction that complies with the increase of entropy principle, that is, $\Delta S_{iso} \geqslant 0$. If $\Delta S_{iso} = 0$, then the process is reversible. If $\Delta S_{iso} < 0$, then it is impossible for the process to occur in fin isolated system.

(2) The increase of entropy principle is suitable for processes occurred in isolated systems. The total entropy of a system is equal to the sum of the entropies of every Parts of the system. A system and its surroundings may constitute an isolated system since they can be enclosed by a sufficiently large arbitrary boundary across which there is no heat, work, or mass transfer. Therefore, a system and its surroundings can be viewed as the two subsystems of the isolated system. And the entropy change of this isolated system during a process is the sum of the entropy changes of the system and its surroundings. The entropy of an isolated system during a process always increases or, in the limiting case of a reversible process, remains constant. However, the entropy change of the two subsystems may increase, decrease or remain constant.

(3) It is pointed out the condition under which a process can take place. It is impossiblc for a process violating this principle to occur in nature. A non-spontaneous process must be accompanied by a process whose entropy increases to occur. As a result, the entropy change of the new isolated system, consisting of the original isolated system

and the accompanying process, is positive. This is the compensation condition that ensures the non-spontaneous process to occur.

(4) Systems in nature have a tendency to change until they attain an equilibrium state. The increase of entropy principle dictates that the entropy of an isolated system increases until the entropy of the system reaches a maximum value. At that point, the system is said to have reached an equilibrium state since the increase of entropy principle prohibits the system from undergoing any change of state that results in a decrease in entropy.

(5) The increase of entropy principle can be used as a quantitative measure of irreversibilities associated with a process in isolated system. The presence of irreversibilities brings about the entropy increase, which is also called entropy generation. It is a measure of the magnitudes of the irreversibilities presented during that process. The greater the extent of irreversibilities, the greater is the entropy generation.

However, the increase of entropy principle does not imply that entropy of system cannot decrease. The entropy change of a system can be negative during a process, though the entropy generation cannot.

5.4.2 Loss of the Conversion Ability of Heat to Work

With the temperature range between T_1 and T_2, Carnot thermal efficiency is the highest. If the system absorbs certain amount of heat Q_1, among which the quantity can be transformed into work is $Q_1(1-T_2/T_1)$. A part of heat not being transformed into work will be released to a low-temperature heat source as useless heat Q_2, $Q_2=Q_1(T_2/T_1)$. From this equation, we can conclude that increasing T_1 or decreasing T_2 will increase the useful energy percentage. Meanwhile, the useless energy percentage will be lowered.

Therefore, for any heat source whose temperature is T, its maximum useful heat which can be transformed into work is $Q(1-T_0/T)$. It is dependant on the temperature of T. The useless energy of $Q(1-T_0/T)$ is called waste heat. This amount of heat energy can not be transformed into work, and they should be released to atmospheric environment, the low-temperature heat source.

In engineering, we consider the conversion of heat to work as an ability to do work which can provide heat Q.

The decrement of the useful heat energy equals to the product of the lowest temperature T_0 and the entropy generation caused by irreversibilities. In another word, we can conclude that the decrement of the useful heat energy is proportional to the total entropy generation.

According to the Second Law of Thermodynamics, even in the ideal Carnot thermal cycle, the conversion of heat to work will cause a certain amount of heat $Q(T_2/T_1)$ released from the high-temperature heat source to a low-temperature one. Though the heat quantity remains constant, the conversion ability of heat to work is decreased due to the heat transfer to a low-temperature heat source. That is, the heat quality is reduced, or the energy level is degraded. Restricted by the environment temperature T_0, such a loss in the conversion ability of heat to work is inevitable.

Reducing the irreversibilities in the conversion can cut the loss to a minimum extent. The irreversibilities cause an isolated thermal system's entropy production, which is proportional to the decrement of heat conversed into work. Therefore, entropy generation call be regarded as a measure to indicate the waste heat transformed from the absorbing heat of the irreversible processes.

Similarly, such conclusion of $L = T_0 \Delta S_{iso}$ can also be drawn in other analysis of irreversible processes. It is a general expression for the loss in conversion of heat to work from irreversibilities.

第6章 水蒸气与湿空气

本章导学

水蒸气是人类使用最早、应用最广的工质。水蒸气由于接近液态，故不宜按理想气体处理，其热力参数的计算较为复杂，因此工程上常采用查图表方法或借助计算机进行求解。含有水蒸气的空气称为湿空气，在通风、空调及干燥工程中，为了满足一定的生活及生产工艺要求，需要对湿空气进行必要的处理。本章主要介绍水蒸气的形成过程、状态参数的确定方法、湿空气的性质以及相应的基本热力过程的能量分析计算。

学习目标

在理解饱和状态概念的基础上，熟悉水蒸气的定压形成过程及其特性，掌握水蒸气在 p-v 图和 T-s 图上的一点、两线、三区和五态。掌握利用水蒸气的热力性质图表确定其状态参数的方法。熟悉水蒸气基本热力过程的分析与简单计算。掌握湿空气的相关概念及性质，掌握未饱和湿空气和饱和湿空气的含义。了解湿空气的简单热力计算。

重点与难点

(1) 实际气体与理想气体。

(2) 正确使用水蒸气热力性质图表分析其热力参数与热力过程。

(3) 干球温度、露点、绝热饱和温度和湿球温度之间的关系。

(4) 含湿量与相对湿度。

(5) 焓湿(h-d)图上湿空气状态参数的确定。

工程上用的气态工质可以分为两类——气体和蒸气，两者之间并无严格的界限。蒸气泛指刚刚脱离液态或比较接近液态的气态物质，在被冷却或被压缩时很容易变回液态。一般地说，蒸气分子间的距离较小，分子间的作用力及分子本身的体积不能忽略，因此，蒸气一般不能作为理想气体处理。工程上常用的蒸气有水蒸气、氨蒸气、氟利昂蒸气等。水蒸气来源丰富，耗资少，无毒无味，比热容大，传热好，有良好的膨胀和载热性能，是热工技术上应用最广泛的一种工质。水蒸气是人类在热力发动机中应用最早的实现热能向机械能转换的工质，至今仍是工业上广泛应用的主要工质。各种蒸气虽然各有特点，但其热力性质及物质变化规律都有许多相似之处。本章以水蒸气为例研究其产生、状态的确定及基本热力过程。

在自然界中，由于江河湖海里水的蒸发，空气中总含有一些水蒸气。这种含有水蒸气的空气称为湿空气。完全不含水蒸气的空气称为干空气。一般认为湿空气是由水蒸气和干空气组成的混合物。由于湿空气中水蒸气的含量极少，在某些情况下往往可以忽略水蒸气的影响。但是，在干燥、空气调节以及精密仪表和电绝缘的防潮等对空气中的水蒸气特别敏感的领域，以及湿空气对干燥过程的速度、对人体舒适感产生影响时，就不能将湿空气视为干空气来进行分析和计算，必须考虑空气中水蒸气的影响。由于干空气状态远离临界点，可视它为一种理想气体。湿空气中含有少量的水蒸气，水蒸气的分压力很低，因而湿空气中的水蒸气也可被认为是理想气体。这样，当水蒸气不发生相变时，湿空气就被认为是由干空气和水蒸气两种理想气体组成的二元混合物。但是，湿空气中的水蒸气常会发生聚集状态的变化，如结露、结霜、蒸发等，因此湿空气又不完全等同于理想气体的混合物。为描述湿空气的特点，本章引入一些反映水蒸气含量和聚集状态变化的专门的状态参数来研究湿空气的特性。

6.1　水蒸气的产生过程及 p-v 图和 T-s 图

工程上应用的水蒸气，通常是在锅炉内对水定压加热产生的。下面用图 6-1 说明水蒸气的产生过程：假设一带有活塞的气缸中盛有 1 kg、0 ℃的水，对活塞施加一定的压力 p，在容器底部对水加热。水蒸气的产生过程一般可以分为以下三个阶段。

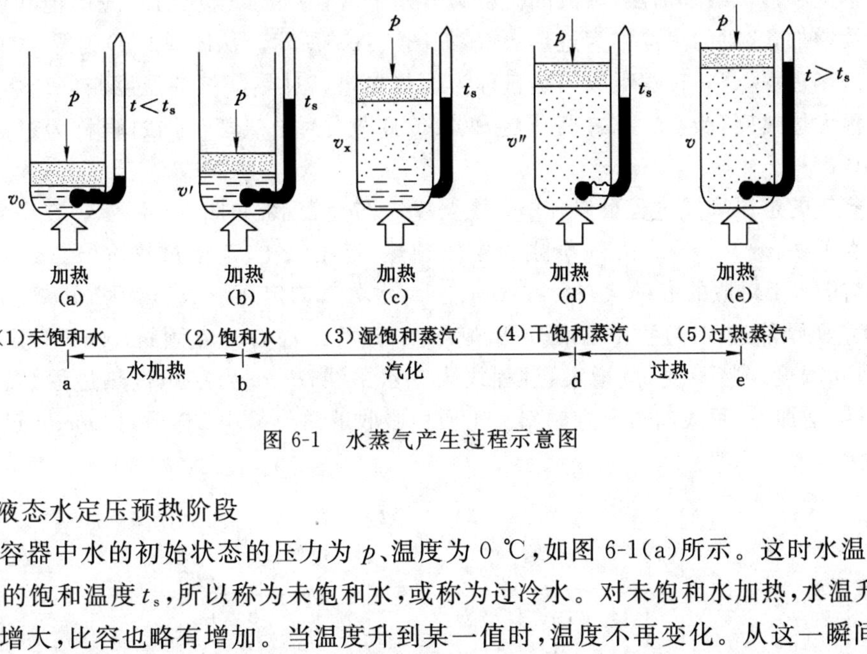

图 6-1　水蒸气产生过程示意图

(1) 液态水定压预热阶段

假设容器中水的初始状态的压力为 p、温度为 0 ℃，如图 6-1(a)所示。这时水温低于压力 p 对应的饱和温度 t_s，所以称为未饱和水，或称为过冷水。对未饱和水加热，水温升高，分子间距离增大，比容也略有增加。当温度升到某一值时，温度不再变化。从这一瞬间起，水开始沸腾，并开始变成蒸汽，这一温度称为该压力(p)下的沸腾温度，即饱和温度，用 t_s表示。此时的压力称为饱和压力，用 p_s表示。此时水达到饱和状态，称为饱和水，如图 6-1(b)所示。水在定压下从未饱和水状态加热到饱和水状态的过程，称为定压预热阶段，即图 6-1 中 a→b。将 1 kg 的未饱和液体从三相点(对于水即冰、水和汽三相平衡共存)状态定压加热成为饱和液体过程中所吸收的热量称为液体热。

(2) 饱和水定压汽化阶段

如对饱和水继续加热，则饱和水开始沸腾，不断地变为蒸汽，水温保持饱和温度 t_s不变。在定温 t_s条件下水沸腾产生大量蒸汽，形成了饱和水和饱和水蒸气的混合物，这种混合物称为湿饱和蒸汽，简称为湿蒸汽，如图 6-1(c)所示。由于湿蒸汽的温度和压力是两个互相依赖的参数，所以给出湿蒸汽的温度和压力并不能确定湿蒸汽的状态。由于湿蒸汽由压力、温度相同的干饱和蒸汽和饱和水按不同的质量比例组成，所以要具体地确定湿蒸汽所处的状态，除了说明它的压力或温度外，还必须指出干饱和蒸汽和饱和水的质量比例。湿蒸汽中所含有的干饱和蒸汽的质量分数，称为湿蒸汽的干度，用 x 表示，即

$$x=\frac{m_{\mathrm{v}}}{m_{\mathrm{w}}+m_{\mathrm{v}}}$$

式中，m_{v}和 m_{w}分别表示湿蒸汽中所含干饱和蒸汽和饱和水的质量。

当对湿蒸汽继续加热直到最后一滴水变为蒸汽时，容器中的蒸汽称为干饱和蒸汽，简称干蒸汽，如图 6-1(d)所示。

将饱和水定压加热为干饱和蒸汽的过程称为汽化阶段，即图 6-1 中 b→d。该阶段的特点是，温度、压力保持不变，所加入的热量用于使水汽化变为蒸汽以及比容增大对外做膨胀功。将 1 kg 饱和液体加热变为同温度的干饱和蒸汽所需的热量，称为汽化潜热，常用 r 表示，其值完全取决于压力或温度。

(3) 干饱和蒸汽定压过热阶段

对干饱和蒸汽继续加热，蒸汽的温度又开始上升，超过了该压力对应的饱和温度，其比容也继续增加，这时的蒸汽称为过热蒸汽，如图 6-1(e)所示。从图 6-1(d)到图 6-1(e)为蒸汽的定压过热过程，这一过程吸收的热量称为过热热量。过热蒸汽的温度与同压力下饱和温度之差称为过热度，$\Delta t=t-t_{\mathrm{s}}$。将干饱和蒸汽加热成为过热蒸汽的过程称为过热阶段，即图 6-1 中 d→e。

水蒸气的定压形成过程可以在 p-v 图与 T-s 图上表示，如图 6-2 所示。在 p-v 图上，它是一条水平线，a—b、b—d、d—e 分别为定压预热、定压汽化、定压过热阶段。a、b、c、d、e 分别表示与图 6-1 对应的五种状态。在 T-s 图上，水蒸气的定压形成过程线 $abcde$ 分为三段：a—b 为定压预热过程，过程中温度升高，熵增大，过程线向右上方倾斜；b—d 为定压汽化过程，压力和温度保持不变，熵增大，过程线为一水平线；d—e 为定压过热过程，温度开始升高，熵继续增加，过程线向右上方倾斜。工质所吸收的总热量由 T-s 图上 $abcde$ 过程线下的面积表示。

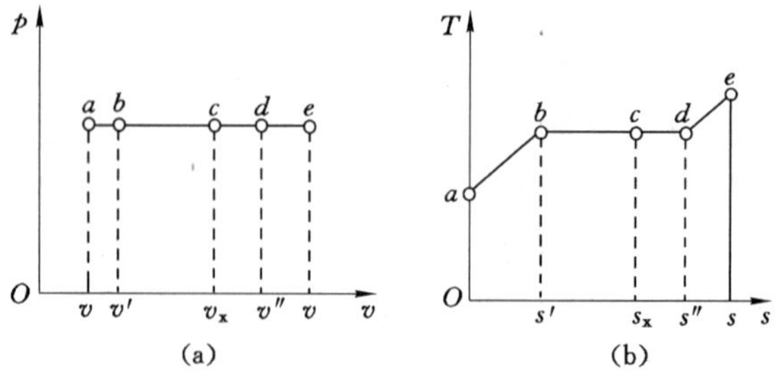

图 6-2　水蒸气定压产生过程的 p-v 图与 T-s 图

如果将不同压力下蒸汽的形成过程表示在 p-v 图与 T-s 图上，并将不同压力下对应的状态点连接起来，就得到了图 6-3 中的 $a_1a_2a_3$…线、$b_1b_2b_3$…线以及 $d_1d_2d_3$…线，它们分别表示各种压力下的 0 ℃的水、饱和水以及干饱和蒸汽状态。$a_1a_2a_3$…线近乎一条垂直线，这是

因为低温时的水几乎不可压缩，压力升高，比容基本不变。$b_1b_2b_3\cdots$线称为饱和水线或下界线，它表示的是不同压力下饱和水的状态。$d_1d_2d_3\cdots$线称为干饱和蒸汽线或上界线，它表示的是不同压力下干饱和蒸汽的状态。

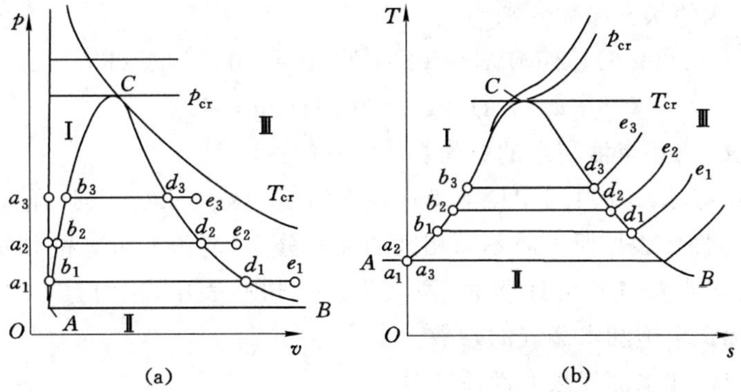

图 6-3　不同压力下水蒸气产生过程的 p-v 图和 T-s 图

由图 6-3 可以清楚地看到，随着压力的增加，饱和水与干饱和蒸汽两点间的距离逐渐缩短。当压力增加到某一临界值时，饱和水与干饱和蒸汽不仅具有相同的压力和比容，而且还具有相同的温度和熵，这时的饱和水与干饱和蒸汽之间的差异已完全消失，在图中由同一点 C 表示。这个点称为临界点，此时饱和水与饱和蒸汽已不再有区别，这样一种特殊的状态称为临界状态。临界参数仅取决于物质种类，每种物质只有一组临界参数。对于水而言，其临界参数为：临界压力 $p_{cr}=22.064$ MPa、临界温度 $t_{cr}=373.99$ ℃、临界比容 $v_{cr}=0.003\ 106$ m^3/kg、$h_{cr}=2\ 085.9$ kJ/kg、$s_{cr}=4.409\ 21$ kJ/(kg·K)。水在临界压力 p_{cr} 下定压加热到临界温度 t_{cr} 时，不存在汽液分界线和汽液共存的汽化过程，再加热就直接成为过热蒸汽，t_{cr} 是最高的饱和水温度。临界点的存在表明，在临界压力下，当温度达到临界温度 t_{cr} 时，液体的汽化过程在瞬间完成。也就是说，压力大于或等于临界压力，就不再存在气液两相共存的湿蒸汽状态。如果工质的温度 t 高于临界温度 t_{cr}，不论压力多高，其状态均为气态；也就是说，当 $t>t_{cr}$ 时，保持温度不变，无论 p 多高也不能使气体液化，因此通常将 $t>t_{cr}$ 的气体称为永久性气体。

可见，临界点以下的未饱和水的定压加热汽化过程在 p-v 图和 T-s 图上的共同特点可以归纳为：

一点：临界点 C；

两线：饱和水线(AC)和饱和蒸汽线(CB)；

三区：未饱和水区(Ⅰ)、湿蒸汽区(Ⅱ)和过热蒸汽区(Ⅲ)；

五态：未饱和水、饱和水、湿饱和蒸汽、干饱和蒸汽和过热蒸汽状态。

6.2　水蒸气性质表及焓熵图

理想气体的状态方程是 $pv=RT$，但是水蒸气不同于理想气体，其状态方程相当复杂，不宜直接用于工程计算。因此，在工程中，常使用基于实验和理论数据绘制成的水蒸气性质表和图。

6.2.1 水蒸气的性质表

对于水和水蒸气，根据1963年国际水蒸气会议的规定，以三相点(即固、液、汽三相共存状态)状态下的饱和水作为基准点，即273.16 K的液相水，规定该状态下液相水的内能和熵为零，三相点液相水的状态参数分别为：

$$p=0.000\ 611\ 2\ \text{MPa}, t=0.01\ ℃, v=0.001\ 000\ 21\ \text{m}^3/\text{kg},$$
$$u=0\ \text{kJ/kg}, s=0\ \text{kJ/(kg·K)}$$

根据焓的定义，三相点液相水的焓为：

$$h=u+pv=0\ \text{kJ/kg}+611.2\ \text{Pa}\times 0.001\ 000\ 21\ \text{m}^3/\text{kg}=0.000\ 61\ \text{kJ/kg}\approx 0\ \text{kJ/kg}$$

水蒸气的热力性质表按照工质不同相区的性质分为"饱和水和饱和水蒸气热力性质表"和"未饱和水和过热水蒸气热力性质表"两大类。在这些表中，用符号′来标识饱和水的参数，而用符号″来标识干饱和水蒸气的参数。

(1) 饱和水与饱和水蒸气热力性质表

由于饱和状态的压力与温度是互相不独立的单值函数，因此，其热力性质表有按温度排列和按压力排列两种形式，详见附录3及附录4。这两种表均给出了饱和水和饱和水蒸气的比容、比焓、比熵及汽化潜热。附表7是以温度为自变量排列的，而附表8则以压力为自变量来排列。因此，如果给定温度，则使用附表7更方便；如果给定压力，则使用附表8更方便。

饱和水与饱和水蒸气的热力性质表具有以下两个作用：

① 对于饱和水与干蒸汽，只要知道压力和温度中的任何一个参数，就可以从饱和水与饱和水蒸气表中直接查得其他参数，利用该表还可以判别工质的状态处于5种状态的哪一种。

② 计算确定湿饱和蒸汽的参数。由于湿蒸汽是由压力、温度相同的饱和水与干蒸汽所组成的混合物，要确定其状态，除了知道它的压力(或温度)外，还必须知道它的干度 x。1 kg湿蒸汽是由 x kg干蒸汽和 $(1-x)$ kg饱和水混合而成的，因此，1 kg湿蒸汽的各有关参数就等于 x kg干蒸汽的相应参数与 $(1-x)$ kg饱和水的相应参数之和，即

湿蒸汽的比容： $v_x = xv'' + (1-x)v' = v' + x(v''-v')$

湿蒸汽的焓： $h_x = xh'' + (1-x)h' = h' + x(h''-h')$

湿蒸汽的熵： $s_x = xs'' + (1-x)s' = s' + x(s''-s')$

(2) 未饱和水与过热水蒸气热力性质表

未饱和水与过热水蒸气是单相物质，它们的各个状态参数均是彼此独立的，由任意两个参数就可以确定蒸汽状态，其表格数据参见相关资料。对于未饱和水和过热蒸汽，已知任何两个状态参数都可以由该表确定出其他状态参数。

6.2.2 水蒸气的焓熵图

在 T-s 图和 p-v 图上虽可用过程曲线与坐标轴所包围的面积来表示过程中工质与外界交换的热量和功量，但是定量地进行能量分析计算却有其不便之处。若采用焓熵图(h-s 图)，则可以直接用线段的长度来表示热量和功量，非常直观，因而焓熵图在工程计算中得到了广泛的应用。

水蒸气的 h-s 图如图6-4所示，h-s 图以焓为纵坐标，熵为横坐标。利用水蒸气性质表的

数据，首先在图上绘制饱和水线 $x=0$、饱和蒸汽线 $x=1$ 及其交点，即临界点 C。饱和水线始于坐标原点。在湿蒸汽区，绘制了等干度线簇，即 $x=$ 常数，它们汇合于临界点 C。湿蒸汽区内等压线是倾斜的直线，且定压线与定温线一致。从干饱和蒸汽线开始，定压线簇与定温线簇分开。在过热蒸汽区内，定压线和定温线的走向如图 6-4 所示，定压线较陡，而定温线较平坦。在低压下，定温线接近水平线，随着压力的增加，定温线的斜率也增加。在 h-s 图上，还绘制了定容线簇，定容线的走向与定压线相近，但比定压线陡，如图中的虚线所示。

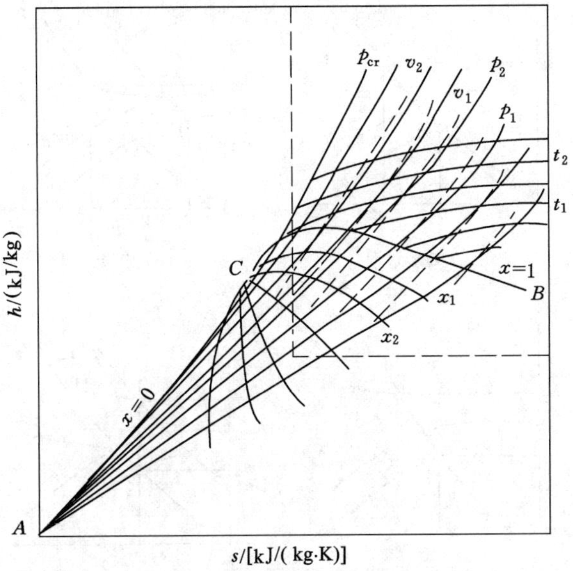

图 6-4　水蒸气的 h-s 图

利用 h-s 图可确定水蒸气的状态参数，其优点是简便、直观，但其读数不够准确，而利用水蒸气性质表确定水蒸气状态参数，优点是读数较准确，但有时需要插值计算。因此，工程上常同时借助水蒸气的 h-s 图和表来分析水蒸气的热力过程，可大大地简化计算。

因为 $x<0.5$ 区域内的图线过于密集，工程中又不经常使用这一区域的数据，通常 h-s 图仅绘制出干度 $x>0.5$ 的湿蒸汽和过热蒸汽部分，工程中应用的水蒸气的 h-s 图可查阅相关资料。

6.3　水蒸气的热力过程

与理想气体一样，分析水蒸气热力过程的目的是了解过程中工质状态的变化规律，确定过程中工质与外界的能量交换。水蒸气热力过程的分析计算不像理想气体一样利用简单的解析式，只能利用热力学第一及第二定律的基本方程，借助水蒸气热力性质图表进行计算。

在热工计算中，经常遇到蒸汽的定压过程和绝热过程。例如，采用水蒸气为工质的换热器均为定压加热或放热过程；水泵、喷管、汽轮机的工作均为绝热过程。蒸汽的形成与凝结过程都是在定压下进行的，这种情况下蒸汽与外界之间无功量交换，与外界交换的热量可以用焓差表示。蒸汽在蒸汽机或汽轮机中的膨胀做功过程，可近似认为是绝热过程，如不考虑摩擦损失，则为可逆绝热过程，在 h-s 图上为一垂直线，过程中与外界之间的功量交换也可用焓差表示。所以，h-s 图用于水蒸气热力过程的定量计算极为方便。

具体分析计算时，一般按下列步骤进行：

(1) 根据初态的给定条件，在 h-s 图上(或水蒸气性质表中)查出初态的其他参数值。

(2) 根据初态和过程的特点以及终态的一个参数值，确定终态，查出终态的其他参数值，并将过程表示在坐标图上，如图 6-5 所示。

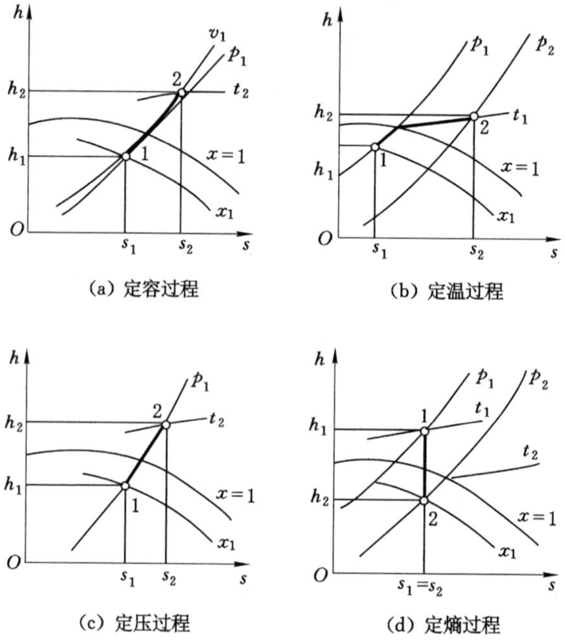

(a) 定容过程　(b) 定温过程

(c) 定压过程　(d) 定熵过程

图 6-5　水蒸气的基本热力过程

(3) 根据热力学第一及第二定律，结合过程的特点，计算过程中工质与外界交换的能量，所采用的计算公式如下。

定容过程：

$$w = 0, w_t = v(p_1 - p_2),$$
$$q = \Delta u = u_2 - u_1 = (h_2 - h_1) - v(p_2 - p_1)$$

定压过程：

$$w = p(v_2 - v_1), \omega_t = -\int_1^2 v\mathrm{d}p = 0, q = h_2 - h_1$$

定温过程：

$$w = q - \Delta u = T(s_2 - s_1) - (h_2 - h_1) + (p_2 v_2 - p_1 v_1),$$
$$w_t = q - \Delta h = T(s_2 - s_1) - (h_2 - h_1),$$
$$q = \int_1^2 T\mathrm{d}s = T(s_2 - s_1)$$

定熵过程：

$$w = -\Delta u = u_1 - u_2 = (h_1 - p_1 v_1) - (h_2 - p_2 v_2),$$
$$w_t = -\Delta h, q = 0$$

6.4　湿空气性质

6.4.1　饱和湿空气和未饱和湿空气

根据道尔顿分压定律，湿空气的总压力 p 等于水蒸气的分压力 p_v 与干空气的分压力 p_a 之和，即

$$p = p_v + p_a \tag{6-1}$$

设湿空气温度为 t，则其中水蒸气在给定的温度下有一对应的饱和分压力 $p_s(t)$。若水蒸气分压力 $p_v(t) < p_s(t)$，则空气中所含的水蒸气处于过热蒸汽状态，如图 6-6 中点 A，此时的湿空气称为未饱和湿空气。未饱和湿空气是干空气和过热水蒸气的混合物。在温度 t 不变的情况下，向未饱和湿空气中加入水蒸气，水蒸气分压力 p_v 增大，如图 6-6 中状态由点 A 向点 C 移动，当达到饱和蒸汽线上点 C 时，$p_v = p_s(t)$，则空气中所含的水蒸气处于干饱和蒸汽状态，此时的湿空气称为饱和湿空气。饱和湿空气是干空气和饱和水蒸气的混合物。由于 $p_s(t)$ 为温度 t 下水蒸气所能达到的最大压力，这时空气中水蒸气的含量也达到最大值，不可能再增加。即使强行加入，也会有液态水珠析出，也就是说饱和湿空气不具有吸水能力。

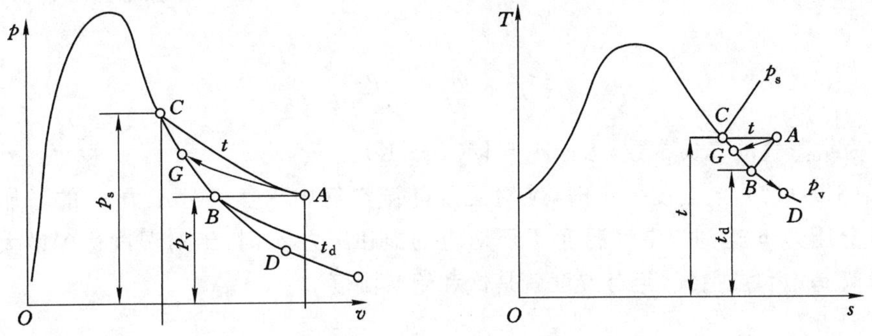

图 6-6　湿空气中水蒸气状态的 p-v 图和 T-s 图

当饱和湿空气的温度升高时，饱和湿空气即变成未饱和湿空气。例如，当湿空气的温度为 20 ℃、水蒸气的分压力 $p_v = 0.002\ 338\ 5$ MPa 时，是饱和湿空气。而在定压下将湿空气的温度提高到 30 ℃时，对应的饱和压力 $p_s = 0.004\ 245\ 1$ MPa。因 $p_s > p_v$，这时的湿空气就是未饱和湿空气。

因此湿空气有两大类：

未饱和湿空气＝干空气＋过热蒸汽

饱和湿空气＝干空气＋干饱和蒸汽

6.4.2　结露和露点

如果将未饱和湿空气（A 点）在定压（p_v 不变）下逐渐冷却，由于湿空气的总压力、干空气和水蒸气的分压力均不变，所以过程中水蒸气的状态将沿着分压力 p_v 不变的过程 A—B 变化。当温度降到与水蒸气分压力相对应的饱和温度（如图 6-6 中的 B 点所示）时，湿空气中的水蒸气便由过热状态变为饱和状态，相应的湿空气也就由未饱和湿空气变为饱和湿空气。

若继续冷却降温，则其中的部分水蒸气将凝结为水，即出现所谓的结露现象。开始结露时的温度称为露点，记作 t_d。换句话说，露点是湿空气在定压条件下被冷却到开始析出水分时的温度，也就是对应于水蒸气分压力的饱和温度，即

$$t_d = t_s(p_v) \tag{6-2}$$

达到露点后若继续冷却，就会有水蒸气凝结成水滴析出，湿空气中的水蒸气状态将沿着饱和水蒸气线变化，如图 6-6 中的 B—D 所示。这时，温度继续降低，水蒸气分压力也随之降低。结露在初秋早晨的草地上最为常见，即使在盛夏，当空气湿度较大时，在自来水管的外表面也会结露。如果露点低于 0 ℃，就会结霜，因此，测定露点还可以预报是否会有霜冻出现。

6.4.3 绝对湿度、相对湿度和含湿量

湿空气的湿度是指湿空气中水蒸气的含量。前面已指出，水蒸气的分压力可以表示湿空气中所含水蒸气的多少。此外，湿空气中水蒸气的含量还可以用绝对湿度、相对湿度和含湿量来表示。

(1) 绝对湿度

1 m^3的湿空气中所含水蒸气的质量称为湿空气的绝对湿度，也就是湿空气中水蒸气的密度 ρ_v，可根据湿空气的温度 T 和水蒸气的分压力 p_v按照理想气体状态方程计算水蒸气的密度 ρ_v：

$$\rho_v = \frac{m_v}{V} = \frac{p_v}{R_v T} \tag{6-3}$$

式中，R_v 为水蒸气的气体常数，为 461.5 J/(kg·K)。

由式(6-3)可知，当温度一定时，绝对湿度随湿空气中水蒸气分压力 p_v的增加而增大。当水蒸气分压力 p_v达到湿空气温度 T 所对应的饱和压力 p_s时，绝对湿度达到最大值，此时的湿空气就是饱和湿空气，绝对湿度就是饱和绝对湿度 ρ_s：

$$\rho_s = \frac{p_s}{R_v T} \tag{6-4}$$

具有相同绝对湿度的湿空气，由于所处的温度不同，吸湿能力也就有所不同，所以绝对湿度的大小还不能完全说明湿空气的吸湿能力，即不能说明湿空气的干燥或潮湿程度。为了说明湿空气的潮湿程度，引进相对湿度的概念。

(2) 相对湿度

湿空气的干湿程度对生产和人体生理的影响，并不单纯地取决于湿空气中含有水蒸气量的多少，而是取决于湿空气接近饱和的程度，相对湿度正是反映这种接近程度的一个状态参数。

湿空气的绝对湿度 ρ_v与同温度下湿空气的最大绝对湿度，即饱和湿空气的饱和绝对湿度 ρ_s的比值称为湿空气的相对湿度，用 φ 表示，即

$$\varphi = \frac{\rho_v}{\rho_s}$$

根据理想气体状态方程 $\rho_v = p_v/(R_v T)$，$\rho_s = p_s/(R_v T)$，可得：

$$\varphi = \frac{\rho_v}{\rho_s} = \frac{p_v}{p_s} \tag{6-5}$$

式中，p_v和 p_s分别为同一温度下未饱和湿空气与饱和湿空气的水蒸气分压力。可见，相对湿度等于湿空气中水蒸气的分压力与同温度下饱和水蒸气分压力的比值。

相对湿度 φ 能反映湿空气接近饱和湿空气的程度。相对湿度的取值范围：$0 \leqslant \varphi \leqslant 1$。$\varphi$ 值越小，湿空气中水蒸气偏离饱和状态越远，空气越干燥，吸湿能力越强，对于干空气，$\varphi = 0$；反之，φ 值越大，则湿空气中的水蒸气越接近饱和状态，空气越潮湿，吸湿能力越弱；当 $\varphi = 1$ 时，湿空气为饱和湿空气，不具有吸湿能力。

(3) 含湿量

在通风空调及干燥工程中，经常要遇到调节湿空气中水蒸气含量的问题，需要确定对湿空气的加湿及减湿的数量。若对湿空气取单位体积或单位质量为基准进行计算，则会由于湿空气在处理过程中体积及质量二者皆随温度及湿度改变而给计算带来麻烦。而湿空气中只有干空气的质量不会随湿空气的温度和水蒸气的含量而改变。因此，在对湿空气的某些参数进行计算时均以 1 kg 干空气作为计算的基准。

含有 1 kg 干空气的湿空气中所包含的水蒸气的质量，称为湿空气的含湿量或比湿度，用 d 表示。

$$d = \frac{m_v}{m_a}, \text{kg/(kg 干空气)}$$

利用理想气体状态方程 $p_a V = m_a R_a T$ 及 $p_v V = m_v R_v T$(式中，V 表示湿空气的体积，也是干空气及水蒸气在各自分压力下所占有的体积，m^3)，干空气及水蒸气的气体常数分别为：

$$R_a = \frac{8\ 314\ \text{J/(kmol·K)}}{28.97\ \text{g/mol}} = 287\ \text{J/(kg·K)};$$

$$R_v = \frac{8\ 314\ \text{J/(kmol·K)}}{18.02\ \text{g/mol}} = 461\ \text{J/(kg·K)}$$

故含湿量可写成：

$$d = 1\ 000\,\frac{R_a}{R_v}\,\frac{p_v}{p_a} = 1\ 000 \times \frac{287}{461}\,\frac{p_v}{p_a} = 622\,\frac{p_v}{p_a} = 622\,\frac{p_v}{B - p_v}, \text{g/(kg 干空气)} \quad (6\text{-}6)$$

相对湿度 φ 能够表示湿空气的饱和程度，但是不能表示湿空气中水蒸气的含量；含湿量 d 与之相反，它能够表示湿空气中水蒸气的含量，但不能表示湿空气接近饱和的程度。

6.4.4 干球温度和湿球温度

干球温度即湿空气的温度，湿空气中干空气和水蒸气的温度是一致的，用“t”表示。图 6-7 示出一个使未饱和空气在绝热情况下稳定流动加湿而达到饱和的物理模型。进入该装置的湿空气是未饱和空气，其温度为 t。如水槽足够长而且绝热，总水量远大于水的蒸发量。空气流与水经过充分的热、质交换之后，达到热湿平衡状态。此时，水槽中水的温度必定会达到一个不变的数值 t_w^*，而出口空气经过绝热加湿后，也达到饱和空气状态，其温度也是 t_w^*。这一稳定的温度值 t_w^* 称为绝热饱和温度，也称热力学湿球温度。热力学湿球温度是湿空气的状态参数，它只决定于进口湿空气的状态。

在工程应用中，要测量绝热饱和温度 t_w^* 是不可能的。因此，常用干、湿球温度计中湿球温度计的读数 t_w 来代替 t_w^*，虽然 t_w 不是一个状态参数，受风速及测量条件的影响，但只要测量方法正确，在风速大于 4 m/s 的情况下，两者相差不大，在一般的工程应用中是完全允许的。

图 6-8 示出用两支水银温度计组成的干、湿球温度计。干球温度计的读数就是湿空气的温度 t。另一支温度计的温包用湿纱布包起来，置于通风良好的湿空气中，当达到热湿平衡状态时，其读数就是湿球温度的读数 t_w。

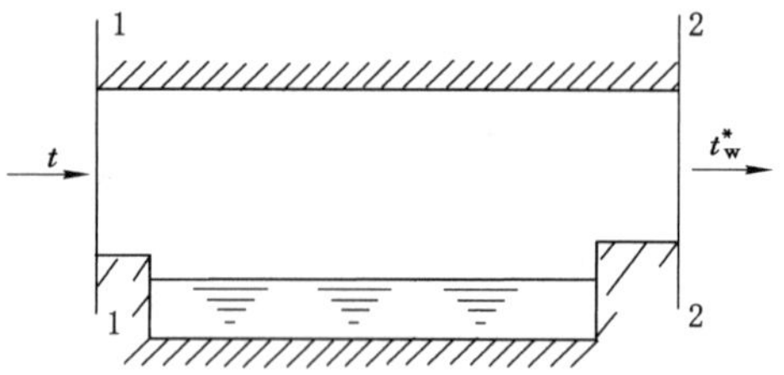

图 6-7　湿空气的绝热饱和温度

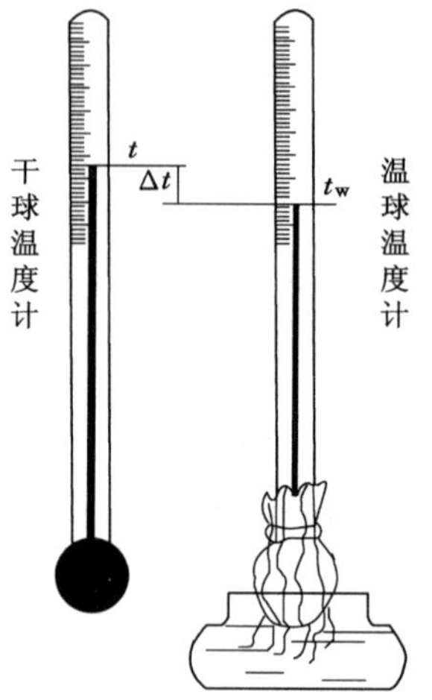

图 6-8　干、湿球温度计

在干、湿球温度计中，如果湿纱布中的水分不蒸发，两支温度计的读数应该是相等的。但由于空气是未饱和空气，湿纱布上的水分将蒸发，水分蒸发所需的热量来自两部分：一部分是降低湿纱布上水分本身的温度而放出热量；另一部分是由于空气温度 t 高于湿纱布表面温度，通过对流换热空气将热量传给湿球。湿纱布上水分不断蒸发的结果，使湿球温度计的读数不断降低。最终，当达到热湿平衡状态时，湿纱布上水分蒸发的热量全部来自空气的对流换热，湿纱布上水分温度不再降低。此时，湿球温度计的读数就是湿球温度 t_w。

由于干、湿球温度计受风速及测量环境的影响，在相同的空气状态下，可能会出现不同的湿球温度的数值。为此，应防止干、湿球温度计与周围环境之间的辐射换热，以及保证 4 m/s以上的风速。这样测得的 t_w 值，才能非常接近绝热饱和温度 t_w^* 的值，否则就会产生较大的误差。

湿空气状态与压力、温度间的关系是这样的：若湿空气为未饱和状态，则具有吸收水分的能力，水蒸气分压力 $p_v < p_s$，那么湿空气温度 t，露点 t_d，湿球温度 t_w之间的关系为$t > t_w > t_d$；若湿空气为饱和状态，则湿空气已没有再吸收水分的能力，则 $p_v = p_s$，$t = t_w = t_d$。

6.4.5　湿空气的密度及比容

湿空气的密度是指 1 m³湿空气中含干空气和水蒸气的质量总和，即

$$\rho = \frac{m}{V} = \frac{m_a + m_v}{V} = \rho_a + \rho_v \tag{6-7}$$

利用理想气体状态方程 $\rho_a = p_a/(R_a T)$ 及 $\rho_v = p_v/(R_v T)$，由式(6-7)得：

$$\rho = \rho_a + \rho_v = p_a/(R_a T) + p_v/(R_v T) = \frac{B - p_v}{R_a T} + \frac{p_v}{R_v T} = \frac{B}{R_a T} - \frac{p_v}{T}\left(\frac{1}{R_v} - \frac{1}{R_a}\right)$$

将 $R_a = 287$ J/(kg·K) 及 $R_v = 461$ J/(kg·K) 代入上式得：

$$\rho = \frac{B}{R_a T} - 0.001\ 315\ \frac{p_v}{T} \tag{6-8}$$

由式(6-8)可见，湿空气的密度比同温度下干空气的密度小，而且温度越高，湿空气的密度越小，即湿空气比干空气轻。在温度和总压力相同的情况下，水蒸气含量越多，湿空气密度越小，即湿空气的密度随着相对湿度的增大而减小。

湿空气的容积是以 1 kg 干空气为基准定义的，它表示在一定温度和总压力下，1 kg 干空气和 0.001d (kg)水蒸气所占有的容积，即含有 1 kg 干空气的湿空气容积，即

$$v = \frac{V}{m_a}, \mathrm{m^3/(kg\ 干空气)}$$

结合湿空气密度的定义可得：$\rho v = 1 + 0.001d$，因此湿空气的比容为：

$$v = \frac{1 + 0.001d}{\rho}, \mathrm{m^3/(kg\ 干空气)}$$

可见，与通常的 $\rho v = 1$ 有所区别。

6.4.6　湿空气的焓

湿空气的焓也是以 1 kg 干空气为基准来表示的，指含有 1 kg 干空气的湿空气的焓，它包括干空气的焓与水蒸气的焓，即

$$h = \frac{H}{m_a} = \frac{m_a h_a + m_v h_v}{m_a} = h_a + 0.001 d h_v$$

湿空气的比焓是指 1+0.001d (kg)湿空气的焓值，单位为 kJ/(kg 干空气)。

选取 0 ℃为焓的计算基准点，0 ℃的干空气的焓为零。因此，温度为 t 的干空气的焓 $h_a = c_{p,a} t = 1.01t$ (kJ/kg)；水蒸气定压质量比热容为 $c_{p,a} = 1.86$ kJ/(kg·K)，0 ℃时水的汽化潜热 $r = 2\,501$ kJ/kg，从 0 ℃的水加热变为温度为 t (℃)的蒸汽，其焓 h_v 为：

$$h_v = r + c_{p,a} t = 2\,501 + 1.86t$$

因此，湿空气的焓为：

$$h = 1.01t + 0.001d(2\,501 + 1.86t) \tag{6-9}$$

6.5　湿空气的焓湿图及热湿比

确定湿空气的状态需要三个彼此独立的状态参数。通常，在湿空气的总压力一定的情况下，只要已知任意另外两个彼此独立的状态参数，如 t 和 φ，就可用解析法确定湿空气的其他状态参数，如 t_d，t_w，h，d，p_v 等，进而对湿空气的热力过程进行分析计算。但解析计算较为烦琐，因此，通常绘制出湿空气的 h-d 图，供工程人员查用。由于 h-d 图都是在一定的大气压力下绘制的，所以只要给出湿空气的另外两个独立参数，就可以利用 h-d 图确定其他参数。

h-d 图的主要结构如图 6-9 所示，图中绘出了在大气压力 $p = 0.1$ MPa 下湿空气的焓 h、含湿量 d、温度 t、相对湿度 φ、水蒸气分压力 p_v 等主要参数的等值线簇，分别介绍如下。

(1) 等焓线簇。为了能够使各种线簇不过于密集，读数方便，将等焓线绘成一组与纵坐标轴(或等 d 线)呈 135°夹角的相互平行的倾斜直线，并取温度 t=0 ℃时的焓值为零。

(2) 等含湿量线簇。等 d 线是一组平行于纵坐标轴的垂直线，因此纵坐标轴即 d=0 的等含湿量线。自左向右，d 值逐渐增加。由式(6-6)可知，在一定的总压力下，水蒸气的分压力 p_v 与 d 值一一对应，因此等 d 线也就是等 p_v 线。另外，湿空气的露点 t_d 仅取决于水蒸气的分压力 p_v，所以等 d 线也是等 t_d 线。为避免图面过大，一般取水平线作为 d 轴。

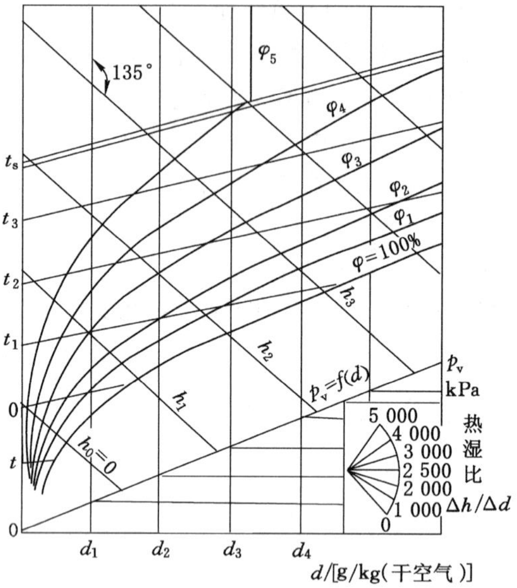

图 6-9　湿空气的 h-d 图

(3) 等温线簇。根据式(6-9)，当 t 为定值时，h 与 d 呈线性关系，其斜率为 2 501＋1.86t。显然，不同的等温线斜率各不相同，t 值越高，斜率越大，但由于 2 501 远大于 1.86t，所以这种差别并不显著。

(4) 等相对湿度线簇。等相对湿度线簇是一组向上弯曲的曲线，随温度的降低，相对湿度增大。因此，$\varphi=1$ 的等相对湿度线处于最下位置，称为饱和湿空气线，线上各点分别代表不同温度下的饱和湿空气。饱和湿空气线将 h-d 图分为上、下两部分，上部是未饱和湿空气。$\varphi=1$ 的等相对湿度线也是不同含湿量时的露点线。$\varphi=1$ 时，$d=0$，即干空气，所以纵坐标轴就是 $\varphi=0$ 的等相对湿度线。

(5) 水蒸气的分压力线。由式(6-6)可知，当大气压力 p 一定时，水蒸气的分压力仅是含湿量 d 的函数，$p_v=f(d)$。由于湿空气中水蒸气的分压力一般很小，所以 $p-p_v$ 近似为一常数，于是 p_v 和 d 的关系接近直线，且 p_v 随 d 的增大而增大。二者之间的函数关系通常绘制在 $\varphi=1$ 的饱和湿空气线下部，并在右边的纵轴上标出水蒸气分压力的数值。

(6) 等热湿比线。在湿空气处理过程中，除了有热交换外还有湿交换，热、湿交换是湿空气重要的能量交换，热、湿交换决定了状态变化的方向，通常用热湿比 ε 表示湿空气状态变化的方向和特征。ε 定义为状态变化前后的焓差与含湿量差之比，即

$$\varepsilon = 1\,000\,\frac{\Delta h}{\Delta d} = 1\,000\,\frac{h_2-h_1}{d_2-d_1}$$

1—2 过程的热湿比 ε 线在 h-d 图上反映了过程线 1—2 的倾斜度，因此也称为角系数。

在 h-d 图上，对于各种过程，无论其初、终状态如何，只要过程的热湿比 ε 值相同，就都是相互平行的。因此，在某些实用的 h-d 图上，在右下方任取一点为基准点，作出一系列等热湿比线，每一斜率的直线对应于一个等 ε 值。

当已知过程的热湿比 ε 时，在 h-d 图上，通过初始状态点 1 作一条平行于热湿比 ε 线的辐射线，即得到通过点 1 的过程线。当知道终状态点 2 的任一参数时，如温度，则等 t_2 线与该过

程线的交点即终状态点 2，从而在 h-d 图上就可确定点 2 的其他未知参数值。因此，在 h-d 图上利用热湿比线来分析和计算过程问题十分方便。

由 $\varepsilon = 1\,000 \dfrac{\Delta h}{\Delta d}$ 可知，在定焓过程中 $\Delta h = 0$，热湿比 $\varepsilon = 0$。在定含湿量过程中，$\Delta d = 0$，如过程吸热，则 $\varepsilon = +\infty$，如过程放热，则 $\varepsilon = -\infty$。因此，定焓线与定含湿量线将 h-d 图分成四个区域，如图 6-10 所示。从两线交点 1 出发，终态点可落在四个不同的区域内，四个区域具有如下特点。

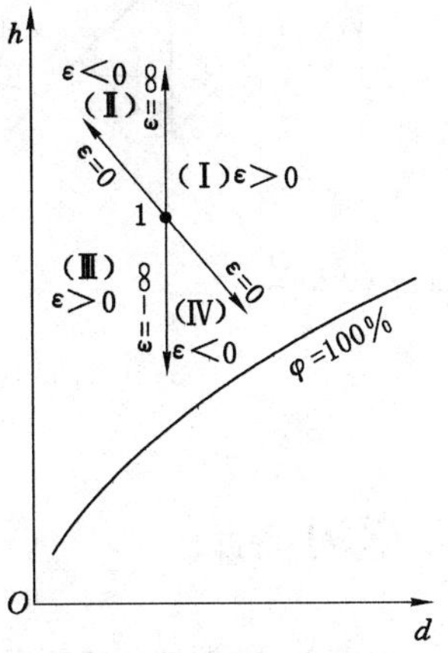

图 6-10　h-d 图四个区域的特征

第 Ⅰ 区域：从初态点 1 出发，落在这一区域内的过程，$\Delta h > 0, \Delta d > 0$，即增焓增湿过程，$\varepsilon > 0$。

第 Ⅱ 区域：从初态点 1 出发，落在这一区域内的过程，$\Delta h > 0, \Delta d < 0$，即增焓减湿过程，$\varepsilon < 0$。

第 Ⅲ 区域：从初态点 1 出发，落在这一区域内的过程，$\Delta h < 0, \Delta d < 0$，即减焓减湿过程，$\varepsilon > 0$。

第 Ⅳ 区域：从初态点 1 出发，落在这一区域内的过程，$\Delta h < 0, \Delta d > 0$，即减焓增湿过程，$\varepsilon < 0$。

6.6　湿空气的热力过程

湿空气处理过程的目的是使湿空气达到一定的温度与湿度，处理过程可以由一个过程或多个过程组合完成。这里介绍两种常见的加热吸湿过程和冷却去湿过程。

6.6.1　加热吸湿过程

工程上经常会遇到干燥处理过程，如谷物、木材、药材、纺织品、食品、造纸等许多加工工艺都需要烘干。如果将被干燥物体放在适当的场地自然干燥，则不仅时间长，而且容易受气候条件的限制，干燥效果也不易控制。所以，工程上一般采用人工干燥的方法，利用未饱和湿空气来吸收被干燥物体的水分，达到干燥的目的。为了提高湿空气的吸湿能力，通常先对湿空气进行加热。由于加热过程中湿空气的含湿量 d 不变，但随着温度 t 升高，相对湿度 φ 降低，湿空气的吸湿能力增强。加热过程一般在加热器中进行，其过程曲线如图 6-11 中的 1—2 所示。

加热后的湿空气送入干燥室，吸收被干燥物体的水分，湿空气的含湿量 d 和相对湿度 φ 皆增加。由于湿空气在吸湿过程中与外界基本绝热，水分蒸发所吸收的潜热完全来自湿空气本身，因此是一绝热加湿过程。在这一过程中，湿空气的焓值基本不变，过程沿着等 h 线向 d 和 φ 增大、t 降低的方向进行，其过程曲线如图 6-11 中的 2—3 所示。

6.6.2　冷却去湿过程

湿空气被冷却时，温度降低。在温度降至露点以前其含湿量保持不变，相对湿度逐渐增加；当相对湿度等于 1 时，再继续冷却，则过程将沿着 $\varphi=1$ 的饱和曲线向含湿量减少、温度降低的方向进行，同时析出水分，如图 6-12 中的 1—2 所示。

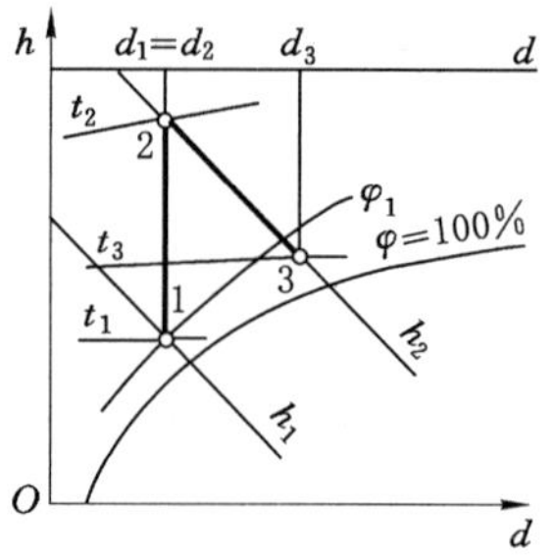

图 6-11　湿空气的加热吸湿过程

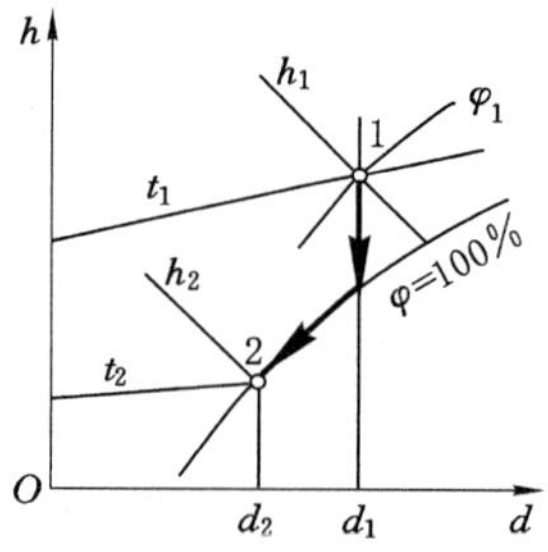

图 6-12　湿空气的冷却去湿过程

本 章 小 结

水蒸气由液态水经汽化产生，它离液态较近；湿空气是指含有水蒸气的空气。这两种气体性质较为复杂，因而不能作为理想气体来处理。本章重点介绍了水蒸气和湿空气的热力性质及基本热力过程的计算。

工业和生活中所用的水蒸气通常是在定压条件下对水加热产生的，水蒸气在定压情况下的产生过程表示在 p-v 图和 T-s 图上可概括为：一点（临界点）、两线（饱和水线和饱和蒸汽线）、三区（未饱和水区、湿蒸汽区和过热蒸汽区）及五态（未饱和水、饱和水、湿饱和蒸汽、干饱和蒸汽和过热蒸汽状态）。由于水蒸气具有复杂的热力性质，工程计算中通常采用一种简易方法，即利用水蒸气的热力性质图表来确定其状态并进行热力过程的功量、热量的计算。

湿空气是一种由氮气、氧气等气体和水蒸气所组成的混合气体，其热力性质可通过一系列的概念来表述，如水蒸气的分压力、饱和压力、相对湿度、含湿量、比焓等。工业中存在两种典型的湿空气热力过程，分别是加热吸湿过程和冷却去湿过程，有时可能是几种热力过程的综合。湿空气的热力计算通常也采用图表的简易算法，最常用的水蒸气热力性质图是 h-d 图。

思考与练习题

(1) 有没有 380 ℃的饱和水，为什么？

(2) 蒸汽锅炉的压力表指示压力为 1.2 MPa，当地大气压为 0.1 MPa，问锅炉内蒸汽温度为多少？

(3) 确定下列参数的状态：① $t=121$ ℃，$p=0.138$ MPa；② $p=0.345$ MPa，$v=0.5\ \mathrm{m^3/kg}$；③ $t=149$ ℃，$v=0.5\ \mathrm{m^3/kg}$；④ $p=0.5$ MPa，$v=0.545\ \mathrm{m^3/kg}$。

(4) 10 kg 130 ℃的水蒸气中含有 1.5 kg 的水，试确定水蒸气的状态及参数。

(5) 水在 $p_1=10$ MPa、$t_1=220$ ℃的状态下送入锅炉，定压加热至 $t_2=550$ ℃生成水蒸气流出锅炉，求每千克水在锅炉中吸收的热量。

(6) 分析下列说法是否正确：

① 当 φ 固定不变时，湿空气温度 t 越高，则含湿量 d 越大；

② 湿空气相对湿度 φ 越高，其含湿量越大；

③ 当 $\varphi=0$ 时不含水蒸气，全为干空气，因此，当 $\varphi=100\%$ 时，湿空气就不含干空气，全为水蒸气。

(7) 已知湿空气温度 $t=20$ ℃，大气压力为 0.1 MPa，水蒸气分压力 $p_v=872.5$ Pa，请判断湿空气是否饱和？什么条件下结露？含湿量如何？

(8) 室内空气压力 $p=0.1$ MPa、$t=30$ ℃，如已知相对湿度 $\varphi=40\%$，试计算空气中水蒸气的分压力、绝对湿度、露点和含湿量。

Chapter 6 Water Vapor and Atmospheric Air

As a kind of working medium, water vapor has many advantages, such as proper thermodynamic properties, non-toxic, odorless, cheap, and so on. It is widely used in steam turbines, steam engines, nuclear power plants and many other places. Water vapor is also used as a heat transfer medium in various heat-exchangers. In thermodynamic systems, water vapor is usually not far away from liquid and often experiences phase changes during working processes. Thus, it can not be treated as an ideal gas. In engineering calculations, the thermodynamic properties of water and water vapor are usually obtained by using water vapor charts and tables.

Air is a mixture of nitrogen, oxygen, and small amounts of some other gases. Air in the atmosphere normally contains some water vapor (or moisture) and is usually referred to as atmospheric air (or moist air). In contrast, air that contains no water vapor is called dry air (air without moisture). It is often convenient to treat atmospheric air as mixture of water vapor and dry air. Although the amount of water vapor is small, it can not be ignored. The amount of water vapor in the air has significant impact on thermal comfort and many processes, such as drying. Therefore, we can not treat moist air simply as dry air for analysis and calculation. As dry air is far away from its critical point, it can be regarded as an ideal gas. Atmospheric air contains a very small amount of water vapor and the partial pressure of water vapor is very low. Thus, the water vapor in the atmospheric air can also be treated as ideal gas That is, atmospheric air can be treated as a binary mixture of ideal gases, dry air and water vapor. However, phase change of water vapor in the atmospheric air often occurs, such as condensing, frosting, snowing, hailing, etc. Thus, it is not exactly the same as an ideal-gas mixture. Therefore, it needs some special state properties in order to describe the characteristics of moist air, such as the phase change of the water vapor and changes of the amount of water vapor.

6.1 Phase Change Process of Water

(1) Per-heat stage: compressed liquid to saturated liquid

Consider, a piston-cylinder device containing liquid water at 0 ℃ and 1 atm pressure. Under this condition, water exists in the liquid phase, and it is called a compressed liquid, or a sub-cooled liquid, meaning that it is not about to vaporize. Heat is now transferred to the water and thus its temperature rises to, say, 60 ℃. As the temperature rises, the liquid water expands slightly, end thus its specific volume increase. To accommodate this expansion, the piston moves up slightly. The pressure in the cylinder remains constant at 1 atm during this process since it is determined by the outside barometric pressure and the weight of the piston, both of which are constant. Water is still a compressed liquid at this state since it has not started to vaporize. As more heat is transferred the temperature keeps rising until it reaches 100 ℃. At this point, water is still a liquid, but any heat addition will cause some of the liquid to vaporize. That is, a phase-change process from liquid to vapor is about to take place. A liquid that is about to vaporize is called a saturated liquid. Therefore, state 2 is the saturated liquid state.

(2) Vaporization stage: saturated liquid to saturated vapor

Once bailing starts, the temperature stops rising until the liquid has completely vaporized. That is, the temperature will remain constant during the entire phase-change process if the pressure is held constant. This can easily be verified by placing a thermometer into pure boiling water on top of a stove. At sea level ($p=1$ atm), the thermometer will always read 100 ℃ if the pan is uncovered or covered with a weightless lid. During a boiling process, the only change observed is a large increase in the volume and a steady decline in the liquid level as a result of more liquid turning to vapor. Midway about the vaporization line, the cylinder contains equal amounts of liquid and vapor. As heat is continuously transferred in, the vaporization process continues until the last drop of liquid is vaporized. At this point, the entire cylinder is filled with vapor that is on the borderline of the liquid phase. Any heat loss from this vapor will cause some of the vapor to condense. A vapor that is about to condense is called a saturated vapor. Therefore, state 4 is a saturated vapor state. A substance at states between 2 and 4 is referred to as a saturated liquid-vapor mixture since the liquid and vapor phases coexist in equilibrium at these states. The amount of energy absorbed during vaporization is called the latent heat of vaporization and is equivalent to the energy released during condensation. Once the phase-change process completes, the water turns back to a single phase region again, this time vapor.

To analyze this mixture of saturated liquid and saturated vapor properly, we need to know the proportions of the liquid and vapor phases in the mixture. This is done by defining a new property called the quality x as the ratio of the mass of vapor to the total mass of the mixture:

$$x=\frac{m_v}{m_w+m_v}$$

Quality has significance for saturated mixtures only. It has no meaning for compressed liquid or superheated vapor. Its value is between 0 and 1. The quality of a system that consists of saturated liquid is 0, and the quality of a system consisting of saturated vapor is 1.

(3) Superheat stage: saturated vapor to superheated vapor

At the saturated vapor state 4, further transfer of heat results in an increase in both the temperature and the specific volume. At state 5, the temperature of the vapor is, let us say, 150 ℃; and if some heat is transferred from the vapor, the temperature may drop by somewhat hot no condensation will take place as long as the temperature remains above 100 ℃ (for $p=1$ atm). A vapor that is not about to condense (i. e., not a saturated vapor) is called a superheated vapor. Therefore, water at state 5 is a superheated vapor. The temperature difference between superheated vapor and saturated vapor is called the degree of superheat, that is $\Delta t=t-t_s$.

The phase-change process of water at 1 atm pressure was described in detail in the previous section. However, the variations of properties during phase-change processes can be more easily studied and understood with the help of property diagrams. Now we repeat the above process at different pressures to develop the p-v and the T-s diagrams for water.

As 1 kg compressed liquid of water is heated at constant pressure, it follows a horizontal line on the p-v diagram, that is, the isobaric line changing from a_1, b_1, d_1 to e_1, which represents the state of unsaturated liquid, saturated liquid, saturated liquid and vapor mixture, saturated vapor and superheated vapor, respectively. Under other pressures, the isobaric lines can be obtained, such as a_2-b_2-d_2-e_2, a_3-b_3-d_3-e_3 ⋯ Connecting saturated liquid state points of b_1, b_2, b_3, ⋯ and saturated vapor states d_1, d_2, d_3, ⋯ at different pressures will form the saturated liquid line ($x=0$) and the saturated vapor line ($x=1$). These two lines meet at the critical point C, forming a dome as shown in Fig. 6-3. All the compressed liquid states are located in the region to the left of the saturated liquid line, called the compressed liquid region. All the superheated vapor states are located to the right of the saturated vapor line, called the superheated vapor region. In these two regions, the substance exists in a single phase, liquid or vapor. All the states that involve both phases in equilibrium are located under the dome called the saturated liquid-vapor mixture region, the wet region.

At the critical point, there is no longer difference between saturated liquid and saturated vapor. The temperature, pressure, and specific volume of a substance at the critical point are called the critical temperature T_C critical pressure p_C and critical specific volume v_C, respectively. At pressures above the critical pressure, there is not a distinct phase change process. Instead, the specific volume of the substance continually increases, and at all times there is only one phase present. Eventually, it resembles a vapor, but we can never tell when the change has occurred. Above the critical state, there is no line that separates the compressed liquid region and the superheated vapor region. However, it is customary to refer to the substance as superheated vapor at temperatures above the critical temperature and as compressed liquid at temperatures below the critical temperature.

Critical properties of a substance are determined by the type of the substances, each substance has only a ground of critical properties. The critical properties of water vapor are $t_C=373.99$ ℃, $p_C=22.064$ MPa and $v_C=0.003\ 106$ m^3/kg.

There are two lines on the diagram, which are the saturated liquid line ($x=0$) and saturated vapor line ($x=1$), and these two lines meet at one point: the critical point; and the two lines divide the attire region into three regions: unsaturated liquid region, wet region and superheated vapor region; during a phase-change process from liquid to vapor, substance will experience five kinds of states: unsaturated liquid stare, saturated liquid state, wet vapor, saturated vapor stare and superheated vapor state.

6.2 Water Vapor Tables and Enthalpy-Entropy Diagram for Steam

The ideal-gas equation of state is $pv=RT$. Water vapor differs from ideal gases and does not follow this equation. The equation of state for water vapor is very complicated and seldom directly used in practical engineering calculations. Therefore, property tables and charts are often compiled and plotted on the basis of experimental and theoretical data for practical application.

6.2.1 Water Vapor Tables

The values of u, h, and s cannot be measured directly, and they are calculated from measurable properties using the relations between thermodynamic properties. However, those relations give the changes in properties, not the values of properties at specified states. Therefore, we need to choose a convenient reference state and assign a value of zero for a convenient property or properties at that state. For water, the state of saturated liquid at 0.01 ℃ is taken as the reference state, and the internal energy and entropy are assigned to be zero values at this state. That is, for saturated water at $t=0.01$ ℃ and $p=0.006\ 117$ MPa, $u=0$ kJ/kg, $s=0$ kJ/(kg · K). At this point, the specific volume of water $v=0.001\ 000\ 21\ \text{m}^3/\text{kg}$ and the enthalpy $h=0.000\ 61$ kJ/kg, which approaches 0 kJ/kg.

Steam tables are divided into a thermodynamic properties table for saturated liquid and saturated vapor and a thermodynamic properties table for unsaturated liquid and superheated vapor. In these tables, the subscript is used to denote the properties of saturated liquid, and the subscript is to denote the properties of saturated vapor.

The properties of saturated liquid and saturated vapor for water are listed in Table 7 and Table 8 in the appendix. It is more convenient to use Table 7 in the appendix when pressure is given and to use Table 8 in the appendix when temperature is given.

Since the compressed liquid region and superheated region are single-phase region (liquid or vapor phase only), temperature and pressure are no longer dependent of each other and they can conveniently be used as the two independent properties. The format of the compressed liquid and superheated vapor table is illustrated in other reference.

A saturated mixture can be treated as a combination of the saturated liquid and the saturated vapor. To analyzc this mixture properly, we need to know the quality x, that is the ratio of the mass of vapor to the total mass of the mixture. For saturated mixtures, quality can serve as one of the two independent intensive properties needed to describe a state. Note $x = m_v/(m_w + m_v)$ that the properties of the saturated liquid are the same whether it exists alone or in a mixture with saturated vapor. Then the properties of this "mixture" will simply be the average properties of the saturated liquid-vapor mixture under consideration.

Thus, at a given pressure, or a given temperature, the properties of a saturated mixture can be calculated by the following formula.

Specific volume of wet vapor: $v_x = xv'' + (1-x)v' = v' + x(v'' - v')$

Enthalpy of wet vapor: $h_x = xh'' + (1-x)h' = h' + x(h'' - h')$

Entropy of wet vapor: $s_x = xs'' + (1-x)s' = s' + x(s'' - s')$

6.2.2 The Enthalpy-Entropy Diagram (*h-s* Diagram) for Steam

On *T-s* and *p-v* diagrams, the area enclosed by process curves and the axes of coordinates indicates the amount of heat addition/rejection and work output/input, but it is inconvenient for use of calculation. If using enthalpy-entropy diagram (*h-s* diagram), the length of lines can represent the amount of heat and work directly. It is very intuitive, and thus it has been widely used in engineering calculations.

The *h-s* diagram for steam is shown in Fig. 6-4. The *h-s* diagram is constructed as follows: the enthalpy is chosen to be the axis of ordinate and the entropy is the abscissa. The origin of coordinates is assumed to be the state of water at the triple point, where $s_0 = 0$ and $h_0 = 0$. Making use of the data taken from the steam tables, the saturated liquid line $x = 0$ and saturated vapor line $x = 1$, meeting at the critical point *C*, are firstly plotted on the diagram. The saturated liquid line initiates from the origin of coordinates. In the wet region the isobars and isotherms are identical and are inclined straight lines. In the superheated region the isobars and isotherms diverge: the isobars run upward and the isotherms turn to approach horizontal lines. In the wet region, lines of constant quality ($x =$ const) are also plotted, which merge at the critical point *C*. Isochors (constant-volume lines) are also plotted on the *h-s* diagram, which run more steeply up wards, compared with the isobars, as shown by the dotted line in Fig. 6-4.

Usually, the entire *h-s* diagram is just partly given out. It is the part of the wet region with quality greater than 0.5 because the lines in the region with quality less than 0.5 is too dense and the data seldom used in projects.

6.3 Thermodynamic Processes of Water Vapor

The main purpose of analyzing thermodynamic processes of water vapor is to determine the final state and state properties of the process and to determine the thermodynamic energy and enthalpy changes of steams to obtain the energy conversion relations, including those between work output and heat addition during the process. In general, steam table and *h-s* diagram are used. The analysis is based on the first and the second laws of thermodynamics. Here we just discuss reversible process. The analyzing procedure is as following:

(1) Based on the given conditions, determine the initial state and related properties.

(2) Based on the characteristics of the process and one of the properties given for the final state, determine the final state and its properties.

(3) Based on the initial and final states, calculate the q, Δu and w during the process.

Isochoric process:

$$w = 0, w_t = v(p_1 - p_2), q = \Delta u = u_2 - u_1 = (h_2 - h_1) - v(p_2 - p_1)$$

Isobaric process:

$$w = p(v_2 - v_1), \omega_t = -\int_1^2 v\mathrm{d}p = 0, q = h_2 - h_1$$

Isothermal process:

$$w = q - \Delta u = T(s_2 - s_1) - (h_2 - h_1) + (p_2 v_2 - p_1 v_1),$$

$$w_t = q - \Delta h = T(s_2 - s_1) - (h_2 - h_1),$$

$$q = \int_1^2 T\mathrm{d}s = T(s_2 - s_1)$$

Isentropic process:

$$w = -\Delta u = u_1 - u_2 = (h_1 - p_1 v_1) - (h_2 - p_2 v_2), w_t = -\Delta h, q = 0$$

6.4 State of Atmospheric Air and Its State Properties

6.4.1 Saturated Air and Unsaturated Air

As the Dalton law of partial pressure stated, the pressure of atmospheric air p is the sum of the partial pressure of the dry air p_a, and the partial pressure of the water vapor p_v that is:

$$p = p_v + p_a \tag{6-1}$$

The partial pressure of water vapor reflects the amount of water vapor in moist air. It is referred to as vapor pressure. The more moisture the atmospheric air contains, the higher the vapor pressure it has. When there is no phase change in water vapor, both the dry air and the water vapor satisfy the ideal-gas relation.

When the temperature of the moist air ist, the corresponding saturated vapor pressure is $p_s(t)$ for the moist air. If $p_v(t) < p_s(t)$, then water vapor in the moist air is superheated vapor at the state A, as shown in Fig. 6-6. The moist air at this point is known as unsaturated air. The unsaturated air is a mixture of dry air and superheated water vapor. Maintaining the temperature t constant, increasing the vapor pressure by adding some water vapor into the air, the state of the moist air changes from A towards C, as shown in Fig. 6-6. When it reaches C on the saturated vapor line, $p_v = p_s(t)$, water vapor in the moist air becomes saturated vapor. At this point, the moist air is called saturated air. Saturated air is a mixture of dry air and saturated water vapor. Due to $p_s(t)$ is the maximal vapor pressure at the temperature t, the amount of water vapor in the moist air reaches the maximum at point C. It can not increase any more, even if water vapor is sprayed into the air, it will drop out as water droplets. That is to say, saturated air is the air that can not hold any more moisture.

6.4.2 Condensation and Dew Point Temperature

If unsaturated moist air at point A is cooled continuously at a fixed water vapor pressure and reaches point B on the saturated vapor line, then the water vapor begins to condense. This phenomenon is known as condensation and the point B is called the dew-point. Temperature at this point is called the dew-point temperature, denoted by t_d. In other words, it is the temperature at which condensation begins when the air is cooled at constant pressure. And the dew point temperature is the saturation temperature corresponding to the vapor pressure, that is:

$$t_d = t_s(p_v) \tag{6-2}$$

When it reaches the dew point, it becomes saturated air. If the temperature of the moist air continues to drop, then condensation is incurred. And the state of the moist air will change along the saturated vapor line, as shown by the curve B-D in Fig. 6-6. At the me time, as the temperature drops, the vapor pressure also decreases.

6.4.3 Absolute and Relative Humidity of Moist Air and Humidity Ratio

At a given temperature, the more water vapor (moisture) the moist air holds, the higher the vapor pressure p_v it possesses. As stated, maintaining the temperature constant, the increase in the amount of moisture will cause the vapor pressure to rise. When it eventually reaches $p_s(t)$, the moist air becomes saturated air. And the amount of moisture reaches the maximum value at the specified temperature.

Generally, the ratio of the amount of moisture the atmospheric air holds at a given temperature to the maximum amount the air can hold at the same temperature is defined as relative humidity, denoted as φ. In terms of this definition and the ideal-gas relation, it is expressed as:

$$\varphi = \frac{\rho_v}{\rho_s} = \frac{p_v}{p_s} \tag{6-3}$$

Where, p_v and p_s are the vapor pressure and the saturated vapor pressure of the atmospheric air at temperature t, respectively. The vapor pressure of saturated air at a given tempers, rare is equal to the saturation pressure of water at that temperature. Relative humidity can also be expressed as the ratio of the vapor pressure to the saturation pressure of water vapor at that temperature.

The relative humidity ranges from 0 for dry air to 1 for saturated air. It reflects the extent the air approaches the saturated air at the same temperature. The lower the value φ, the less water vapor it holds, the more water vapor it can absorb, and the drier it is. Note that the amount of moisture the air can hold depends on its temperature. Therefore, the relative humidity of air changes with temperature even when the amount of moisture it holds remains constant.

In ventilation and air conditioning or drying processes, it is often required to adjust the amount of water vapor to the moist air. If a unit mass or a unit volume of moist air is chosen as the datum, it may bring some trouble because the mass or the volume will change during a process in which the temperature and the amount of moisture in the moist air changes as well. However, the amount of dry air does not change during energy and mass exchange processes. Therefore, the calculation of the properties of moist air is based on per unit mass of dry air.

The amount of water vapor in the air can be specified directly by the mass of water vapor present in per unit mass of dry air. This is called specific humidity (also called humidity ratio) denoted as d.

$$d = \frac{m_v}{m_a} = \frac{p_v}{p_a} \text{ (kg water vapor/kg dry air)}$$

Since there are $p_a V = m_a R_a T$ and $p_v V = m_v R_v T$, yields the calculation equation of the humidity ratio:

$$d = 1\,000 \frac{R_a}{R_v}\frac{p_v}{p_a} = 1\,000 \times \frac{287}{461}\frac{p_v}{p_a} = 622\frac{p_v}{p_a} = 622\frac{p_v}{B - p_v} \text{ (g water vapor/kg dry air)} \tag{6-4}$$

6.4.4 Dry-Bulb Temperature and Wet-Bulb Temperature

Dry-bulb temperature is the ordinary temperature of the moist air. The dry-bulb temperature of water vapor and dry air are the same. It is denoted as "t". As shown in Fig. 6-7, when the unsaturated moist air with a certain amount of moisture at temperature t_1, flows through a long insulated channel containing a pool of water, water may evaporate into the moist air. During this process, the moisture contained by the air stream increases and its temperature decreases as the result of providing the latent heat of vaporization to the water when it evaporates into the air. If the channel is long enough, the air stream may exit as saturated air (φ=100%) at temperature t_2, which is called the adiabatic saturation temperature. This device is called adiabatic saturator. During the evaporation process, water absorbs heat so that the air temperature drops. In general, the adiabatic saturation temperature is lower than the dry-bulb temperature of the moist air. The adiabatic saturation temperature is a state property. It only depends on the state of the inlet moist air.

The measurement of relative humidity and specific humidity is not only necessary to determine the properties of moist air, but also of great significance for industrial and agricultural measurement and control, weather forecasting, aerospace, microenvironment and other fields. However, they can not be measured directly. It requires along channel in adiabatic saturator to measure the adiabatic saturation temperature indirectly. But it is difficult to realize. In practice, the wet-bulb temperature t_w is taken as a substitution for the adiabatic saturation temperature. Thus, the dry-bulb and wet-bulb thermometer are used to measure the relative humidity of the moist air indirectly.

Wet-bulb temperature is measured by using a thermometer whose bulb is covered with a cotton wick saturated into water and letting the air blow over the wick, as shown in Fig. 6-8. The temperature measured in this manner is called the wet-bulb temperature t_w, and it is commonly used for air-conditioning applications.

The principle involved is as following: when unsaturated air passes over the wet wick, some of the water in the wick evaporates. As a result, the temperature of the water drops, creating a temperature difference between the air and the water. After a while, the heat loss from the water by evaporation equals to the heat gain from the air, and the water temperature tends to be stable. The reading of the thermometer at this point is the wet-bulb temperature. Typically, when the wind speed is greater than 4 m/s, the difference between the wet-bulb temperature measured and the adiabatic saturation temperature is very slight.

6.4.5 Density and Specific Volume of Moist Air

The density of moist air is the ratio of its total mass to its total volume, that is:

$$\rho = \frac{m}{V} = \frac{m_a + m_v}{V} = \rho_a + \rho_v \tag{6-5}$$

According to the ideal-gas equation of state, substituting these formulas above into Eq. (6-5), we can obtain:

$$\rho = \frac{B}{R_a T} - 0.001\,315\,\frac{p_v}{T} \tag{6-6}$$

It indicates that the density of moist air is less than that of dry air. And the higher the temperature. the lighter the moist air is. At a fixed atmospheric pressure, the density of moist air is related to temperature and the specific humidity.

The specific volume of moisture air is defined as:

$$v = \frac{1 + 0.001d}{\rho}, \text{m}^3/(\text{kg dry air}) \tag{6-7}$$

Obviously, it is different from $\rho v = 1$.

6.4.6 Enthalpy of Moist Air

In most practical applications, the amount of dry air in the air-water vapor mixture remains constant, but the amount of water vapor changes. Therefore, the enthalpy of the atmospheric air is expressed per unit mass of dry air instead of per unit mass of moist air. It is the sum of the enthalpies of dry air and the water vapor.

$$h = \frac{H}{m_a} = \frac{m_a h_a + m_v h_v}{m_a} = h_a + 0.001 d h_v$$

Selecting the enthalpy at 0 ℃ as the datum point, the dry air enthalpy at 0 ℃ is zero and thus the enthalpy of the dry air at the temperature t is $h_a = c_{p,a} t = 1.01$ kJ/kg; For water vapor, the specific heat at constant pressure $c_{p,a} = 1.86$ kJ/(kg · K) and the latent heat of vaporization at 0 ℃ is $r = 2\,501$ kJ/kg. Thus, the enthalpy of the water vapor at the temperature t is:

$$h_v = r + c_{p,a}t = 2\,501 + 1.86t$$

Therefore, the total enthalpy of moist air per unit mass dry air is:

$$h = 1.01t + 0.001d(2\,501 + 1.86t) \tag{6-8}$$

6.5 Psychrometric Chart (*h-d* Chart) and Heat Humidity Ratio

It needs three separate properties to determine the state of moist air. Typically, the state of the atmospheric air at a given pressure is completely specified by two independent properties. Other properties can be calculated easily in terms of the relations introduced above. However, it may involve tedious calculations and incur errors. Therefore, there is a clear motivation to do these calculations once and to present the data in the form of easily readable charts. Such charts ere called psychrometric chart (*h-d* chart). It is a plot of specific humidity versus enthalpy. It shows lines of constant dry-bulb temperature, relative humidity, etc., for the atmospheric air. They are widely used in air-conditioning engineering.

The *h-d* chart of the moist air for a fixed pressure is shown in Fig. 6-9. It is usually plotted for moist air at barometric pressure of 101 325 Pa, but it can be used with sufficient accuracy for other pressures close to the basic one. On this chart, enthalpy h, kJ/(kg dry air), is chosen to be the vertical axis, and the specific humidity d, g/(kg dry air) to be the horizontal axis. For the convenience of arranging various lines on the chart, the axes of coordinates are arranged at an angle of 135 ℃.

On the chart, there is a curve called the saturation line. All the states of saturated air are located on this curve. Therefore, it is also the curve of 100 percent relative humidity. There are the following lines illustrated on the *h-d* chart.

(1) Lines of constant specific humidity and lines of constant enthalpy (Isenthalpic lines)

Constant specific humidity lines are straight lines vertical to the horizontal axis and isenthalpic lines (lines of constant enthalpy) are parallels might lines in an angle of 135° with the horizontal axis.

(2) Lines of constant dry-bulb temperature (Isotherm lines)

Enthalpy of the moist air $h = 1.01t + 0.001d(2\,501 + 1.86t)$ kJ/(kg dry air) shows that at a certain dry-bulb temperature, h is linear with d. The slopes of the constant dry-bulb temperature lines are slightly different from each other. The higher the temperature, the larger the intercept and the slope of constant dry-bulb temperature lines have. It means that constant dry-bulb temperature lines are not parallel. However, since $1.86t \ll 2\,501$, they are nearly parallel.

(3) Constant relative humidity lines (φ lines)

The relative humidity at a certain total pressure is a function of d and t. The relative humidity line $\varphi = 0$ is corresponding to the longitudinal axis (vertical axis), it indicates the state of the dry air ($d=0$); the relative humidity $\varphi=100\%$ line is of the saturated air line. Other constant relative-humidity curves generally have the same shape as that of the saturated air. The unsaturated air situates between the vertical axis and saturated moist air line, as shown in Fig. 6-9. For moist air with the same specific humidity, the higher the dry-bulb temperature, the smaller is the value of φ. Thus, the closer the constant relative humidity line approaches to the vertical axis.

(4) Constant vapor pressure lines

Eq. (6-6) shows that at a fixed total pressure, when $p_v \ll p$, there is a one-to-one correspondence relation between the vapor pressure p_v and the specific humidity d. Typically, on the h-d chart, another horizontal axis on the top of the horizontal axis of d is given, as shown in Fig. 6-9, or a transformation line between constant d and constant p_v is given.

(5) Heat humidity ratio line

The psychrometric chart also serves as a valuable aid in visualizing the air conditioning process. During air conditioning processes of moist air, the heat and moisture exchange processes occur simultaneously and uniformly. The process 1-2 processes along s line connecting the initial scare point 1 with the final state point 2 on the h-d chart. The slope of the line is referred to as the ratio of the change in enthalpy to that of the moisture, denoted by ε:

$$\varepsilon = 1\ 000\frac{\Delta h}{\Delta d} = 1\ 000\frac{h_2 - h_1}{d_2 - d_1}$$

It is called heat humidity ratio. It reflects the inclination extent of the line 1-2, so it is also called shape factor. It shows the direction and characteristics of the heat and moisture exchange process.

On the h-d chart, for all kinds of processes, regardless of its initial and final states, the processes lines are parallel with each other as long as the heat humidity ratios are the same. Therefore, on some h-d chart, in the lower part on the right side, series of lines of constant heat humidity ratio are drawn with slopes corresponding to a starting from a randomly chosen initial point.

When the heat humidity ratio, ε, of the process is known, the process fine can be drawn from the initial state point 1 parallel to the line of a on the h-d chart. When any property of the final state is given, for example, the dry-bulb temperature, then the intersection point of the isotherm line t_2 and the process line is the final state point 2. And then, it is easy to determine other properties of the final state. Therefore, it is very convenient to make use of the heat humidity ratio a line to analyze thermodynamic processes of moist air.

Various air-conditioning processes can be illustrated on the psychrometric chart. The isenthalpic lines and constant specific humidity lines divide the h-d chart into four regions, as shown in Fig. 6-10. If the initial state of moist air is indicated with the point 1, the final state point may fall into the following four regions. The characteristics of the process vary with which region the final state point is in:

If in region Ⅰ, then both the enthalpy and the humidity ratio of the moist air increase during the process, that is $\Delta h>0$, $\Delta d>0$;

If in region Ⅱ, then the enthalpy increases, bur the humidity ratio decreases during the process, that is $\Delta h>0$, $\Delta d<0$;

If in region Ⅲ, then both the enthalpy and the humidity ratio of the moist air decrease during the process, that is $\Delta h<0$, $\Delta d<0$;

If in region Ⅳ, then the enthalpy decreases but the humidity ratio increases during the process, that is $\Delta h<0$, $\Delta d>0$.

6.6 Thermodynamic Process of Atmospheric Air

Maintaining a living space or an industrial facility at a desired temperature and humidity requires some processes called air-conditioning processes. They include simple heating process, simple cooling process, humidifying process (adding moisture), and dehumidifying process (removing moisture), etc. Sometimes two or more of these processes are needed so as to bring the air to a desired temperature and humidity level. Therefore, some processes commonly encountered in air-conditioning engineering are presented.

6.6.1 Heating Process

When moist air flows over a heating coil in which hot water flows, the temperature and the enthalpy of the air increase, but the relative humidity decreases, the specific humidity remains constant. Notice that the relative humidity of the air decreases during a hearing process even if the specific humidity d remains constant. This is because the relative humidity is the ratio of the amount of moisture at a given temperature to the maximum moisture it can hold at the same temperature, and moisture capacity increases with temperature.

6.6.2 Cooling with Dehumidification

Hot, moist air enters a cooling section at state 1. As it passes through the cooling coils, its temperature decreases and the relative humidity increases at constant specific humidity. If the cooling section is long enough, air reaches its dew point (state A, saturated air). Further cooling of air will result in the condensation of the moisture. Air remains saturated during the entire condensation process. And the process follows the saturated air line to proceed until it reaches the final state (state 2).

This process can achieve the purposes of cooling the air with dehumidification. It can only occur when the surface temperature of the cooling coil is below the dew point temperature of the moist air.

第7章　动力循环与制冷循环

本章导学

要实现热能和机械能之间连续不断地转换，必须依靠一定的热力装置来完成工质的热力循环。正如前面所述，循环按方向分为正循环和逆循环，按目的来分有动力循环(正循环)和制冷(或制热)循环(逆循环)。动力循环能够将热能转换为机械能，而制冷(或制热)循环能将机械能转变为热能。动力装置、制冷装置和热泵统称为热力装置。动力装置的任务是将热量通过能量的传递和转换，转变成人们所需要的功。制冷装置的任务是将热量不断地从系统排向环境以使系统温度降到所要求的某一低于环境温度的水平，并使该系统温度保持不变。热泵装置的任务则相反，是将热量不断地传给系统以使系统温度提高到所要求的某一高于环境温度的水平，并使该系统温度保持不变。研究热力装置循环是为了通过循环分析，理解热能利用和转换的基本定律，同时也为现有的热力装置提出提高热效率的途径。本章主要讲述蒸汽动力循环、内燃机循环两种动力循环和空气压缩制冷循环、蒸汽压缩制冷循环两种制冷循环的设备、流程及热力过程的组成，循环中热量和功量转换以及热效率的分析计算，分析影响热效率的各种因素，指出提高热力装置循环热效率的途径。

学习目标

了解各种热力装置循环的实施设备及工作流程。掌握将实际循环抽象和简化为理想循环的一般方法，并会分析各种循环的热力过程组成。了解各种热力循环的吸热量、放热量、做功量及热效率等能量分析和计算的方法。会分析影响各种循环热效率的主要因素。了解提高各种热力循环能量利用经济性的具体方法和途径。

重点与难点

(1) 动力循环与制冷循环的工作原理。

(2) 理解影响动力循环与制冷循环热效率的主要因素。

(3) 提高各种循环能量利用经济性的具体方法和途径。

7.1　概　　述

动力循环设备能够将热能转换为机械能，称为热能动力装置或热力发动机，简称热机。

在热机中，热能连续地转换为机械能是通过工质的热力循环过程来实现的。根据工质的不同，动力循环可分为蒸汽动力循环（如蒸汽机、蒸汽轮机的工作循环）和燃气动力循环（如内燃机、燃气轮机装置的工作循环）两大类。

对物体进行冷却，使其温度低于周围环境的温度，并维持低温的过程称为制冷。为了获得或保持低温，必须从需要制冷的物体或空间提取热量，制冷装置便是以消耗能量（功量或热量）为代价来实现这一任务的设备，它是一种将热量从低温介质转移到高温介质的装置。为了把热量不断排向外部热源（通常就是大气环境），就必须构成循环，使制冷装置能够连续运转。制冷循环所采用的工质称为"制冷剂"，如空气、水、氨、氟利昂等。

压缩制冷装置是目前使用广泛的一种制冷装置，绝大多数家用冰箱、空调机、冷柜等都是采用压缩式制冷。如果制冷工质（即制冷剂）在循环过程中一直处于气态，则称制冷循环为气体压缩制冷循环。如果制冷工质的状态变化跨越液、气两态，则制冷循环称为蒸气压缩制冷循环。除此之外，还有吸收式制冷循环、吸附式制冷循环、蒸汽喷射式制冷循环以及半导体制冷循环等。

所有实际循环过程都是十分复杂的，并且是不可逆的。因此，在进行循环分析时，首先建立实际循环的简化热力学模型，用简单、典型的可逆过程和循环来近似实际复杂的不可逆过程和循环，通过热力学分析和计算，找出其基本特性和规律。只要这种简化的热力学模型是合理的、接近实际的，那么分析和计算的结果就具有理论上的指导意义。必要时还可以进一步考虑各种不可逆因素的影响，进行必要的修正，以提高精度。

7.2 蒸汽动力循环

利用固体、液体或气体燃料燃烧放热产生动力发电的工厂称为热力发电厂（或称为火力发电厂）。现代大型热力发电厂都是由锅炉、汽轮机和发电机等主要设备构成的，其中由锅炉、汽轮机、凝汽器、水泵等主要热力设备组成的整套装置称为蒸汽动力装置。像热力发电厂这样以水蒸气为工质的蒸汽动力装置工作循环称为蒸汽动力循环。

7.2.1 蒸汽卡诺循环

根据热力学第二定律，在一定温度范围内工作的一切循环，以卡诺循环的热效率最高。那么，采用饱和蒸汽为工质的蒸汽动力循环，是否可以实现卡诺循环呢？

图7-1是饱和蒸汽卡诺循环的 p-v 图和 T-s 图，它们说明了饱和蒸汽卡诺循环的构成。4—1为工质在锅炉内的定温（定压）吸热过程，1—2为工质在汽轮机内绝热膨胀的做功过程；2—3为工质在凝汽器内的定温（定压）放热过程，3—4是工质在压气机内的绝热压缩过程。上述四个过程中，前三个过程可近似地实现，而最后一个过程却难以实现。这是因为压缩的是湿蒸汽，比容大，需要制造结构庞大的压缩机，技术上存在很大困难，也需要消耗很大的压缩功，使循环的净功量大为减少，而且压缩湿蒸汽对压缩机是极为不利的。

此外，饱和蒸汽卡诺循环的上限温度只能在其临界温度以下，而下限温度也只能高于环境温度，其间温差不大，即使实现了饱和蒸汽的卡诺循环，其热效率也不是很高。

7.2.2 朗肯循环

鉴于上述原因，难以采用蒸汽卡诺循环。为了改进上述压缩过程，使乏汽完全凝结为

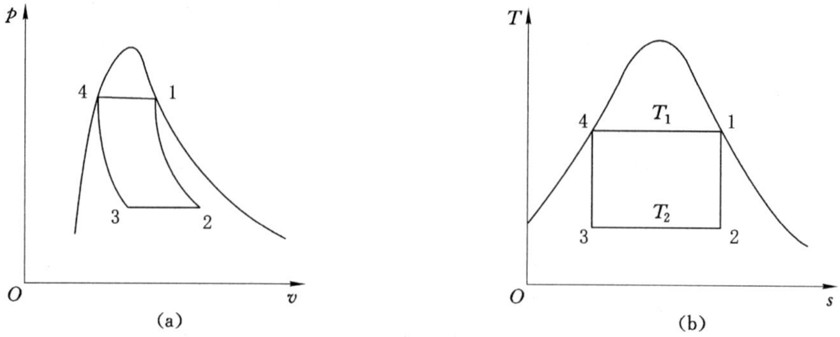

图 7-1　饱和蒸汽卡诺循环

水，可采用结构较小的水泵。同时，采用定压吸热来提高工质的上限温度，用高于临界温度的过热蒸汽来代替饱和蒸汽。这样改进的结果就构成了一个切实可行的蒸汽动力循环，这个循环即蒸汽动力的基本循环——朗肯循环。朗肯循环已成为现代火力发电厂蒸汽动力装置普遍采用的一个基本循环。尽管目前火力发电厂蒸汽动力装置所实施的实际循环非常复杂，但它们都是在朗肯循环的基础上加以演变和改进而得到的。

7.2.2.1　朗肯循环的工作原理

朗肯循环系统由锅炉、汽轮机、冷凝器和给水泵等设备组成。图 7-2(a)给出了系统的装置图。循环可理想化为 4 个可逆过程(两个定压过程和两个定熵过程)：

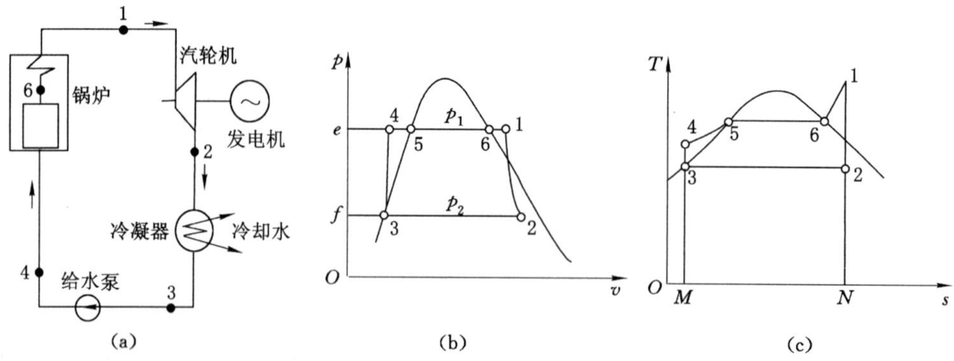

图 7-2　朗肯循环系统装置图及朗肯循环的 p-v 图和 T-s 图

(1) 水在给水泵中的可逆绝热压缩过程 3—4：水经给水泵增压后再送入锅炉，此过程要消耗功。

(2) 水与水蒸气在锅炉中的可逆定压加热过程 4—5—6—1：该过程中工质由未饱和水被加热成为过热水蒸气。

(3) 水蒸气在汽轮机中的可逆绝热膨胀过程 1—2：蒸汽在汽轮机中膨胀并对外输出轴功，在汽轮机出口处，蒸汽已处于低压湿蒸汽状态，通常称其为乏汽。

(4) 乏汽在冷凝器中的定压放热过程 2—3：乏汽在冷凝器内向冷却水放热而凝结，湿蒸汽凝结成乏汽压力下的饱和水，完成一个循环。

朗肯循环在 p-v 图和 T-s 图上的表示分别如图 7-2(b)和图 7-2(c)所示。

由于水的压缩性很小，经过给水泵定熵压缩而产生的温升极小，在状态图上，可认为点

4 与点 3 重合。简单蒸汽动力的朗肯循环在状态图上就可简单表示为 123561。

7.2.2.2　朗肯循环的净功及热效率

在朗肯循环中，每千克蒸汽对外所做的净功 w_{net} 应等于蒸汽流过汽轮机所做的功 $w_{s,1-2}$ 与给水在给水泵内被绝热压缩所消耗的功 $w_{s,3-4}$ 之差。根据稳定流动能量方程：

$$w_{s,1-2} = h_1 - h_2, w_{s,3-4} = h_4 - h_3$$

于是：

$$w_{net} = (h_1 - h_2) - (h_4 - h_3)$$

锅炉中每千克蒸汽的定压吸热量为：

$$q_1 = h_1 - h_4$$

冷凝器中，每千克蒸汽的定压放热量为：

$$q_2 = h_2 - h_3$$

朗肯循环的热效率：

$$\eta_t = \frac{w_{net}}{q_1} = \frac{(h_1 - h_2) - (h_4 - h_3)}{h_1 - h_4} \tag{7-1}$$

由于给水泵消耗的轴功与汽轮机输出的轴功相比很小，可以忽略，即 $h_4 \approx h_3$，式(7-1)可简化为：

$$\eta_t = \frac{w_{net}}{q_1} = \frac{h_1 - h_2}{h_1 - h_3} \tag{7-2}$$

除了热效率以外，汽耗率也是衡量蒸汽动力装置性能优劣的重要经济指标之一。汽耗率表示蒸汽动力装置每输出 1 kW·h 的功(等于 3 600 kJ)需要消耗的蒸汽量，用 d' 表示，单位为 kg/(kW·h)。因为 1 kg 蒸汽在一个循环中所做的功 $w_{s,t} = h_1 - h_2$，所以，汽耗率为：

$$d' = \frac{3\ 600}{w_{s,t}} = \frac{3\ 600}{h_1 - h_2} \tag{7-3}$$

7.2.2.3　提高朗肯循环热效率的基本途径与方法

(1) 提高蒸汽初温度、初压力和降低排汽压力。提高蒸汽初温度 t_1、初压力 p_1 可以提高蒸汽动力循环的平均吸热温度 T_H，降低排汽压力 p_2 主要可降低循环的平均放热温度 T_L。因此，提高蒸汽初温度、初压力和降低排汽压力是提高蒸汽动力循环热效率的有效措施。在强度和排汽干度允许的条件下应尽量提高 t_1 和 p_1，在环境温度允许的条件下尽量降低 p_2。

(2) 再热循环。提高蒸汽的初压力可以提高朗肯循环的热效率。但如果蒸汽的初温度不能同时提高，则蒸汽在汽轮机内膨胀终了时的干度降低，影响汽轮机的安全运行。因此，为了提高蒸汽的初压又不致使乏汽的干度过低，常采用蒸汽中间再过热的方法。如图 7-3 所示，新蒸汽在高压汽轮机中膨胀做功到某一中间压力以后，全部抽出导入锅炉中的再热器，吸收烟气放出的热量，然后再导入低压汽轮机继续膨胀做功到终压力 p_2。这种循环称为蒸汽再热循环，简称再热循环。蒸汽经中间再过热以后，其乏汽的干度明显

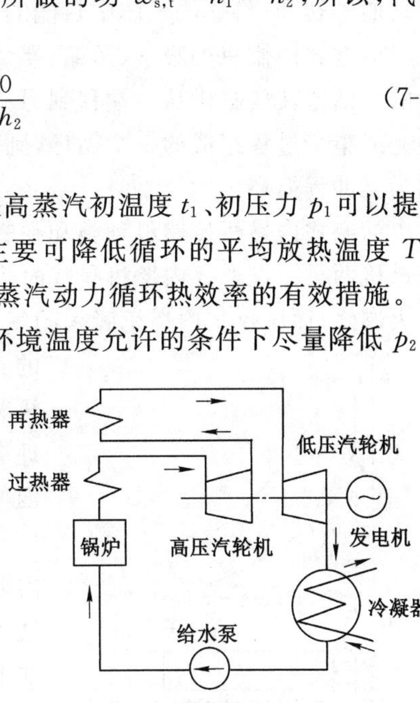

图 7-3　再热循环系统示意图

提高了。再热循环的吸热平均温度高于基本的朗肯循环，故再热循环的热效率有所提高。

(3) 抽汽回热循环。朗肯循环平均吸热温度不高的主要原因在于，由未饱和水加热到饱和水的过程 4—5 是吸热温度最低的一段。若能预热锅炉给水，使其温度升高后再送入锅炉，就能提高水在锅炉内的平均吸热温度，从而提高循环效率。但是，利用乏汽加热是不可能的，因为乏汽的温度与给水的温度相等。目前采用的一种切实可行的方案是从汽轮机中间抽出部分已做过功但压力尚不太低的少量蒸汽来加热进入锅炉之前的低温给水。这种方法称为给水回热。有给水回热的蒸汽循环称为抽汽回热循环，在现代蒸汽动力循环中普遍采用，可以有效地提高循环热效率。

(4) 热电联产。在现代火力发电厂中，尽管采用了高参数、再热、回热等措施，但循环热效率仍低于 50%，即燃料所放出的热量中有 50%以上没有得到利用。其中大部分被排放于冷却水或大气中，这部分热能数量虽大，但因温度接近环境温度，故不能用来转换为机械能。而在日常生活及工业生产的很多工艺过程中，又需要大量的温度不太高的供热蒸汽或热水。因此，如果蒸汽动力装置在生产电能的同时，把做过功的蒸汽的余热用来供给需要用热的场合，就可有效地利用冷凝过程中所放出的热量。这种既发电又供热的动力循环称为热电联产，既发电又供热的电厂，通常称为热电厂。它是目前我国发展集中供热的方向之一。

7.3　活塞式内燃机循环

在燃气动力装置中，燃料在气缸内燃烧，则称为内燃机循环。内燃机一般都是活塞式(或称往复式)的，燃料直接在气缸里燃烧，用燃烧的产物作为工质推动活塞做功，再由连杆带动曲轴转动，将燃烧气体中热能的一部分直接转换为机械能。

活塞式内燃机的吸气、压缩、燃烧、膨胀和排气过程都是在同一个带有活塞的气缸中进行的。活塞在气缸中从一端移到另一端的距离称为“冲程”，也称为行程。如果在四个冲程中完成整个过程组成的一个循环，则称为四冲程循环；如果在两个冲程中完成一个循环，就称为二冲程循环。

活塞式内燃机按燃料种类和在气缸内燃烧方式的不同，可分为点燃式内燃机和压燃式内燃机两类。点燃式内燃机吸气时吸入的气体是燃料和空气的混合物，经压缩后，由电火花点火燃烧；而压燃式内燃机吸入的气体则仅仅是空气，经压缩后使空气温度上升到燃料自燃的温度，而后喷入燃料燃烧。典型的点燃式内燃机为汽油机和煤气机，而典型的压燃式内燃机则为柴油机。对应的理论循环可分为定容加热循环和定压加热循环。下面以四冲程内燃机为例介绍其工作原理和循环过程。

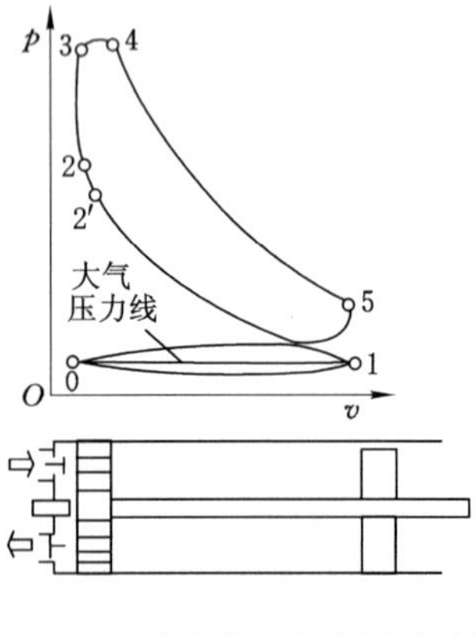

图 7-4　四冲程柴油机实际示功图

现以四冲程柴油机为例分析其实际工作循环。这种内燃机的每个工作循环有四个冲程，即每个循环活塞在气缸内往返两次。柴油机工作循环的四个冲程(图 7-4)分别如下：

(1) 吸气冲程 0—1：活塞从气缸上死点下行，进气阀开启，吸入空气。由于进气阀的节流作用，气缸内气体的压力低于大气压力。

(2) 压缩冲程 1—2：活塞到达下死点 1 时，进气阀关闭，活塞上行，压缩空气，当行至上死点 2 前的点 2′时，空气的压力可达 3～5 MPa，温度达 600～800 ℃，大大超过了柴油的自燃温度(3 MPa 时柴油的自燃温度约为 205 ℃)。这时柴油经高压雾化喷嘴喷入气缸。由于柴油有一个滞燃期，加之柴油机的转速较高，柴油实际上在活塞接近上死点 2 时才开始燃烧。

(3) 动力冲程 2—3—4—5：活塞到达上死点 2 时，气缸内已有相当数量的柴油，一旦燃烧就十分迅猛，压力迅速上升至 5～9 MPa，而活塞向下移动甚微，所以这一燃烧过程接近定容过程，如图 7-4 中的 2—3 所示。随着活塞下行，喷油和燃烧继续进行，此时缸内压力变化不大，这段燃烧过程接近定压，如图 7-4 中的 3—4 所示。活塞到达点 4 时喷油结束，此时气体温度可达 1 700～1 800 ℃。高温高压气体膨胀做功，压力、温度下降，活塞到达点 5 时气体的压力下降到 0.3～0.5 MPa，温度约 500 ℃。

(4) 排气冲程 5—0：活塞移动到点 5 时，排气阀突然打开，部分废气排入大气，气缸中气体的压力迅速下降，而活塞移动极微，接近定容降压过程。当气体压力降至略高于大气压力时，活塞开始上行，将气缸中剩余的气体排出，至此完成了一个实际循环。

四冲程汽油机与柴油机循环过程几乎相同，不同在于，汽油机吸气冲程中吸入气缸的是汽油与空气在气缸外充分混合后的混合物，再就是着火方式不同，汽油机由于汽油的自燃温度(约 415 ℃)较高，所以在压缩冲程终了时靠电火花点燃，而不是压燃。

实际的内燃机循环并非闭合的，也并非可逆循环。为了便于热力学分析，需要对实际循环加以合理的抽象和简化，使之成为闭口的、可逆的理想循环，且工质为空气。经合理抽象和简化，理想化的内燃机循环分为三类：定容加热循环、定压加热循环及混合加热循环。为了说明内燃机的工作过程对循环热效率的影响，首先引入内燃机的 3 个特性参数：

压缩比 $\varepsilon' = v_1/v_2$，表示压缩过程中工质体积被压缩的程度。

升压比 $\lambda' = p_3/p_2$，表示定容加热过程中工质压力升高的程度。

预胀比 $\rho' = v_4/v_3$，表示定压加热时工质体积膨胀的程度。

7.3.1　定容加热循环

在汽油机中，吸气过程吸入的是汽油与空气的混合物，经活塞压缩到上死点时由火花塞点火而迅速燃烧，几乎在瞬间完成，在此过程中活塞位移极小，可以认为是定容燃烧加热过程。此种情况下图 7-4 中的 4 与 3 重合为一点，定压预胀比 $\rho'=1$。如图 7-5 所示的定容加热循环，又称奥图(Otto)循环，它是汽油机和煤气机的理想循环。

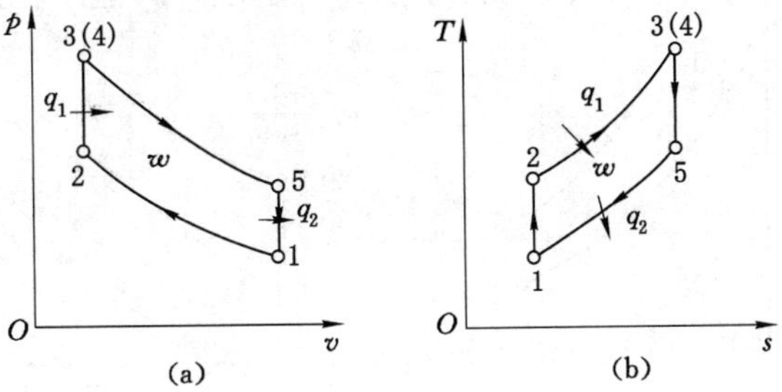

图 7-5　定容加热循环

根据图 7-5,可以很方便地进行热力学分析。以 1 kg 空气为工质,对定容加热循环进行分析,可以求出该循环的热效率。

吸热量:
$$q_1 = c_v(T_3 - T_2)$$

放热量:
$$q_2 = c_v(T_5 - T_1)$$

循环对外输出的净功:

$$w_s = q_1 - q_2 = c_v(T_3 - T_2) - c_v(T_5 - T_1)$$

循环的热效率:

$$\eta_t = \frac{w_s}{q_1} = 1 - \frac{q_2}{q_1} = 1 - \frac{T_5 - T_1}{T_3 - T_2}$$

由于 1—2,3—5 是定熵过程,则有:

$$\frac{T_2}{T_1} = \left(\frac{v_1}{v_2}\right)^{\kappa-1}, \frac{T_3}{T_5} = \left(\frac{v_5}{v_3}\right)^{\kappa-1}$$

而 2—3,5—1 是定容过程,则有:$v_2 = v_3$,$v_1 = v_5$。

代入热效率的定义式,整理可得:

$$\eta_t = 1 - \frac{T_1}{T_2} = 1 - \frac{1}{\left(\frac{v_1}{v_2}\right)^{\kappa-1}} = 1 - \frac{1}{\varepsilon'^{\kappa-1}} \tag{7-4}$$

从热效率的计算式中可以发现,汽油机的热效率仅取决于压缩比 ε'(κ 值一定时),并随着 ε' 的增大而升高。在实际过程中,压缩比要依据汽油的品质来确定,一般 ε' 为 5 ~ 9,不能随意增大。这主要是因为,气缸中压缩的是汽油和空气的混合物,若压缩比过高,压缩终了时混合物的温度接近或超过汽油的自燃温度,将发生爆燃(猛烈的燃烧过程),其结果不仅会使热效率降低,而且对机器寿命也有不良影响。

7.3.2 定压加热循环

由于汽油机的压缩比不能太高,因而限制了其热效率的提高。为了提高压缩比,发展了将空气和燃料分开送入气缸内的压燃式内燃机。这种内燃机以柴油为燃料,空气在气缸外加压,利用压缩空气将柴油射入气缸中,使之分散成雾状,被高温空气加热后进行燃烧,随喷随燃,活塞向右移动,缸内压力变化不大,可理想化为定压加热过程,故压燃式内燃机循环又称为定压加热循环,也称狄塞尔(Diesel)循环。此种情况下图 7-4 中的 3 与 2 重合为一点,升压比 $\lambda'=1$,成为图 7-6 所示的定压加热循环。早期的低速柴油机就属于这种情况,柴油喷入气缸燃烧的同时,活塞向下移动,气缸内的压力变化很小,近于在定压下燃烧。

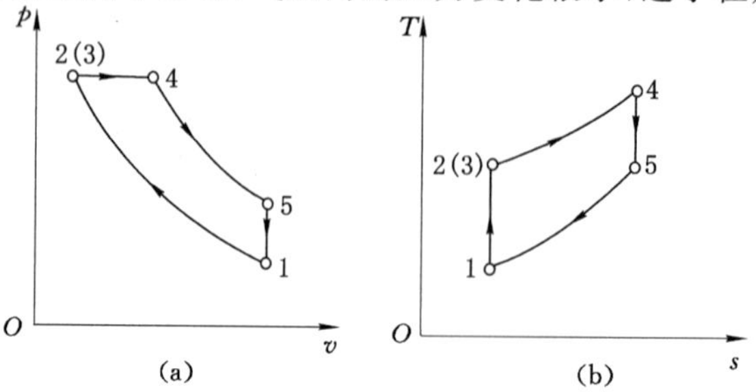

图 7-6 定压加热循环

对循环进行分析，同样可求得其热效率：

$$\eta_t = 1 - \frac{c_V(T_5 - T_1)}{c_p(T_4 - T_2)} = 1 - \frac{\rho'^{\kappa} - 1}{\kappa(\rho' - 1)\varepsilon'^{\kappa-1}} \tag{7-5}$$

由式(7-5)可知，热效率随压缩比 ε' 的增加而增大，随定压预胀比 ρ' 的增加而降低。柴油机吸入的是单纯空气，压缩后不会发生爆燃，所以压缩比可以比汽油机大，同时也只有提高压缩比，才有可能使压缩终了的空气温度高于燃料的燃点。而减小定压预胀比 ρ' 可以提高热效率，但 ρ' 的大小取决于喷入气缸的燃料量。对于柴油机而言，喷入的燃料量又决定于机器负荷的大小。所以，负荷大时，喷入的油量大，ρ' 也增大，热效率则降低。

这种柴油机由高压空气将柴油喷入气缸进行雾化。尽管其喷油雾化好，但由于转速低，活塞移动慢，同时还需附带喷油用的压气机，整个设备从外形到质量都较庞大，现已被按混合加热循环工作的柴油机所代替。

7.3.3　混合加热循环

现代高速柴油机并非单纯按定压加热循环工作，燃烧过程基本上包含定容燃烧和定压燃烧两个阶段，此循环也称为混合加热循环，或萨巴德循环。

经推导可得混合加热循环的热效率：

$$\eta_t = 1 - \frac{c_V(T_5 - T_1)}{c_V(T_2' - T_2) + c_p(T_3 - T_2')} = 1 - \frac{\lambda'\rho'^{\kappa} - 1}{[(\lambda' - 1) + \kappa\lambda'(\rho' - 1)]\varepsilon'^{\kappa-1}} \tag{7-6}$$

由式(7-6)可知，当 $\lambda' = 1$ 时，混合加热循环即为定压加热循环；当 $\rho' = 1$ 时，则为定容加热循环。

由三个循环热效率的计算式可以看出，影响循环热效率的主要因素是压缩比 ε'。无论哪种内燃机循环，提高压缩比 ε'，总是可以提高循环热效率。

7.4　空气压缩制冷循环

第 5 章已提到逆向卡诺循环是在一定温度范围内最有效的制冷循环，即逆向卡诺循环的制冷系数最大。逆向卡诺循环由两个可逆定温过程和两个定熵过程组成。按照逆向卡诺循环运行的制冷机称为卡诺制冷机。制冷装置一般由四大件组成，即压缩机、冷凝器、膨胀机和蒸发器(换热器)。虽然逆向卡诺循环的制冷系数最高，但受工质性质的限制，逆向卡诺循环在实际工程中不易实现，比如空气的定温加热和定温排热不易实现。实际的制冷循环不能按逆向卡诺循环工作，而是根据制冷剂的性质采用不同的循环方式。

由第 4 章的内容可知，当空气绝热膨胀做功时，其温度随之降低。空气压缩制冷装置就是利用这一原理获得低温的。

从冷藏室换热器出来的空气被压缩机吸入并进行压缩，提高压力和温度后进入冷凝器，被冷却后进入膨胀机膨胀做功，压力和温度大幅度下降，低温低压空气进入冷藏室吸取热量，从而达到维持冷藏室低温(即制冷)的目的。吸热升温后的空气再次被吸入压缩机进行下一个循环。由于空气定温加热和定温排热不易实现，故不能按逆向卡诺循环运行。在这种装置中，可以将两个定温过程近似为两个定压过程。

若空气的比热容为定值，则单位质量空气在冷凝器中放出的热量为：

$$q_1 = h_2 - h_3 = c_p(T_2 - T_3)$$

在冷藏室中，空气从低温物体吸收的热量为：

$$q_2 = h_1 - h_4 = c_p(T_1 - T_4)$$

于是，空气压缩制冷循环的制冷系数为：

$$\varepsilon = \frac{q_2}{q_1 - q_2} = \frac{T_1 - T_4}{(T_2 - T_3) - (T_1 - T_4)} = \frac{1}{\dfrac{T_2 - T_3}{T_1 - T_4} - 1}$$

对于可逆绝热过程 1—2 及 3—4，有：

$$\frac{T_2}{T_1} = \left(\frac{p_2}{p_1}\right)^{\frac{\kappa-1}{\kappa}}, \frac{T_3}{T_4} = \left(\frac{p_3}{p_4}\right)^{\frac{\kappa-1}{\kappa}}$$

因为 $p_2 = p_3$，$p_1 = p_4$，所以：

$$\frac{T_2}{T_1} = \frac{T_3}{T_4} = \frac{T_2 - T_3}{T_1 - T_4}$$

于是可得空气压缩制冷循环的制冷系数：

$$\varepsilon = \frac{1}{(T_2/T_1) - 1} = \frac{1}{(p_2/p_1)^{\frac{\kappa-1}{\kappa}} - 1} \tag{7-7}$$

相同温度范围内的逆向卡诺循环如图 7-7(c)中循环 1—3′—3—1′—1 所示。相同温度范围是指冷藏室温度 T_1(即制冷剂在换热器盘管出口的温度)至冷却水温度 T_3(即制冷剂在冷凝器出口能够达到的环境温度)，该逆向卡诺循环的制冷系数为：

$$\varepsilon_c = \frac{T_1}{T_3 - T_1} = \frac{1}{(T_3/T_1) - 1} \tag{7-8}$$

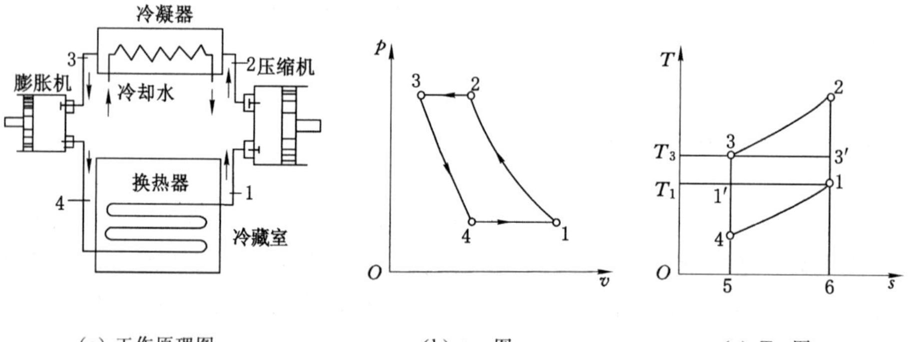

(a) 工作原理图　(b) p-v 图　(c) T-s 图

图 7-7　空气压缩制冷循环

比较上述两种制冷循环在相同温度范围内的制冷系数，由图 7-7(c)显见，$T_3 < T_2$，所以空气压缩制冷循环的制冷系数要比逆向卡诺循环的制冷系数小。由图 7-7(c)还可看出，空气压缩制冷循环 1—2—3—4—1 所消耗的功量(面积 12341)大于逆向卡诺循环所消耗的功量(面积 13′31′1)，但其制冷量却比后者少(二者相差面积 411′4)，所以前者的制冷系数小于后者。

空气以其容易获得、成本低、无毒安全等优点，最早被用作制冷工质。由于空气压缩制冷循环不易实现定温吸热和放热，同时空气的比热容较小，而它在冷藏室中的温升($T_1 - T_4$)又不宜太大[由图 7-7(c)可知，若要使 $T_1 - T_4$ 增大，压力比 p_2/p_1 就要增大，而空气压缩制冷循环的制冷系数 ε 偏离 ε_c 则越大]。由于空气的液化温度低，在制冷循环中不出现液气转变，不能利用汽化潜热，加上空气的比热容又较小，所以空气压缩制冷循环中单位工质的制冷能力较低。由于活塞式压缩机一次吸入的空气不多，否则压缩机将很大、很笨重，所以空气压缩制冷装置单位时间的制冷量受到限制。因此在常规制冷范围内（冷库温度不低于 －50 ℃），除了飞机上空调等特殊用途以外，空气压缩制冷循环现今很少应用。但近年来采用低压力比、大流量的叶轮压气机和有回热措施的装置，以及环保的需要，使空气压缩制冷循环又有了应用前景。

7.5　蒸汽压缩制冷循环

如上一节所述，空气的热物性决定了空气压缩制冷循环的制冷系数低和单位工质的制冷能力小。如果采用低沸点的物质作为工质，利用该种工质在定温定压下液化和气化的相变性质，可以实现定温定压吸热或放热过程（在湿蒸汽区），因而原则上可实现逆向卡诺循环。蒸汽压缩制冷循环采用在工作温度范围内发生相变的蒸汽作为工质，依靠相变潜热来实现制冷，单位质量制冷量大，并且在湿蒸汽区可实现定温吸热和定温放热过程。蒸汽压缩制冷在冷库、冰箱、空气调节等家用电器和工业领域都有应用，是目前应用最广泛的制冷方式。

图 7-8 是蒸汽压缩制冷装置及其理想制冷循环示意图。该制冷装置主要由压缩机、冷凝器、膨胀阀（或称节流阀）和蒸发器四大部件组成，蒸汽压缩制冷循环由四个过程组成：在压缩机中的定熵压缩过程 1—2，在冷凝器中的定压放热过程 2—3，在膨胀装置中的节流过程 3—4，在蒸发器中的定压吸热过程 4—1。其工作循环如下：从蒸发器出来的状态 1 的干饱和蒸汽被吸入压缩机进行绝热压缩过程 1—2，升压、升温至过热蒸汽状态 2；然后进入冷凝器，进行定压过程 2—3，先从过热蒸汽状态 2 定压下冷却成为干饱和蒸汽 $2'$，然后继续在定压、定温下凝结为饱和液体 3；从冷凝器出来的饱和液体经过膨胀阀绝热节流，使部分液体蒸发，降压降温至湿蒸汽状态 4；干度比较小的湿蒸汽进入蒸发器（冷库），进行定压吸热过程 4—1，离开蒸发器时已成为干饱和蒸汽，从而完成了一个循环 12341。图 7-8(b) 与图 7-8(c) 是蒸汽压缩制冷理想循环的 T-s 图及 $\lg p$-h 图（又称压焓图，由于循环中包含两个定压换热过程，因此用以压力为纵坐标、焓为横坐标所绘成的制冷剂的压焓图进行制冷循环的热力计算非常方便），每种制冷剂都有其相应的 T-s 图及 $\lg p$-h 图。其中，绝热节流过程 3—4 是不可逆过程，所以用虚线表示，这是使用结构简单、造价低廉的节流阀代替膨胀机造成的，采用节流阀将可逆绝热膨胀变成不可逆绝热节流膨胀，虽然损失了一点可回收的功，但是从实用观点来看，节流阀结构简单、体积小、造价低、调节方便，极易控制所需的制冷温度，而采用膨胀机取代膨胀阀并不现实，附加的收益不能抵偿增加设备所引起的费用增加，而且膨胀机结构更复杂。

蒸汽压缩制冷循环装置处于稳定运行状态时，组成循环的四个过程能够被看作稳定流动过程来分析。制冷剂的动能和势能变化相对功量和所传递的热量而言很小，因此，单位质量制冷剂的稳定流动过程能量方程可以简化为：$q = \Delta h + w_s$。

冷凝器和蒸发器不涉及任何功，压缩机被近似认为是绝热的。据此，可从制冷剂的 $\lg p$-h 图上方便地获得如下结果：

制冷量：
$$q_2 = h_1 - h_4$$

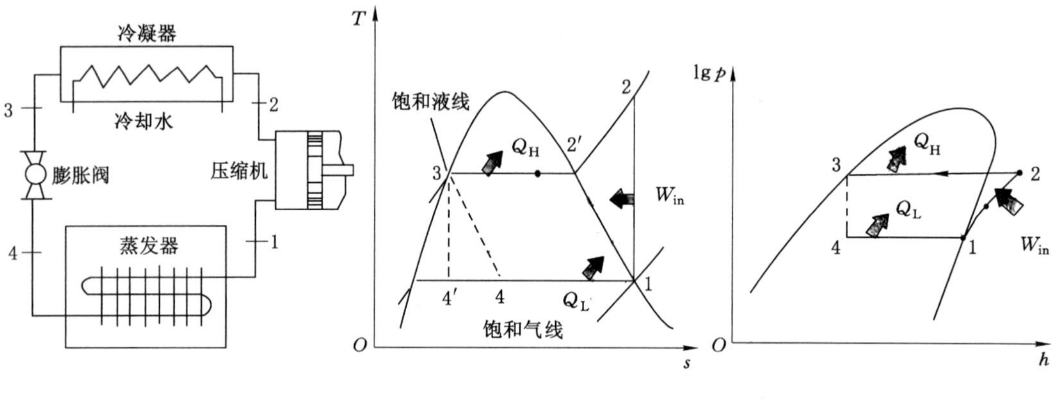

（a）工作原理图　　（b）理想制冷循环 T-s 图　　（c）理想制冷循环 lg p-h 图

图 7-8　蒸汽压缩制冷循环

冷凝放热量：　　$q_1 = h_2 - h_3$

循环消耗的净功量：　　$w_0 = h_2 - h_1$

此蒸汽压缩制冷循环的制冷系数为：

$$\varepsilon = \frac{q_2}{w_0} = \frac{h_1 - h_4}{h_2 - h_1} \tag{7-9}$$

式中　h_1——压力 p_1 下的饱和制冷剂蒸汽的焓值，kJ/kg；

h_4——压力 p_2 下的饱和制冷剂液体的焓值，kJ/kg；

h_2——被压缩后的制冷剂的焓值，kJ/kg。

制冷剂的质量流量：

$$\dot{m} = Q_2/q_2, \text{kg/s}$$

式中　Q_2——制冷装置的供冷能力，kW。

压缩机的功率为：　　$P = \dot{m} w_0$

冷凝器的热负荷为：　　$Q_1 = \dot{m} q_1$

本章小结

将热能转换成机械能的设备称为热机。根据循环介质不同热机主要分为两种形式：蒸汽动力装置和气体动力装置。本章主要介绍了动力装置循环的基本构成及热力学计算。通过热力计算，分析动力循环的能量转换的经济性，寻求提高热效率的方向及途径。

实际循环都是复杂的不可逆的，为使分析简化，通常将实际循环抽象概括成可逆的理论循环，通过理论循环分析，找出影响循环热效率的主要因素，从而获得提高热效率的有效措施。朗肯循环是基本的蒸汽动力装置循环，通过理论循环的热力学分析，得出提高循环的热效率主要有两种途径：一是改变循环初参数，即提高蒸汽的初温度、初压力及降低乏汽压力；二是改变循环的方式，即采用回热、再热循环及热电联产。活塞式内燃机循环和燃气轮机循环是两种典型的气体动力装置循环，前者根据工质不同可分为煤气机、汽油机和柴油机，根据循环方式不同又可分为混合加热循环、定压加热循环和定容加热循环。通过柴油机的理论循环分析得出,提高循环的压缩比、升压比及降低预胀比均可提高循环的热效率。

制冷循环是一种不完全逆卡诺循环,它通过消耗机械能或外界驱动热源实现了热量从低温物体向高温物体的传递,是一种重要的热力循环。蒸汽压缩制冷循环依靠相变潜热制冷,单位质量制冷剂的制冷量较大,因而应用最为广泛。热泵循环也是逆循环,其不同于制冷循环之处在于它的目的是向高温热源释放热量。由于热泵装置的供暖系数永远大于 1,故在节能方面优于其他供热方式。但热泵循环的上限温度为被加热物体的温度,下限温度为环境温度,因而它的应用会受到一定限制。

思考与练习题

(1) 朗肯循环效率低的主要原因是什么?

(2) 卡诺循环优于相同温度范围的其他循环,为什么蒸汽动力循环不采用卡诺循环?

(3) 简述蒸汽动力循环装置采用再热技术与回热技术的理由。

(4) 既然利用抽汽回热可以提高蒸汽动力装置循环的热效率,能否将全部蒸汽抽出来用于回热?为什么回热能提高热效率?

(5) 蒸汽动力装置循环热效率不高的原因是冷凝器放热损失太大,如取消冷凝器而用压缩机将乏汽直接升压送回锅炉是否可以?

(6) 影响内燃机定压加热理想循环效率的因素有哪些?

(7) 柴油机的热效率高于汽油机的热效率主要原因是什么?

(8) 为什么空气压缩制冷循环不采用逆向卡诺循环?

(9) 为何蒸汽压缩制冷循环采用节流阀来代替膨胀机?空气压缩制冷机能否采用节流阀?

(10) 蒸汽压缩制冷循环中对制冷剂有何基本要求?常用的制冷剂有哪几种?

Chapter 7 Power Cycle and Refrigeration Cycles

7.1 Overview

A power cycle device that converts thermal energy into mechanical energy is called a thermal power device or a thermal engine. In a heat engine, the continuous conversion of thermal energy to mechanical energy is achieved through the thermal cycle of the working medium. According to the different working medium, the power cycle can be categorized into steam power cycle (such as the working cycle of steam engine and steam turbine) and gas power cycle (such as the working cycle of internal combustion engine and gas turbine device).

The process of cooling an object so that its temperature is lower than that of the surrounding environment and maintained at that temperature is called refrigeration. In order to obtain or maintain a low temperature, heat must be extracted from the object or space that needs to be cooled. A refrigeration device is a device that achieves this task at the cost of energy (work or heat). It is a device that transfers heat from a low-temperature medium to a high-temperature medium. In order to continuously discharge heat to an external source (usually the atmosphere), it is necessary to form a cycle that allows the refrigeration unit to operate continuously. The refrigerant used in the refrigeration cycle is called "refrigerant", such as air, water, ammonia, Freon and so on.

Compression refrigeration device is a widely used refrigeration device at present, the vast majority of household refrigerators, air conditioners, refrigerators and so on are using compression refrigeration. If the refrigerant (that is, the refrigerant) is always in the gaseous phase during the cycle, the refrigeration cycle is called the gas compression refrigeration cycle. If the state change of the refrigerant medium spans the liquid and gas states, the refrigeration cycle is called the vapor compression refrigeration cycle. In addition, there are absorption refrigeration cycle, adsorption refrigeration cycle, steam injection refrigeration cycle and semiconductor refrigeration.

All real cycles are complex and irreversible. Therefore, in the cycle analysis, the simplified thermodynamic model of the actual cycle is firstly established, and the simple and typical reversible process and cycle are used to approximate the actual complex irreversible process and cycle. Through thermodynamic analysis and calculation, the basic characteristics and laws are found out. As long as the simplified thermodynamic model is reasonable and close to reality, the results of analysis and calculation have theoretical guiding significance. If necessary, the influence of various irreversible factors can be further considered and necessary corrections can be introduced to improve its accuracy.

7.2 Steam Power Cycle

7.2.1 Steam Carnot Cycle

Consider a steady-flow Carnot cycle executed within the saturation dome of pure water, as shown in Fig. 7-1. The fluid is heated reversibly and isothermally in a boiler (process 4-1), and expands isentropically in a turbine (process 1-2). Then it is condensed reversibly and isothermally in a condenser (process 2-3), and compressed isentropically by a compressor to the initial state (process 3-4).

Isothermal heat transfer to or from a two-phase system is not difficult to achieve in practice since maintaining a constant pressure in the device automatically fixes the temperature at the saturation value. Therefore, processes 4-1 and 2-3 can be approached closely in actual boilers and condensers. However, the quality of the steam decreases during the process 1-2, as shown on the *T-s* diagram in Fig. 7-1. Thus the turbine has to handle steam with low quality, that is, steam with a large amount of moisture. The impingement of liquid droplets on the turbine blades causes erosion and is a major source of wear. The isentropic compression process 3-4 involves the compression of a liquid-vapor mixture to a saturated liquid. It is not easy to control the condensation process so precisely as to end up with the desired quality at state 3. On the other hand, it is not practical to design a compressor that handles two phases.

Thus we conclude that the Carnot cycle cannot be approximated in actual devices and is not a realistic model for vapor power cycles. Even the Carnot cycle can be accomplished, the thermal efficiency is limited by the maximum temperature in the cycle (it has to remain under the critical-point value, which is 374 ℃ for water).

7.2.2 Rankine Cycle

The impracticalities associated with the Carnot cycle can be eliminated by superheating the steam in the boiler and condensing it completely in the condenser, as shown schematically on the diagrams in Fig. 7-2. The system consists of four parts: a boiler, a turbine, a condenser and a pump. The cycle that result is the Rankine cycle. It is an ideal cycle for vapor power plants. It does not involve any internal irreversibilities and widely used in modern thermal power plants. The ideal Rankine cycle consists of the following two constant pressure processes and two isentropic processes.

3-4: Isentropic compression process in the pump;

4-5-6-1: Constant pressure heat addition process in the boiler;

1-2: Isentropic expansion process in the turbine;

2-3: Constant pressure heat rejection process in the condenser.

In the process 3-4, water enters the pump at state 3 as saturated liquid and is cornpressed isentropically to the operating pressure of the boiler. During this process, the pump consumes some work and pumps the water into the boiler. During the process 4-5-6-1, water enters the boiler as a compressed liquid at state 4 and leaves as a superheated vapor at state 1. The boiler is basically a large heat exchanger where the heat originating from combustion gases, nuclear reactors, or other sources is transferred to the water essentially at constant pressure.

The superheated vapor at state 1 enters the turbine, where it expands isentropically and outputs work by rotating the shaft connected to an electric generator or to other work consuming devices. The pressure and the temperature of the steam drop during this process to the values at state 2, where steam enters the condenser. At this state, steam is usually a saturated liquid-vapor mixture with a high quality, which is called exhaust vapor.

Steam is condensed at constant pressure in the condenser, which is basically a large heat exchanger, by rejecting heat to cooling medium, such as the atmosphere or cooling water. Steam leaves the condenser as saturated liquid and enters the pump, completing the cycle.

The Rankine cycle can be analyzed as steady-flow processes. Based on steady-flow energy equation per unit mass of steam, the heat absorption in the process 4-5-6-1 is:

$$q_1 = h_1 - h_4$$

The shaft work output in the isentropic expansion process 1-2 in the turbine is:

$$w_{s,1-2} = h_1 - h_2$$

The heat rejection during the constant pressure process 2-3 in the condenser is:

$$q_2 = h_2 - h_3$$

The shaft work consumed by the water pump in the isentropic compression process 3-4 is:

$$w_{s,3-4} = h_4 - h_3$$

Thus, during the whole cycle, the net work output is:

$$w_{net} = (h_1 - h_2) - (h_4 - h_3)$$

Since the water in the system undergoes a cycle, the change in internal energy is 0, $\Delta u=0$. From the First Law of Thermodynamics, we know $q_1 - q_2 = w_{net}$.

Therefore the thermal efficiency of the ideal Rankine cycle can be determined by:

$$\eta_t = \frac{w_{net}}{q_1} = \frac{(h_1 - h_2) - (h_4 - h_3)}{h_1 - h_4} \tag{7-1}$$

The work consumed by the pump is comparatively small and can be neglected, thus:

$$\eta_t = \frac{w_{net}}{q_1} = \frac{h_1 - h_2}{h_1 - h_3} \tag{7-2}$$

7.3 Gas Power Cycle with Internal Combustion Engines

In gas power cycles, the working fluid remains as gas throughout the entire cycle. Spark ignition engines, diesel engines, and conventional gas turbines are familiar examples of devices that operate on gas cycles. In all these engines, energy is provided by burning a fuel within the system boundaries. That is, they are internal combustion engines. Because of this combustion process, the composition of the working fluid changes from air and fuel to combustion products during the course of the cycle. However, considering that air is predominantly nitrogen that hardly undergoes any chemical reactions in the combustion chamber, the working fluid closely resembles air at all times.

In gas power cycle, reciprocating engines are always adopted. And the internal combustion engines are classified as spark-ignition (SD) engines or compression-ignition (CI) engines, depending on how the combustion process in the cylinder is initiated. In SI engines, the combustion of the air-fuel mixture is initiated by a spark plug. In CI engines, the air-fuel mixture is self-ignited as a result of compressing the mixture above its self-ignition temperature.

7.3.1 Ideal Cycle for Spark-Ignition Engines: Otto Cycle

The Otto cycle is the ideal cycle for spark-ignition reciprocating engines. Gasoline is the fuel used in SI engine. In most spark-ignition engines, the piston executes four complete strokes within the cylinder, and the crankshaft completes two revolutions for every thermodynamic cycle. These engines are called four-stroke internal combustion engines.

The schematic of the actual Otto cycle is illustrated in Fig. 7-4. Initially, both the intake and the exhaust valves are closed, and the piston is at its lowest position (BDC). During the intake stroke, the intake valve is opened and fresh air-fuel mixture is draw in through the intake valve; then during the compression stroke, the piston moves upward, compressing the air-fuel mixture. Shortly before the piston reaches its highest position (TDC), the spark plug fires and the mixture is ignited, increasing the pressure and temperature of the system. And then, during the expansion or power stroke, the high-pressure gases force the piston down, which in turn forces the crankshaft to rotate, producing a useful work output. At the end of this stroke, the piston is at the lowest position, and the cylinder is filled with combustion products. Continuously, during the exhaust stroke, the piston moves upward one more time, purging the exhaust gases through the exhaust valve, and down a second time, drawing in fresh air-fuel mixture through the intake valve again. Notice that the pressure in the cylinder is slightly above the atmospheric value during the exhaust stroke and slightly below during the intake stroke.

The thermodynamic analysis of the actual four-stroke or two-stroke cycles described is not a simple task. However, the analysis can be simplified significantly if the air-standard assumptions are utilized. The resulting cycle, which closely resembles the actual operating conditions, is the ideal Otto cycle. The idealized Otto cycle together with p-v and T-s diagrams is illustrated in Fig. 7-5. It consists of four internally reversible processes:

1-2: Isentropic compression;

2-3: Constant-volume heat addition;

3-4: Isentropic expansion;

4-1: Constant-volume heat rejection.

No work is involved during the two heat transfer processes for both take place at constant volume. Heat transferred to and from the working fluid can be expressed as:

$$q_1 = c_v(T_3 - T_2)$$

and

$$q_2 = c_v(T_5 - T_1)$$

The net work output is:

$$w_s = q_1 - q_2 = c_v(T_3 - T_2) - c_v(T_5 - T_1)$$

Thermal efficiency of the Otto cycle is:

$$\eta_t = \frac{w_s}{q_1} = 1 - \frac{q_2}{q_1} = 1 - \frac{T_5 - T_1}{T_3 - T_2}$$

Since process 1-2 and 3-5 are isentropic, $v_2 = v_3$ and $v_1 = v_5$:

$$\frac{T_2}{T_1} = \left(\frac{v_1}{v_2}\right)^{\kappa-1} = \left(\frac{v_5}{v_3}\right)^{\kappa-1} = \frac{T_3}{T_5} \tag{7-3}$$

Substituting these equations into the thermal efficiency relation and simplifying it gives:

$$\eta_t == 1 - \frac{T_1}{T_2} = 1 - \frac{1}{\left(\frac{v_1}{v_2}\right)^{\kappa-1}} = 1 - \frac{1}{\varepsilon'^{\kappa-1}} \tag{7-4}$$

Eq. (7-4) shows that under the air-standard assumptions, the thermal efficiency of an ideal Otto cycle depends on the compression ratio of the engine and the specific heat ratio of the working fluid. The thermal efficiency of the ideal Otto cycle increases with both the compression ratio and the specific heat ratio. However, the increase in thermal efficiency with the compression ratio is not as pronounced at high compression ratios. Also, when high compression ratios are used, the temperature of the air-fuel mixture rises above the auto-ignition temperature of the fuel (the temperature at which the fuel may ignite without the help of a spark) during the combustion process, causing an early and rapid burn of the fuel at some point or points ahead of the flame front, followed by almost instantaneous inflammation of the gases.

Improvement of the thermal efficiency of gasoline engines by utilizing higher compression ratios (up to about 12) without facing auto-ignition problem has been made possible by using gasoline blends that have good antiknock characteristics, such as gasoline mined with tetraethyl lead. However, this measure has brought a very undesirable side effect and the combustion compounds are hazardous to health and pollute the environment.

7.3.2 Diesel Cycle: the Ideal Cycle for Compression-Ignition Engine

The Diesel cycle is the ideal cycle for CI reciprocating engines. In CI engines (also known as diesel engines), the air is compressed to a temperature higher than the auto-ignition temperature of the fuel, and combustion starts as the fuel is injected into this hot air. Therefore, the spark plug and carburetor are replaced by a fuel injector in diesel engines.

In diesel engines, only air is compressed during the compression stroke, eliminating the possibility of auto-ignition. Therefore, diesel engines can be designed to operate at much higher compression ratios, typically between 12 and 24. Not having to deal with the problem of auto-ignition, it has another benefit: many of the stringent requirements placed on the gasoline can now be removed, and fuels that are less refined (thus less expensive) can be used in diesel engines.

The fuel injection process in diesel engines starts when the piston approaches TDC and continues during the first part of the power stroke. Therefore, the combustion process in these engines takes place over a longer interval. Because of this longer duration, the combustion process in the ideal Diesel cycle is approximated as a constant-pressure heat-addition process. In fact, this is the only process where the Otto and the Diesel cycles differ. The remaining three processes are the same for both ideal cycles. The similarity between the two cycles is also apparent from the p-v and T-s diagrams of the Diesel cycle, as shown in Fig. 7-6.

Similarly, we can get the thermal efficiency for Diesel engine:

$$\eta_t = 1 - \frac{c_v(T_5 - T_1)}{c_p(T_4 - T_2)} = 1 - \frac{\rho'^{\kappa} - 1}{\kappa(\rho' - 1)\varepsilon'^{\kappa-1}} \tag{7-5}$$

The efficiency of a Diesel cycle differs from that of an Otto cycle by the quantity in the brackets. This quantity is always greater than 1. Also, as the cutoff ratio decreases, the efficiency of the Diesel cycle increases. For the limiting case of $\rho' = 1$, the quantity in the brackets becomes unity and the efficiencies of the Otto and Diesel cycles become identical. Remember that diesel engines operate at much higher compression ratios and thus are usually more efficient than the spark-ignition (gasoline) engines. Thermal efficiencies of large diesel engines range from about 35% to 40%.

7.3.3 The Dual Cycle

Approximating the combustion processin internal combustion engines as a constant-volume or a constant-pressure heat-addition process is overly simplistic and not quite realistic. A better approach is to model the combustion process in both gasoline and diesel engines as a combination of two heat-transfer processes, one at constant volume and the other at constant pressure. The ideal cycle based on this concept is called the dual cycle. The relative amounts of heat transferred during each process can be adjusted to approximate the actual cycle more closely.

The thermal efficiency of the dual cycle is:

$$\eta_t = 1 - \frac{c_v(T_5 - T_1)}{c_v(T_2{}' - T_2) + c_p(T_3 - T_2{}')} = 1 - \frac{\lambda' \rho'^{\kappa} - 1}{[(\lambda' - 1) + \kappa\lambda'(\rho' - 1)]\varepsilon'^{\kappa-1}} \quad (7\text{-}6)$$

Note that both the Otto and the Diesel cycles can be treated as special cases of the dual cycle. It can be found that as $\lambda' = 1$, the thermal efficiency of the dual cycle becomes that of the Diesel cycles; further as $\rho' = 1$, it is the Otto cycle.

7.4 Air-Compression Refrigeration Cycles

The reversed Carnot cycle has the maximal coefficient of performance within given temperature limits and it is an ideal cycle of refrigerators. In practice, the two constant temperature processes can be obtained easily. In the evaporator and the condenser, the evaporation process and the condensation process of the refrigerant at constant pressure are in wet vapor region. And we can obtain constant temperature processes approximately because of the constant saturated temperatures. It seems these problems can be overcome by the reversed Carnot cycle operating outside the saturation region, however, it is difficult to realize the isothermal heat-absorption and heat-rejection processes. Therefore the actual device using the reversed Carnot cycle is unrealistic, but the reversed Carnot cycle can be taken as a datum for comparison of coefficient of performance of refrigeration cycles.

Actual refrigeration cycles using air as working substance are not based on the reversed Carnot cycle because of the difficulties involved in carrying out the isothermal heat-absorption and heat-rejection processes. The schematic diagram of an air-compression refrigerating installation is shown in Fig. 7-7. The cycle adopts two constant pressure processes instead of two constant temperature processes in the reversed Carnot cycle. The air from the cooling chamber 1 is compressed in the compressor 2. As a result, its temperature rises. Then the compressed air is discharged at a constant pressure into the heat exchanger 3, in which it is cooled by water to the temperature of the ambient medium. And then the compressed air passes into the expansion cylinder or the expander 4, in which it expands to the initial pressure. As the compressed air expands, its temperature drops. Then the chilled air absorbs heat from the refrigerated space and its temperature rises. It enters the compressor again to start another cycle.

The ideal cycle of an air-compression refrigerating installation is plotted on the p-v and T-s diagrams shown in Fig. 7-7. In the process represented by line 1-2, air is compressed adiabatically from pressure p_1 to p_2. During the isobaric process 2-3, heat is rejected from the air to an external source and its temperature decreases from T_2 to T_3. Through adiabatic expansion process represented by line 3-4, the air is further cooled to the temperature T_4. And then, the chilled air absorbs heat during the isobaric process 4-1 to remove thermal energy from the cooling chamber. As a result, the air temperature rises from T_4 to T_1.

The work needed to accomplish an air-compression cycle is equal to the difference between the amounts of heat released q_1 and heat absorbed q_2. Assuming the specific heat to be constant, we obtain:

$$q_1 = h_2 - h_3 = c_p(T_2 - T_3)$$

$$q_2 = h_1 - h_4 = c_p(T_1 - T_4)$$

Thus, the coefficient of performance of the cycle is equal to:

$$\varepsilon = \frac{q_2}{q_1 - q_2} = \frac{T_1 - T_4}{(T_2 - T_3) - (T_1 - T_4)} = \frac{1}{\dfrac{T_2 - T_3}{T_1 - T_4} - 1}$$

During isentropic processes 1-2 and 3-4, we know that:

$$\frac{T_2}{T_1} = \left(\frac{p_2}{p_1}\right)^{\frac{\kappa-1}{\kappa}} \text{ and } \frac{T_3}{T_4} = \left(\frac{p_3}{p_4}\right)^{\frac{\kappa-1}{\kappa}}$$

Since $p_2 = p_3$ and $p_1 = p_4$, the relation among temperatures is:

$$\frac{T_2}{T_1} = \frac{T_3}{T_4} = \frac{T_2 - T_3}{T_1 - T_4}$$

Finally we obtain:

$$\varepsilon = \frac{1}{(T_2/T_1) - 1} = \frac{1}{(p_2/p_1)^{\frac{\kappa-1}{\kappa}} - 1} \tag{7-7}$$

Where T_1 is temperature of the room being cooled, the same as the air temperature drawn into the compressor; T_2 is temperature of the compressed air.

During earlier period, piston compressor and piston expander were used in air refrigeration cycles. And soon air refrigeration cycles were eliminated for its small refrigerating capacity. However, the adoption of heat regenerator, centrifugal compression and expander provide the possibility to improve the performance of the air compression refrigeration cycle, and make it retrieved in industry application. At present, the air compression refrigeration cycle is mainly used for cooling cabin, and it also can be used for the cryogenic engineering and gas liquefaction, etc.

7.5 The Vapor-Compression Refrigeration Cycle

The vapor-compression refrigeration cycle is the most widely used in refrigerators and air-conditioning systems.

Schematic and *T-s* diagram of the ideal vapor-compression refrigeration cycle are shown in Fig. 7-8. It consists of four processes:

1-2: Isentropic compression in the compressor;

2-3: Constant-pressure heat rejection in the condenser;

3-4: Throttling in the expansion device;

4-1: Constant-pressure heat absorption in the evaporator.

In an ideal vapor-compression refrigeration cycle, the refrigerant enters the compressor at state 1 as saturated vapor and is compressed isentropically to the condenser pressure, during which the temperature of the refrigerant increases and it becomes superheated vapor. Then the superheated refrigerant vapor at state 2 enters the condenser and leaves as saturated liquid at state 3 as a result of heat rejection to the surroundings. The temperature of the refrigerant at this state is still above the temperature of the surroundings. The saturated liquid refrigerated at state 3 is throttled to the evaporator pressure by passing through an expansion valve or capillary tubes. The refrigerant temperature drops below the temperature of the refrigerated space during this process. It enters the evaporator at state 4 as a low-quality saturated liquid-vapor mixture, and it completely evaporates by absorbing heat from the refrigerated space. The refrigerant leaves the evaporator as saturated vapor and enters the compressor again, completing a cycle.

Remember that the area under the process curve on a T-s diagram represents the heat transferred during internally reversible processes. Thus, the area under the process curve 4-1 represents the heat absorbed by the refrigerant in the evaporator, and the area under the process curve 2-3 represents the heat rejected in the condenser. Another diagram frequently used in the analysis of vapor-compression refrigeration cycles is the lg p-h diagram, as shown in Fig. 7-8(c).

All the four components associated with the vapor-compression refrigeration cycle are steady-flow devices, and thus all four processes that snake up the cycle can be analyzed as steady-flow processes. The kinetic and potential energy changes of the refrigerant are usually small compared with the work input and the heat transferred. Therefore, they can be neglected. Then the steady flow energy equation on a unit-mass basis reduces to:

$$q = \Delta h + w_s$$

The condenser and the evaporator do not involve any work, and the compressor can be approximated as adiabatic. Accordingly, we can obtain the following results easily:

Refrigerating capacity is:

$$q_2 = h_1 - h_4$$

Heat released to the cooling medium in the condenser is:

$$q_1 = h_2 - h_3$$

Net work consumption is:

$$w_0 = h_2 - h_1$$

Thus the coefficient of performance of refrigerators operating on the vapor-compression refrigeration cycle can be expressed as:

$$\varepsilon = \frac{q_2}{w_0} = \frac{h_1 - h_4}{h_2 - h_1} \tag{7-8}$$

Where h_1 and h_4 are the enthalpies of the saturated vapor and saturated liquid refrigerant at pressure p_1, respectively, kJ/kg; h_2 is the enthalpy of the superheated refrigerant leaving the compressor, kJ/kg.

The mass flow rate of the refrigerant is:

$$\dot{m} = Q_2/q_2$$

Where Q_2 is the cooling load of the refrigerator, kW.

Compression work:

$$P = \dot{m}w_0$$

Condenser heat load:

$$Q_1 = \dot{m}q_1$$

第8章 导　热

本章导学

物体各部分之间不发生相对位移时,依靠分子、原子及自由电子等微观粒子的热运动而产生的热量传递称为导热。学习导热的目标主要有两个:(1) 传导了多少热量?(2) 固体内部的温度怎么分布?前者由傅立叶导热定律决定,后者主要由导热微分方程进行求解。

本章首先介绍了导热机理和傅立叶定律,其次根据能量守恒推导出了导热微分方程,并阐述了相应的定解条件。在此基础上,针对典型的一维平壁、圆筒壁导热,肋片导热以及二维稳态导热进行求解,以获得物体中的温度分布和热流量。最后阐述了非稳态导热的特点、零维问题的集总参数法以及集总参数法的适用范围。

学习目标

掌握导热基本定律——傅立叶定律的物理意义、数学表达式及应用;理解温度场、等温面及温度梯度的意义和特点;掌握平壁、圆筒壁一维稳态导热问题的温度场和导热量求解计算;掌握导热系数的物理意义及影响因素;掌握热阻的概念及不同情况下导热热阻的分析和计算;理解非稳态导热过程的基本特点及非稳态导热问题的求解方法。

重点与难点

(1) 建立导热微分方程的基本方法。

(2) 一维稳态导热问题温度场和导热量的计算。

(3) 导热系数与导温系数。

(4) 导热热阻。

8.1 导热概述

8.1.1 导热机理

提及传导,应该立刻想到原子和分子的运动,因为导热就是原子和分子的热运动造成的。导热可看作微观粒子之间的相互作用:能量较大的粒子向能量小的粒子传输能量。通过讨论气体的导热并利用大家熟悉的热力学的基本概念可以很容易地说明导热的物理机理。设想在一种气体中存在温度梯度，并假定气体没有宏观运动。两个不同温度的表面之间充满了气体，如图 8-1 所示。温度越高，分子的能量越大，当邻近的分子相互碰撞时（这总是在不断发生的），能量较大的分子必然会向能量较小的分子传输能量。当存在温度梯度时，通过导热的能量传输总是向温度降低的方向进行。这种传输可由图 8-1 容易得到：由于分子的随机运动，不断地有分子从上面或下面越过 x_0 处的假想平面。由于 x_0 平面上方的温度比 x_0 平面下方的高，所以净的能量传输必然沿 x 正方向，这种基于分子随机运动的净能量传输即能量的扩散。

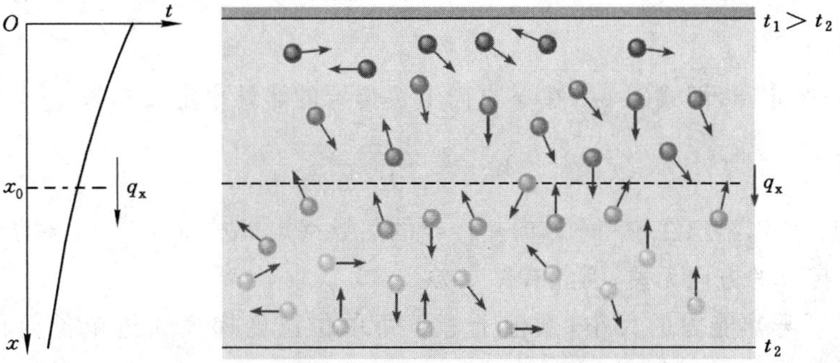

图 8-1 由分子活动导致的与能量扩散相联系的热传导

8.1.2 傅立叶定律

傅立叶定律是现象学定律，就是说它是由观察到的现象得到的，并非由热力学第一定律推得。傅立叶定律是基于大量实验结果的归纳而得到的普遍规律，其表述式为：

$$\Phi = -\lambda A \frac{\partial t}{\partial x} \tag{8-1}$$

式中，Φ 为热流量，表示在单位时间内通过某一给定传热面积 A 传递的热量，单位为 W。

在单位时间内，通过单位面积传递的热量，称为热流密度或热流通量，用 q 表示，单位为 W/m^2。Φ 与 q 的关系为 $\Phi = Aq$。热流密度的表达式：

$$q = \frac{\Phi}{A} = -\lambda \frac{\partial t}{\partial x} \tag{8-2}$$

注意，由于热能总是向温度降低的方向传输，所以式(8-2)中必须有负号。

由于热流密度是一个向量，可将热传导方程（傅立叶定律）写成下述更通用的表达式：

$$q = -\lambda \nabla t = -\lambda \left(\mathbf{i} \frac{\partial t}{\partial x} + \mathbf{j} \frac{\partial t}{\partial y} + \mathbf{k} \frac{\partial t}{\partial z} \right) \tag{8-3}$$

式中，∇为三维倒三角算子；$t(x,y,z)$ 为标量温度场。

式(8-3)说明热流密度方向是垂直于等温面的。所以，傅立叶定律的另一种表达式是：

$$q = q_n \mathbf{n} = -\lambda \frac{\partial t}{\partial n} \mathbf{n} \tag{8-4}$$

式中，q_n为 $\mathbf{n}$ 方向上的热流密度，它垂直于等温面。

传热过程是由沿 $\mathbf{n}$ 方向的温度梯度维持的。还要注意的是，可以将热流密度向量分解为几个分量，在直角坐标系中，q_n的通用表达式为：

$$q_n = \mathbf{i} q_x + \mathbf{j} q_y + \mathbf{k} q_z \tag{8-5}$$

由式(8-3)可得：

$$q_x = -\lambda \frac{\partial t}{\partial x}, q_y = -\lambda \frac{\partial t}{\partial y}, q_z = -\lambda \frac{\partial t}{\partial z} \tag{8-6}$$

傅立叶定律是热传导的基础，其关键要点可总结如下：它不是可以由热力学第一定律导出的一个表达式，而是基于实验结果的归纳。这也是定义材料的一个重要物性——热导率的一个表达式。此外，傅立叶定律是一个向量表达式，它指出热流密度是垂直于等温面的，并且是沿温度降低的方向。另外需要指出的是，傅立叶定律适用于所有物质，不管它处于什么状态(固体、液体或气体)。

8.1.3 导热系数

根据傅立叶定律，即式(8-6)，在 x 方向上热传导的热导率定义式为：

$$\lambda_x = -\frac{q_x}{(\partial t/\partial x)}$$

与此类似，可以给出在 y 和 z 方向热传导的热导率定义式(λ_y，λ_z)，但对于各向同性的材料，热导率与传导的方向无关，因此有 $\lambda_x = \lambda_y = \lambda_z$。

λ 表征物质导热能力的大小，数值上等于每单位温度梯度通过单位面积所传递的热流量。

由上式可知，对于规定的温度梯度，传导热流密度是随热导率的增大而增大的。根据导热的物理机理可知，一般来说，固体的热导率比液体的大，而后者的又比气体的大。固体的热导率可以比气体的大 4 个数量级以上。这种差异在很大程度上是这两种状态分子间距的不同所导致的。

8.2 导热问题的数学描述

8.2.1 导热微分方程

讨论一个内部无整体运动(平流)的均匀介质，温度分布 $t(x,y,z)$ 用直角坐标系表示。根据内能量守恒的原理，首先定义一个如图 8-2 所示的无限小的(微元)控制体积 $\mathrm{d}x\mathrm{d}y\mathrm{d}z$，选用适用于瞬时的热力学第一定律表达式。其次讨论与这个控制体积有关的能量过程。如果存在温度梯度，热传导会通过控制体积的各个表面发生。垂直于坐标位置分别为 x，y 和 z 轴的各控制表面的热流量分别记作 Φ_x，Φ_y 和 Φ_z。利用泰勒级数展开并忽略高次项，可得通过每个相对的控制表面的热流量：

$$\Phi_{x+\mathrm{d}x} = \Phi_x + \frac{\partial \Phi_x}{\partial x}\mathrm{d}x \tag{8-7a}$$

$$\Phi_{y+\mathrm{d}y} = \Phi_y + \frac{\partial \Phi_y}{\partial y}\mathrm{d}y \tag{8-7b}$$

$$\Phi_{z+\mathrm{d}z} = \Phi_z + \frac{\partial \Phi_z}{\partial z}\mathrm{d}z \tag{8-7c}$$

具体地说，方程式(8-7)只是说明：在 $x+\mathrm{d}x$ 处的 x 热流量分量等于它在 x 处的分量加上它相对 x 方向的变化率乘 $\mathrm{d}x$。

基于能量守恒的通用形式为：

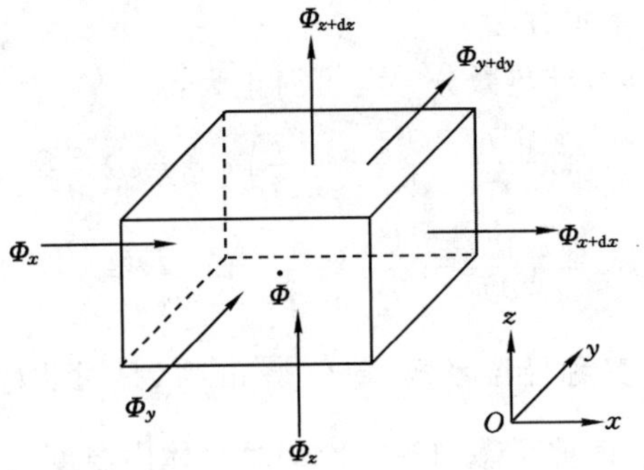

图 8-2　直角坐标系中进行导热分析的微元控制体积

$$\dot{E}_{in} + \dot{E}_g - \dot{E}_{out} = \dot{E}_{st} \tag{8-8}$$

在介质内可能有与不断产生热能有关的能源项，此项由式(8-9)表示：

$$\dot{E}_g = \dot{q}dxdydz \tag{8-9}$$

式中，$\dot{q}$ 是单位体积介质的热流密度，W/m^3。

另外，控制体积内物质所贮存的总的热能也可能发生变化。对于固体和不可压缩流体，定压比热 c_p 和定容比热 c_v 相差很小，$c_p = c_v = c$。如果物质不发生相变，则不存在潜热的影响，能量贮存项可表示为：

$$\dot{E}_{st} = \rho c \frac{\partial t}{\partial \tau}dxdydz \tag{8-10}$$

式中，$\rho c \dfrac{\partial t}{\partial \tau}$ 是单位体积介质的显热能的时间变化率。

需要注意的是，$\dot{E}_g$ 和 $\dot{E}_{st}$ 这两项代表了不同的物理过程。产能项 $\dot{E}_g$ 表示某种能量转换过程，涉及这种过程的一方是热能，另一方可以是化学能、电能或核能。如果在介质内靠消耗某种其他形式的能量而产生热能，此项为正（源）；如果消耗的是热能，此项为负（汇）。与此不同，贮能项 $\dot{E}_{st}$ 指物质所贮存的热能的变化速率。

因此，注意能量输入 $\dot{E}_{in}$ 和输出 $\dot{E}_{out}$ 是由导热速率项构成的，将式(8-9)和式(8-10)代入式(8-8)可得：

$$\Phi_x + \Phi_y + \Phi_z + \dot{q}dxdydz - \Phi_{x+dx} - \Phi_{y+dy} - \Phi_{z+dz} = \rho c \frac{\partial t}{\partial \tau}dxdydz \tag{8-11}$$

将式(8-7)代入式(8-11)，有：

$$-\frac{\partial \Phi_x}{\partial x}dx - \frac{\partial \Phi_y}{\partial y}dy - \frac{\partial \Phi_z}{\partial z}dz + \dot{q}dxdydz = \rho c \frac{\partial t}{\partial \tau}dxdydz \tag{8-12}$$

热流量可据傅立叶定律计算：

$$\Phi_x = -\lambda dydz \frac{\partial t}{\partial x}, \Phi_y = -\lambda dxdz \frac{\partial t}{\partial y}, \Phi_z = -\lambda dxdy \frac{\partial t}{\partial z} \tag{8-13}$$

将式(8-13)代入式(8-12)，并除以控制体积 dxdydz，可得：

$$\frac{\partial}{\partial x}\left(\lambda \frac{\partial t}{\partial x}\right) + \frac{\partial}{\partial y}\left(\lambda \frac{\partial t}{\partial y}\right) + \frac{\partial}{\partial z}\left(\lambda \frac{\partial t}{\partial z}\right) + \dot{q} = \rho c \frac{\partial t}{\partial \tau} \tag{8-14}$$

式(8-14)是直角坐标系中热扩散方程的普通形式。这个方程通常称为导热微分方程，它提供了分析导热过程的基本方法。

当导热系数 λ 为常数时，导热微分方程可简化为：

$$\lambda\left(\frac{\partial^2 t}{\partial x^2}+\frac{\partial^2 t}{\partial y^2}+\frac{\partial^2 t}{\partial z^2}\right)+\dot{q}=\rho c\,\frac{\partial t}{\partial \tau}$$

$$\frac{\lambda}{\rho c}\left(\frac{\partial^2 t}{\partial x^2}+\frac{\partial^2 t}{\partial y^2}+\frac{\partial^2 t}{\partial z^2}\right)+\frac{\dot{q}}{\rho c}=\frac{\partial t}{\partial \tau} \tag{8-15}$$

$a=\dfrac{\lambda}{\rho c}$称为热扩散率，又称导温系数，单位是 m^2/s，是对非稳态导热过程有重要影响的热物性参数，其大小反映物体被瞬态加热或冷却时物体内温度变化的快慢，即热量在物体中的扩散速率。a 越大，热量传递越快，物体内部温度趋于平衡的速度越快，物体内温度分布趋向于一致的能力越大。

由热扩散率的定义 $a=\lambda/\rho c$ 可知：① 分子 λ 是物体的导热系数，λ 越大，在相同的温度梯度下可以传导更多的热量；② 分母 ρc 是单位体积的物体温度升高 1 ℃所需的热量(热容量或热惰性)，ρc 越小，温度上升 1 ℃所吸收的热量越少，可以剩下更多的热量继续向物体内部传递，使物体内各点的温度更快地随界面温度的升高而升高。导热系数 λ 越大，且单位容积的热容量 c 越小(本身蓄热或放热的能量越小)的材料，扩散热量的能力越大，热扩散率越大。在非稳态导热过程中，热扩散率大的材料温度变化快，或整块材料温度比较均匀，所以热扩散率又称导温系数。例如，一般木材的热扩散率约为 $a=1.5\times10^{-7}\ m^2/s$，紫铜(纯铜)的热扩散率约为 $a=5.33\times10^{-5}\ m^2/s$，是木材的 355 倍。如果两手分别握住同样长短、粗细的木棒和紫铜棒，同时将另一端伸到灼热的火炉中，则当拿紫铜棒的手感到很烫时，拿木棒的手尚无热的感觉。这说明，紫铜棒中温度的变化要比木棒中快得多。

导热方程也可以用圆柱坐标系和球坐标系来表示。其表述式如下：

圆柱坐标系：

$$\frac{1}{r}\frac{\partial}{\partial r}\left(\lambda r\frac{\partial t}{\partial r}\right)+\frac{1}{r^2}\frac{\partial}{\partial \varphi}\left(\lambda\frac{\partial t}{\partial \varphi}\right)+\frac{\partial}{\partial z}\left(\lambda\frac{\partial t}{\partial z}\right)+\dot{q}=\rho c\,\frac{\partial t}{\partial \tau} \tag{8-16}$$

球坐标系：

$$\frac{1}{r^2}\frac{\partial}{\partial r}\left(\lambda r^2\frac{\partial t}{\partial r}\right)+\frac{1}{r^2\sin^2\theta}\frac{\partial}{\partial \varphi}\left(\lambda\frac{\partial t}{\partial \varphi}\right)+\frac{1}{r^2\sin\theta}\frac{\partial}{\partial \theta}\left(\lambda\sin\theta\frac{\partial t}{\partial \theta}\right)+\dot{q}=\rho c\,\frac{\partial t}{\partial \tau} \tag{8-17}$$

8.2.2 导热微分方程的定解条件

导热微分方程是描写导热过程共性的数学表达式。求解导热问题，实质是求解上述导热微分方程。为了得到某一具体导热问题的解，必须给出表征该特定问题的附加条件，这些附件条件称为定解条件。对于非稳态问题，定解条件包括两方面，即给出初始时刻温度分布的初始条件，以及导热物体边界上的温度分布或换热情况，即边界条件。对于稳态问题，定解条件仅需要边界条件，不需要初始条件。导热问题常见的边界条件可归纳为以下三类：

(1) 第一类边界条件：规定了边界上的温度分布，在非稳态导热问题中，应给出其随时间的变化规律，即 $\tau>0$ 时，$t_w=f(x,y,z,\tau)$。此类边界条件最简单的例子是边界表面温度保持恒定不变，即 $t_w=$ 常数，简称恒壁温边界条件。

(2) 第二类边界条件：给出物体边界上的热流密度分布随时间的变化规律，即 $\tau>0$ 时，$q_w=-\lambda\dfrac{\partial t}{\partial n}\Big|_w=f(x,y,z,\tau)$。此类边界条件最简单的例子是边界上热流密度保持为常数，

即 $q_w=$ 常数，简称恒热流边界条件，其中 $q_w=0$ 称为绝热边界条件。

(3) 第三类边界条件：给出物体边界上的换热状态，即对流换热表面传热系数 h 和周围流体的温度 t_f，一般写成 $-\lambda \frac{\partial t}{\partial n}\big|_w = h(t_w - t_f)$。

8.3 稳态导热

8.3.1 通过平壁的导热

(1) 单层平壁

一维稳态无内热源的平壁，如图 8-3 所示，平壁厚度为 δ，两个表面的温度分别为 t_1 和 t_2。其导热问题的数学描述为：

$$\frac{d^2 t}{dx^2} = 0 \tag{8-18}$$

定解条件为：

$$x=0, t=t_1; x=\delta, t=t_2 \tag{a}$$

对式(8-18)进行两次积分，得其通解：

$$t = c_1 x + c_2 \tag{b}$$

式中，c_1 和 c_2 为积分常数，代入定解条件公式(b)可以得到：

$$c_1 = \frac{t_2 - t_1}{\delta}, c_2 = t_1 \tag{c}$$

将式(c)代入式(b)得到温度分布：

$$t = \frac{t_2 - t_1}{\delta} x + t_1 \tag{8-19}$$

流过平壁的热流密度为：

$$q = -\lambda \frac{t_2 - t_1}{\delta} = \lambda \frac{t_1 - t_2}{\delta} = \frac{\lambda}{\delta} \Delta t \tag{8-20}$$

流过平壁的热流量为：

$$\Phi = A \frac{\lambda}{\delta} \Delta t \tag{8-21}$$

由式(8-19)可以看出，对于一维无内热源的平壁而言，其温度分布为线性分布。对式(8-21)进行改写变出如下形式：

$$\Phi = \frac{\Delta t}{\delta/(A\lambda)} = \frac{\Delta t}{R} \tag{8-22}$$

式中，热流量 Φ 为导热过程的转移量；温差 Δt 为转移过程中的动力；分母 $\delta/(A\lambda)$ 为热量转移中的阻力，令 $R=\delta/(A\lambda)$，则称为导热热阻。对单位面积的平壁而言，其导热热阻 $R_A=\delta/\lambda$。

热阻的概念类似于电路里面的欧姆定律，热阻概念的建立给复杂热传递过程的分析带来了很大的便利。譬如，可以借用比较熟悉的电路里的串、并联电阻的计算公式来计算多层平壁或者兼有导热、对流、辐射等的复杂传热过程中的总热阻。

(2) 多层平壁

所谓多层平壁，就是由几层不同材料叠在一起组成的复合壁。例如，采用耐火砖、保温砖层和普通砖层叠合形成的锅炉炉墙便为多层平壁。下面以三层平壁为例来讨论。如图 8-4 所示,已知各层的厚度分别为 $\delta_1,\delta_2,\delta_3$,各层的导热系数分别为 $\lambda_1,\lambda_2,\lambda_3$,左右两边的温度分别为 t_1,t_4。需要确定通过多层平壁的热流密度以及各层的层间温度。

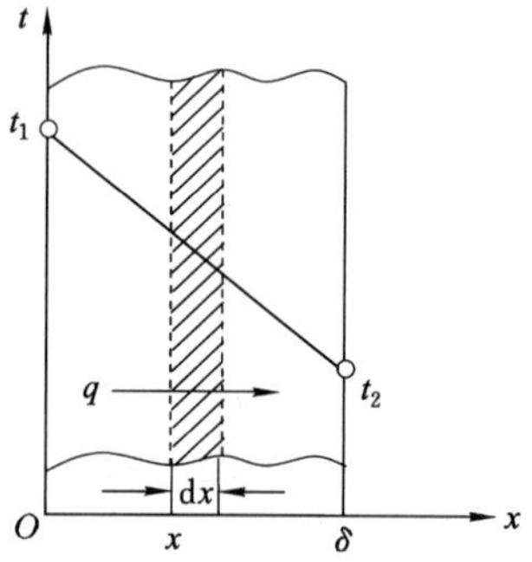

图 8-3　通过单层平壁的导热

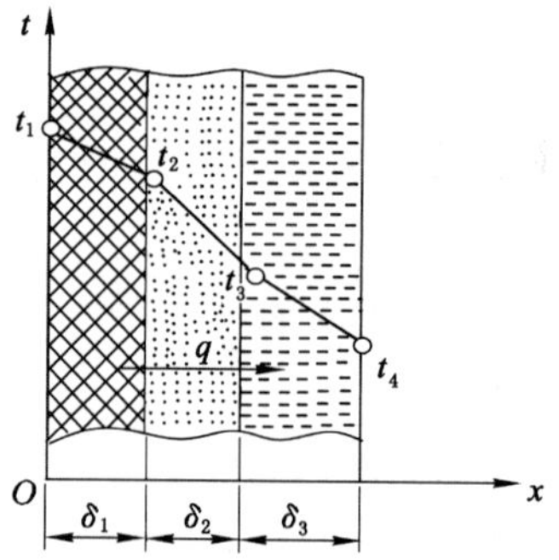

图 8-4　通过多层平壁的导热

通过三层平壁的导热可以采用上述热阻的概念。各层应用热阻的表达式如下:

$$\frac{t_1-t_2}{q}=\frac{\delta_1}{\lambda_1},\frac{t_2-t_3}{q}=\frac{\delta_2}{\lambda_2},\frac{t_3-t_4}{q}=\frac{\delta_3}{\lambda_3} \tag{8-23}$$

应用串联过程的总热阻等于各分热阻的总和,即所谓的串联热阻叠加原则,将各层热阻叠加即可以得到多层平壁的总热阻:

$$\frac{t_1-t_4}{q}=\frac{\delta_1}{\lambda_1}+\frac{\delta_2}{\lambda_2}+\frac{\delta_3}{\lambda_3} \tag{8-24}$$

因此,可以得到通过多层平壁的热流密度:

$$q=\frac{t_1-t_4}{\dfrac{\delta_1}{\lambda_1}+\dfrac{\delta_2}{\lambda_2}+\dfrac{\delta_3}{\lambda_3}} \tag{8-25}$$

计算得到热流密度之后,层间分界面上的未知温度 t_2,t_3 就可以利用式(8-22)求出。例如:

$$t_2=t_1-q\frac{\delta_1}{\lambda_1} \tag{8-26}$$

【例 8-1】 已知一炉壁由三层材料组成:第一层是耐火砖,导热系数 $\lambda_1=1.85$ W/(m·K);第二层是绝热砖,导热系数 $\lambda_2=0.45$ W/(m·K),允许使用的最高温度为 1 300 ℃;第三层是钢板,厚度 $\delta_3=5$ mm,导热系数 $\lambda_3=0.45$ W/(m·K)。炉墙内侧和外侧的表面温度分别为 $t_{w_1}=1\,600$ ℃和 $t_{w_4}=80$ ℃。稳定时,通过炉壁的热量 $q=2\,000$ W/m^2。试计算各层壁厚为多少才能使炉壁总厚度最小。

【解】 此题为多层平壁导热过程,为使炉壁总厚度最小,应取 $t_{w_2}=1\,300$ ℃。

第一层耐火砖的厚度至少为:

$$\delta_1=\frac{\lambda_1(t_{w_1}-t_{w_2})}{q}=\frac{1.85\times(1\,600-1\,300)}{2\,000}=0.277\,5\ (\text{m})$$

第二层耐火砖的厚度至少为:

$$\delta_2=\lambda_2\left(\frac{t_{w_2}-t_{w_4}}{q}-\frac{\delta_3}{\lambda_3}\right)=0.45\times\left(\frac{1\,300-80}{2\,000}-\frac{0.005}{0.45}\right)=0.269\,5\ (\text{m})$$

8.3.2 通过圆筒壁的导热

(1) 单层圆筒壁

设有一长度为 l，内、外直径各为 d_1 与 d_2 的圆筒壁，且 $l \gg d_2$，内、外壁面温度各为 t_{w_1}, t_{w_2}，且 $t_{w_1} > t_{w_2}$，如图 8-5 所示。若用圆柱坐标系，该问题就成为沿半径方向的一维导热问题，即其温度场是一维的，故各等温面都是彼此同心的圆柱面。为了便于分析，假设材料的导热系数为定值 λ。

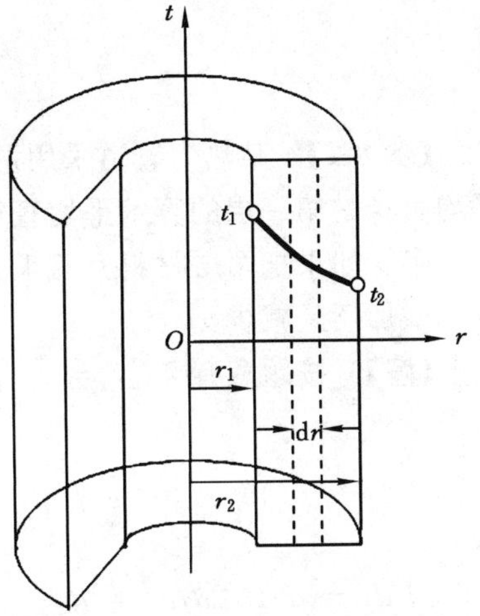

图 8-5 通过单层圆筒壁的导热

单层圆筒壁导热微分方程及边界条件如下：

$$\frac{\mathrm{d}}{\mathrm{d}r}\left(r\frac{\mathrm{d}t}{\mathrm{d}r}\right) = 0 \tag{8-27}$$

$$r = r_1, t = t_1 \tag{d}$$

$$r = r_2, t = t_2 \tag{e}$$

对式(8-27)连续积分两次，得其通解：

$$t = c_1 \ln r + c_2 \tag{f}$$

将边界条件式(d)和式(e)代入式(f)，求解得：

$$c_1 = \frac{t_2 - t_1}{\ln(r_2/r_1)}, c_2 = t_1 - \ln r_1 \frac{t_2 - t_1}{\ln(r_2/r_1)} \tag{g}$$

将式(g)代入式(f)，得到温度分布：

$$t = t_1 + \frac{t_2 - t_1}{\ln(r_2/r_1)}\ln(r/r_1) \tag{8-28}$$

由此可见，与平壁中的线性温度分布不同，圆筒壁中的温度分布呈对数曲线。对式(8-28)求导得到：

$$\frac{\mathrm{d}t}{\mathrm{d}r} = \frac{1}{r}\frac{t_2 - t_1}{\ln(r_2/r_1)}$$

代入傅立叶定律得：

$$q = -\lambda\frac{\mathrm{d}t}{\mathrm{d}r} = \frac{\lambda}{r}\frac{t_1 - t_2}{\ln(r_2/r_1)} \tag{8-29}$$

由式(8-29)可见，圆筒壁的稳态导热中，不同半径处的热流密度与半径呈反比。但是，通过整个圆筒壁的热流量 Φ 是恒定的，不随半径而变。将式(8-29)两边同时乘 $2\pi rl$ 可得热流量：

$$\Phi = 2\pi rlq = \frac{2\pi\lambda l(t_1 - t_2)}{\ln(r_2/r_1)} \tag{8-30}$$

根据热阻的定义可以得到通过圆筒壁的导热热阻：

$$R = \frac{\Delta t}{\Phi} = \frac{\ln(r_2/r_1)}{2\pi\lambda l} \tag{8-31}$$

(2) 多层圆筒壁

多层圆筒壁的导热计算公式与多层平壁一样，可利用串联热阻叠加的原则，得到多层圆筒壁的导热热流量：

$$\Phi = \frac{2\pi l(t_{w_1} - t_{w_2})}{\sum_{i=1}^{n} \frac{1}{2\pi\lambda_i}\ln\frac{d_{i+1}}{d_i}} \tag{8-32}$$

【例 8-2】 某热力管道采用两种不同的材料组成保温层，两层厚度相等，第二层的算术平均直径是第一层的算术平均直径两倍，而第二层材料的导热系数为第一层材料导热系数的一半。如果把两层材料互相调换，其他情况都保持不变，问每米长管道热损失改变了多少？增加还是减少？

【解】 按题意有 $2d_{m1} = d_{m2}$，$\delta_1 = \delta_2$，则有：

$$2\,\frac{d_1 + d_2}{2} = \frac{d_2 + d_3}{2}$$

$$d_3 = 2d_1 + d_2$$

因 $d_2 = d_1 + 2\delta$，$d_3 = d_2 + 2\delta$，代入上式，得 $\delta = d_1$，则 $\frac{d_2}{d_1} = 3$，$\frac{d_3}{d_2} = \frac{5}{3}$。

未调换时：

$$R_{\lambda,l} = R_{\lambda_1} + R_{\lambda_2} = \frac{1}{2\pi\lambda_1}\ln\frac{d_2}{d_1} + \frac{1}{2\pi\lambda_2}\ln\frac{d_3}{d_2}$$

$$= \frac{1}{2\pi\lambda_1}\ln\frac{d_2}{d_1} + \frac{1}{2\pi\lambda_1/2}\ln\frac{d_3}{d_2} = \frac{1}{\pi\lambda_1}\left(\frac{1}{2}\ln\frac{d_2}{d_1} + \ln\frac{d_3}{d_2}\right)$$

$$q = \frac{\Delta t}{R_{\lambda,l}} = \frac{\pi\lambda_1\Delta t}{\frac{1}{2}\ln\frac{d_2}{d_1} + \ln\frac{d_3}{d_2}}$$

调换后：

$$R_{\lambda,l}{}' = R_{\lambda_1}{}' + R_{\lambda_2}{}' = \frac{1}{2\pi\lambda_1}\ln\frac{d_2}{d_1} + \frac{1}{2\pi\lambda_2}\ln\frac{d_3}{d_2}$$

$$= \frac{1}{2\pi\lambda_1/2}\ln\frac{d_2}{d_1} + \frac{1}{2\pi\lambda_1}\ln\frac{d_3}{d_2} = \frac{1}{\pi\lambda_1}\left(\ln\frac{d_2}{d_1} + \frac{1}{2}\ln\frac{d_3}{d_2}\right)$$

$$q' = \frac{\Delta t}{R_{\lambda,l}{}'} = \frac{\pi\lambda_1\Delta t}{\ln\frac{d_2}{d_1} + \frac{1}{2}\ln\frac{d_3}{d_2}}$$

每米长管道热损失变化为：

$$\frac{q' - q}{q} = \frac{\frac{1}{2}\ln\frac{d_2}{d_1} + \ln\frac{d_3}{d_2}}{\ln\frac{d_2}{d_1} + \frac{1}{2}\ln\frac{d_3}{d_2}} - 1 = \frac{\frac{1}{2}\ln 3 + \ln\frac{5}{3}}{\ln 3 + \frac{1}{2}\ln\frac{5}{3}} - 1$$

$$= 0.783 - 1 = -21.7\%$$

由此可知，用两种不同导热系数的保温材料对管道进行保温时，若两保温层厚度相等，为减少散热损失，应将导热系数小的材料放在内层。

8.3.3 肋片的导热

为了强化换热或满足其他要求，可在换热表面上附加各种形状的肋片，以增加对流传热面积，减小对流换热热阻，起到增强传热的作用。图 8-6 画出了不同形式的肋片。与平壁连接的扩展面都是直肋片，它可以是等截面的，也可以是变截面直肋片，即截面随离平壁的距离 x 变化而变化。与圆柱在圆周面接触的为环肋片，它的截面随离圆柱中心线的径向距离变化而变化。针肋片(也称柱肋片)则不同,它是横截面为圆形的扩展表面。针肋片可以是等截面的,也可以是变截面的。

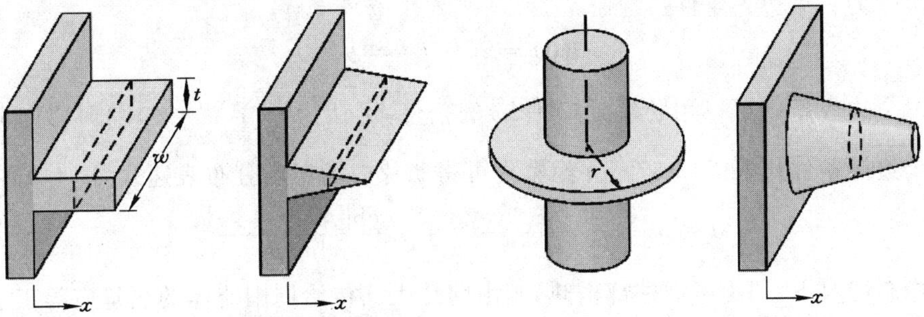

(a) 等截面直肋片　　(b) 变截面直肋片　　(c) 环肋片　　(d) 针肋片

图 8-6　肋片的形状

肋片导热分析的主要任务是确定沿肋片高度方向的温度分布和肋片散热量。这里讨论最简单的情况,即等截面的矩形肋片或针肋片,如图 8-7 所示。每个肋片都与温度 $t(0) = t_0$ 的基面连接,并延伸到温度为 t_∞ 的流体中。

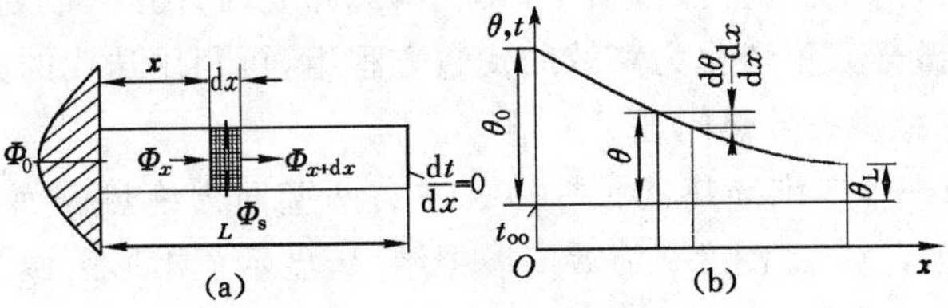

图 8-7　等截面直肋片

肋片一般用金属制造,导热系数较大,可认为任一截面上的温度分布几乎均匀一致,即可假定只沿 x 方向(即肋高方向)温度才有显著变化。同时,把肋片表面向周围流体的换热量当作冷源吸热。因而,肋片内的温度场可视为具有负内热源的一维稳态温度场。对给定的肋片,A_c 为常数,$A_s = Px$,此处的 A_s 是从肋基到 x 处的表面积,P 是肋片的周长。等截面直肋片的温度场微分方程可写成:

$$\frac{d^2 t}{dx^2} - \frac{hP}{\lambda A_c}(t - t_\infty) = 0 \tag{8-33}$$

为简化式(8-33),定义一个过余温度 θ 来替代因变量:

$$\theta(x) = t(x) - t_\infty \tag{8-34}$$

由于式(8-34)中的 t_∞ 是常数,所以 $d\theta/dx = dt/dx$。将式(8-34)代入式(8-33),得:

$$\frac{d^2 \theta}{dx^2} - m^2 \theta = 0 \tag{8-35}$$

式中:

$$m^2 = \frac{hP}{\lambda A_c} \tag{8-36}$$

式(8-35)为线性、齐次、常系数二阶微分方程,其通解的形式为:

$$\theta(x) = C_1 e^{mx} + C_2 e^{-mx} \tag{8-37}$$

将式(8-37)代入式(8-35)，很容易证明它确实是式(8-35)的一个解。

为确定式(8-37)中的常数 C_1 和 C_2，必须具体地给出合适的边界条件。肋基($x=0$)处的温度可作为一个边界条件：

$$\theta(0)=t_0-t_\infty=\theta_0 \tag{8-38}$$

第二个边界条件可以用肋端处($x=L$)的条件，可认为 $\left.\frac{\mathrm{d}\theta}{\mathrm{d}x}\right|_{x=L}=0$，即肋端部不散热。将此两个边界条件代入式(8-37)，并经整理可得肋片内的温度分布表达式：

$$\theta=\theta_0\,\frac{\mathrm{e}^{-mx}+\mathrm{e}^{-2mL}\mathrm{e}^{mx}}{1+\mathrm{e}^{-2mL}}=\theta_0\,\frac{\mathrm{ch}[m(L-x)]}{\mathrm{ch}(mL)} \tag{8-39}$$

在稳态情况下，如不考虑端部散热，则由肋片表面传给周围流体的热量应等于通过肋基导入的热量：

$$\Phi=\Phi_0=\frac{hP}{m}\theta_0\,\mathrm{th}(mL) \tag{8-40}$$

8.3.4 二维稳态导热

前面讨论的内容均为只在一个坐标方向上有温度梯度的导热问题，但在很多情况下，用一维来处理，会使某些问题因过于简化而不能反映实际情况，因而必须考虑多维效应。多维问题的求解主要借助计算机进行数值模拟，其基本思路是对计算区域划分网格，将其进行离散化，从而将连续的微分方程转化为离散的代数方程组，再利用迭代的方法进行求解。多维问题里最简单的是二维稳态导热问题。

如图 8-8 所示为一个二维导热的长柱体。左右两壁面为绝热表面，上下两个表面保持不同的温度 t_1，t_2，且 $t_1>t_2$。图 8-9 为数值模拟结果，显示了矩形区域二维稳态导热的温度分布和热流密度分布。从图中可以看出，在 y 方向，物体中的温度梯度随 y 的增加而增加。固体中的局部热流密度是一个处处垂直于等温线的矢量。其方向由图 8-9 中的热流线箭头方向表示。值得注意的是，在上部两个顶角处，所有的等温线都汇集在一起，物理上是不合理的。这主要是给定的边界条件中四个顶点都有两个温度所导致的。

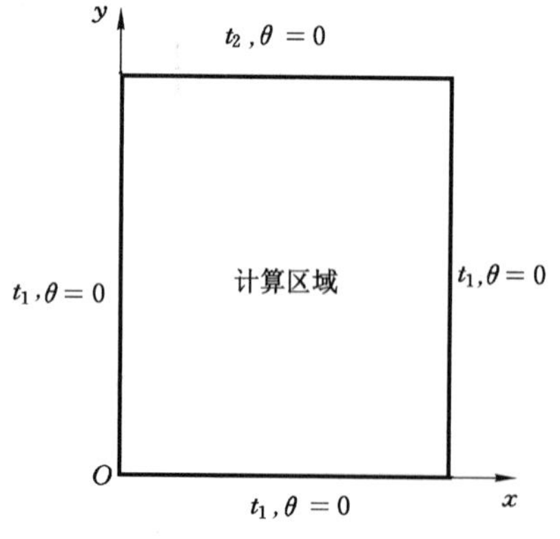

图 8-8 矩形区域的二维稳态导热

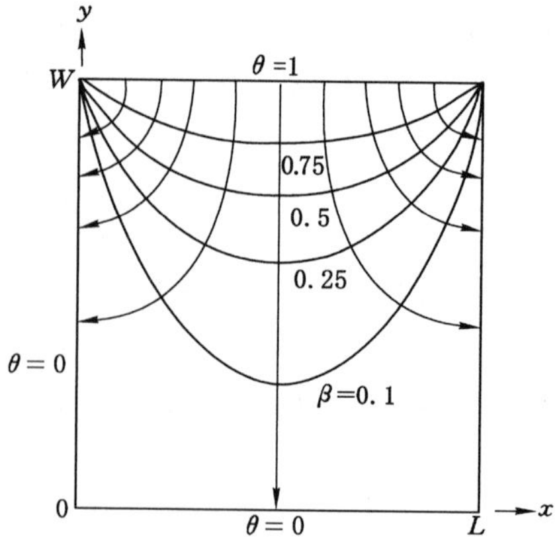

图 8-9 矩形区域二维稳态导热温度和热流密度分布

8.4　非稳态导热

8.4.1　非稳态导热的特点

各点温度随时间而变动的温度场称为非稳态温度场。物体处于非稳态温度场内的导热过程，称为非稳态导热。它是由于物体受热或冷却所引起的。工程上经常遇到的，如热力设备的启动、停机及变负荷，零部件的热处理等都属于非稳态导热现象。

为说明非稳态导热过程的特点，以平壁为例，如图 8-10 所示。假定初始时刻，即 $\tau = \tau_0$ 时，平壁置于空气中，整个平壁温度均为 t_0；$\tau > \tau_0$ 时，平壁左边受热温度升至 t_1，此时，平壁最右侧暴露于空气中温度仍为 t_0。随着时间的推移，靠近平壁左侧的板内温度依次升高，图 8-10(a)示意性地表示了这种变化过程。最终达到稳态时，壁内各层的温度不随时间而变化，温度分布呈线性。图 8-10(b)示意性地表示了非稳态导热的热流量变化。Φ_1 表示从左侧表面传入的热量，Φ_2 表示右侧表面向四周的散热。从图中容易看出，当 $\tau < \tau_4$ 时，随着时间的推移，Φ_1 逐渐降低，由于平壁右侧的温度始终等于 t_0，不向空气传递热量，即 $\Phi_2 = 0$；当 $\tau > \tau_4$ 时，Φ_1 仍然降低，而 Φ_2 逐渐升高。当 $\tau = \tau_5$ 时，得热等于失热，即 $\Phi_1 = \Phi_2$，此时，平壁内最终建立起了稳态的传热工况。

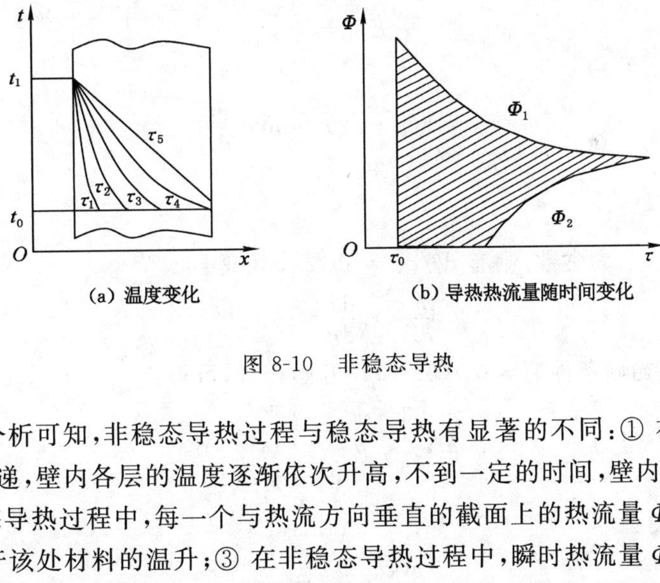

图 8-10　非稳态导热

由上面的分析可知，非稳态导热过程与稳态导热有显著的不同：① 在非稳态导热过程中，随着热量传递，壁内各层的温度逐渐依次升高，不到一定的时间，壁内某层的温度不会升高；② 在非稳态导热过程中，每一个与热流方向垂直的截面上的热流量 Φ 是处处不等的，Φ 减少的部分用于该处材料的温升；③ 在非稳态导热过程中，瞬时热流量 Φ_1 和 Φ_2 不是定值，而是时间 τ 的函数，图 8-10(b)所示的阴影面积表示非稳态导热过程中材料所积蓄的内能。

8.4.2　集总参数法

(1) 集总参数法温度场的分析解

一个固体的热环境发生突然的变化是简单而常见的导热问题。假设一个初始均匀温度为 t_i 的金属锻件，将其浸入温度较低（$t_\infty < t_0$）的液体中进行淬火，如图 8-11 所示。如果令淬火的起始时刻为 $\tau = 0$，在 $\tau > 0$ 后固体的温度将降低，直到最终达到 t_∞。这种温度降低是固-液交界面上发生对流换热的结果。集总参数法的实质是假定在瞬态过程中的任何时刻固体中的温度在空间上是均匀的。这就意味着固体中的温度梯度可以忽略不计。根据傅立叶定律，没有温度梯度的导热意味着存在无限大的热导率。这显然是不可能的。然而，如果固体中的导热热阻与固体和其环境之间的换热热阻相比很小的话，这就是一种很精确的近似。

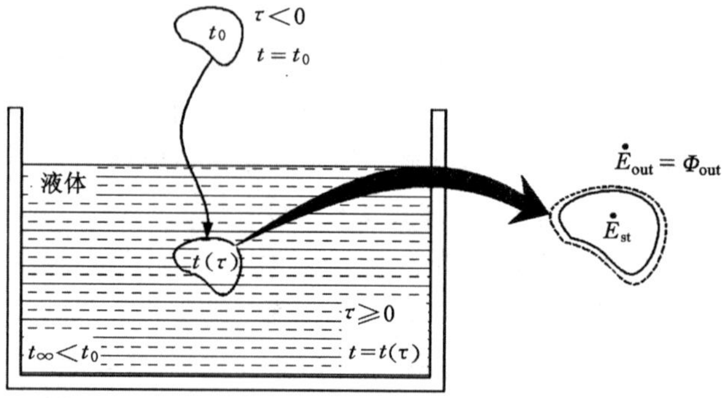

图 8-11　高温金属锻件淬火冷却

忽略了固体中的温度梯度，就不能再在导热方程的框架内来讨论问题，取代方法是对固体写出总的能量平衡关系来确定瞬态温度响应。这个平衡关系必须建立固体表面的热损失速率与固体内能的变化速率之间的联系。图 8-11 中的控制体热平衡要求的形式为：

$$-\dot{E}_{out}=\dot{E}_{st} \tag{8-41}$$

或

$$-hA(t-t_\infty)=\rho Vc\frac{dt}{d\tau} \tag{8-42}$$

引入温度差：

$$\theta=t-t_\infty \tag{8-43}$$

并注意到如果 t_∞ 为常数，则有 $d\theta/d\tau=dt/d\tau$，可得：

$$\frac{\rho Vc}{hA}\times\frac{d\theta}{d\tau}=-\theta$$

分离变量并从初始条件 $\tau=0, t(0)=t_0$ 开始积分，可得：

$$\frac{\rho Vc}{hA}\int_{\theta_0}^{\theta}\frac{d\theta}{\theta}=-\int_0^{\tau}d\tau$$

式中，$\theta_0=t_0-t_\infty$。

计算积分式，得：

$$\frac{\rho Vc}{hA}\ln\frac{\theta_0}{\theta}=\tau \tag{8-44}$$

或

$$\frac{\theta}{\theta_0}=\frac{t-t_\infty}{t_0-t_\infty}=\exp\left[-\left(\frac{hA}{\rho Vc}\right)\tau\right] \tag{8-45}$$

式(8-43)可用来确定固体达到某个温度 t 所需的时间，而式(8-45)则可用来计算固体在某个时间 τ 达到的温度。上述结果表明，当时间 τ 趋近无限大时，固体与流体之间的温差必定按指数函数规律衰减到零。上述过程中传递的总热量为：

$$Q=\rho Vc\theta_0\left[1-\exp\left(-\frac{hA}{\rho cV}\tau\right)\right] \tag{8-46}$$

将式(8-45)右边的指数项做如下变化：

$$\frac{hA}{\rho Vc}\tau = \frac{hl}{\lambda}\frac{\lambda}{\rho c}\frac{\tau}{l^2} = \frac{hl}{\lambda}\frac{\alpha\tau}{l^2} = Bi \cdot Fo \tag{8-47}$$

$$Bi = \frac{hl}{\lambda} = \frac{l/\lambda}{1/h}, Fo = \frac{\alpha\tau}{l^2} \tag{8-48}$$

式中，Bi 是以 l 为特征长度的毕渥数；Fo 称为傅立叶数，同样以 l 为特征长度。

这样式(8-45)可以写成：

$$\frac{\theta}{\theta_0} = \exp(-Bi \cdot Fo) \tag{8-49}$$

(2) 集总参数法的适用范围

式(8-49)表明非稳态导热温度分布受 Fo 和 Bi 两个参数的影响。其中，傅立叶数为表征非稳态导热过程进行深度的无量纲时间，其数值越大，热扰动就越深地传播到物体内部，因而物体内部各点的温度就越接近周围介质的温度。毕渥数为物体内部导热热阻与外部对流换热热阻之比，即无量纲热阻。Bi 表达式中的 λ 是指固体材料的导热系数。Bi 决定物体内部温度分布的特征。当 $Bi \to \infty$ 时，意味着外部热阻趋近零，此时内部导热热阻起主导作用，在传热过程中温度降低主要发生在物体内部；当 $Bi \to 0$ 时，意味着内部导热热阻趋近零，物体内部温度均匀，此时外部热阻起主导作用。理论计算表明，当 $Bi < 0.1$ 时，若采用集总参数法进行简化计算，其误差小于5%。因此，工程上应用集总参数法的判别式为：

$$Bi = \frac{hV/A}{\lambda} < 0.1M \tag{8-50}$$

式中，M 是与几何形状有关的量纲为一的数，对于无限大平板、无限长圆柱和球，分别取 $M=1, M=1/2, M=1/3$。

【例8-3】 测温热电偶球形热接点的直径为0.5 mm，密度 $\rho = 8\,930\ \mathrm{kg/m^3}$，比热容 $c=400\ \mathrm{J/(kg \cdot K)}$，导热系数 $\lambda = 20\ \mathrm{W/(m \cdot K)}$，初始温度为20 ℃，将它突然插入表面传热系数 $h = 50\ \mathrm{W/(m^2 \cdot K)}$，温度为200 ℃的气流中，试求：① 热电偶指示的过余温度与初始过余温度之比达到1%时所需时间；② 如果要求热电偶在15 s内达到 $\theta/\theta_0=1\%$，那么热电偶接点的直径应为多少？

【解】 ① 计算 Bi 数

对于球体，其体积与表面积之比为：

$$\frac{V}{A} = \frac{4}{3}\frac{\pi r^3}{4\pi r^2} = \frac{r}{3} = \frac{d}{6}$$

$$Bi = \frac{hV/A}{\lambda} = \frac{50 \times 0.000\,5/6}{20} = 2.083 \times 10^{-4} < 0.1M$$

对于球体，M 取1/3，故可以采用集总参数法。

根据公式：
$$\tau = \frac{\rho Vc}{hA}\ln\frac{\theta_0}{\theta}$$

达到 $\theta/\theta_0=1\%$ 时所需时间为：

$$\tau = -\frac{\rho cV}{hA}\ln 0.01 = -\frac{8\,930 \times 400 \times 0.000\,5}{50 \times 6}\ln 0.01 = 27.4\ (\mathrm{s})$$

② 计算15 s内使 $\theta/\theta_0=1\%$ 所需接点直径

$$d = -\frac{6\tau h}{c\rho \ln 0.01} = -\frac{6 \times 15 \times 50}{8\ 930 \times 400 \times \ln 0.01} = 2.74 \times 10^{-4}(\text{m})$$

验证：

$$Bi = \frac{hV/A}{\lambda} = \frac{50 \times 2.74 \times 10^{-4}}{20 \times 6} = 1.142 \times 10^{-4}$$

可以采用集总参数法。

本章小结

本章首先讲述了导热机理和傅立叶定律，其次根据能量守恒推导出了导热微分方程，并阐述了相应的定解条件。在此基础上，针对典型的一维平壁、圆筒壁导热，肋片导热以及二维稳态导热进行求解，以获得物体中的温度分布和热流量。最后阐述了非稳态导热过程的基本特点与非稳态导热问题的求解方法，以及零维问题的集总参数法和集总参数法的适用范围。同时介绍了热阻的概念及不同情况下导热热阻的分析和计算。

思考与练习题

(1) 试写出傅立叶定律的一般表达式，说明其适用条件及式中各符号的物理意义。

(2) 在寒冷的北方地区，建房用砖采用实心砖还是多孔的空心砖好？为什么？

(3) 试说明得出导热微分方程所依据的基本定律。

(4) 导热问题的边界条件有几类？如何用数学式描述？

(5) 何谓导热问题的定解条件？它包含哪些内容？

(6) 试说明在什么条件下平壁和圆筒壁的导热可以按一维导热处理。

(7) 试用传热学观点说明为什么冰箱要定期除霜。

(8) 什么叫热阻？平壁、圆筒壁的导热热阻如何表示？

(9) 某固体的导热系数 $\lambda = \lambda_0 + at$，$\lambda_0 > 0$，为常数，试分析 $a > 0$，$a = 0$ 和 $a < 0$ 三种情况下，由此种材料做成的平壁内的温度分布。

(10) 非稳态导热的热量传递有何特点？Fo 和 Bi 的物理意义是什么？

(11) 集总参数法的适用条件是什么？满足集总参数法的物体，其内部温度分布有何特点？

(12) 一平壁，厚度为 δ，两侧表面的温度分别为 t_{w_1} 和 t_{w_2}，且 $t_{w_1} > t_{w_2}$，平壁材料的导热系数随温度变化的关系式为 $\lambda = \lambda_0(1 + bt)$。试求热流密度和平壁内温度分布的表达式。

(13) 一导热系数 $\lambda_1 = 1.3$ W/(m·K)，厚 2 cm 的无限大平壁，外覆盖一层导热系数 $\lambda_2 = 0.35$ W/(m·K) 的保温材料以减少热损失。当组合壁的内、外温度分别为1 300 ℃和30 ℃时，欲使稳态导热时热损失不超过 1 830 W/m²，保温材料的厚度应为多少？

(14) 已知一内外径为 r_1 和 r_2 的圆筒壁，它的密度 ρ 和比热容 c 均为常数，无内热源。圆筒壁内外侧温度分别为 t_{w_1} 和 t_{w_2}。试求圆筒壁稳态导热时壁内的温度分布表达式。

(15) 一根直径为 3 mm 的铜导线，每米长的电阻为 2.22×10^{-3} Ω。导线外包有厚 1 mm 导热系数为 0.15 W/(m·K)的绝缘层。限定绝缘层的最高温度为 65 ℃，环境温度为0 ℃，空气与绝缘层外表面的表面传热系数为 20 W/(m^2·K)，试确定在这种条件下导线中允许通过的最大电流。

(16) 将温度为 t_0 的固体(密度为 ρ、比热容为 c_p、体积为 V、表面积为 A)突然置于温度为 t_∞($t_\infty < t_0$)的冷流体中，固体与流体间的对流换热表面传热系数为 h。利用集总参数法，求物体温度随时间的变化关系。

知识拓展：传热学发展简史

知识拓展：导热系数与导温系数

Chapter 8 Heat Conduction

When the relative displacement among every parts of object does not occur, heat transfer generated only by the thermal motion of molecules, atoms, free electrons and other particles is called conduction. Study conduction has two objectives: ① How much is heat transfer? ② How distribute the temperature inside the solid? The former is determined by the Fourier′s law, and the latter are mainly solved by the differential equation of heat conduction.

This chapter first describes the thermal mechanism and Fourier′s law, secondly, according to energy conservation, derive the differential equation of heat conduction, and states the corresponding definite conditions. Based on the differential equation, to solve the typical one-dimensional flat, cylindrical wall, as well as fin conduction and two-dimensional steady-state conduction, so as to obtain the heat flow rate and temperature distribution of the object. Finally elaborated the characteristics of the unsteady conduction, lumped parameter method of the zero-dimensional problem and the scope of the lumped parameter method.

8.1 Introduction to Heat Conduction

8.1.1 Conduction Mechanism

At mention of the word conduction, we should immediately conjure up concepts of atomic and molecular activity because processes at these levels sustain this mode of heat transfer. Conduction may be viewed as the transfer of energy from the more energetic to the less energetic particles of a substance due to interactions between the particles. The physical mechanism of conduction is most easily explained by considering a gas and using ideas familiar from your thermodynamics background. Consider a gas in which a temperature gradient exists, and assume that there is no bulk, or macroscopic, motion. The gas may occupy the space between two surfaces that are maintained at different temperatures, as shown in Fig. 8-1. Higher temperatures are associated with higher molecular energies. When neighboring molecules collide, as they are constantly doing, a transfer of energy from the more energetic to the less energetic molecules must occur. In the presence of a temperature gradient, energy transfer by conduction must then occur in the direction of decreasing temperature. This would be true even in the absence of collisions, as is evident from Fig. 8-1. The hypothetical plan x_0 is constantly being crossed by molecules from above and below due to their random motion. However, molecules from above are associated with a higher temperature than those from below, in which case there must be a net transfer of energy in the positive x-direction. Collisions between molecules enhance this energy transfer. We may speak of the net transfer of energy by random molecular motion as a diffusion of energy.

8.1.2 Fourier's Law

Fourier's law is phenomenological, that is, it is developed from observed phenomena rather than being derived from the first law. Fourier's law is the universal law based on the induction of a large number of experimental results, and its expression is as follows:

$$\Phi = -\lambda A \frac{\partial t}{\partial x} \tag{8-1}$$

Or for the heat flux expression:

$$q = \frac{\Phi}{A} = -\lambda \frac{\partial t}{\partial x} \tag{8-2}$$

Recall that the minus sign is necessary because heat is always transferred in the direction of decreasing temperature.

Recognizing that the heat flux is a vector quantity, we can write a more general statement of the conduction rate equation (Fourier's law) as follows:

$$q = -\lambda \nabla t = -\lambda\left(\mathbf{i}\frac{\partial t}{\partial x} + \mathbf{j}\frac{\partial t}{\partial y} + \mathbf{k}\frac{\partial t}{\partial z}\right) \tag{8-3}$$

Where ∇ is the three-dimensional del-operator and $T(x, y, z)$ is the scalar temperature. It is implicit in Eq. (8-3) that the heat flux vector is in a direction perpendicular to the isothermal surfaces. Therefore an alternative form of Fourier's law is:

$$q = q_n \mathbf{n} = -\lambda \frac{\partial t}{\partial n}\mathbf{n} \tag{8-4}$$

Where q_n is the heat flux in a direction $\mathbf{n}$, Which is perpendicular to the plane of the isothermal, and $\mathbf{n}$ is the unit normal vector in that direction. Note also that the heat flux vector can be resolved into components such that, in Cartesian coordinates, the general expression for q_n is:

$$q_n = \mathbf{i}q_x + \mathbf{j}q_y + \mathbf{k}q_z \tag{8-5}$$

Where, from Eq. (8-3), it follows that:

$$q_x = -\lambda \frac{\partial t}{\partial x}, q_y = -\lambda \frac{\partial t}{\partial y}, q_z = -\lambda \frac{\partial t}{\partial z} \tag{8-6}$$

Fourier's law is the cornerstone of conduction heat transfer, and its key features are summarized as follows. It is not an expression that may be derived from the first law; it is instead a generalization based on experimental evidence. It is an expression that defines an important material property, the thermal conductivity. In addition, Fourier's law is a vector expression indicating that the heat flux is normal to an isotherm and in the direction of decreasing temperature. Finally, note that Fourier's law applies for all matter, regardless of its state (solid, liquid, or gas).

8.1.3 Thermal Conductivity

From Fourier's law, Eq. (8-6), the thermal conductivity associated with conduction in the x-direction is defined as:

$$\lambda_x = -\frac{q_x}{(\partial t/\partial x)}$$

Similar definitions are associated with thermal conductivities in the y- and z-directions (λ_y, λ_z), but for an isotropic material the thermal conductivity is independent of the direction of transfer, $\lambda_x = \lambda_y = \lambda_z = \lambda$.

From the foregoing equation, it follows that, for a prescribed temperature gradient, the conduction heat flux increases with increasing thermal conductivity. In general, the thermal conductivity of a solid is larger than that of a liquid, which is larger than that of a gas. The thermal conductivity of a solid may be more than four orders of magnitude larger than that of a gas. This trend is due largely to differences in intermolecular spacing for the two states.

8.2 Mathematical Description of Heat Conduction

8.2.1 Thermal Differential Equation

Consider a homogeneous medium within which there is no bulk motion (advection) and the temperature distribution $t(x, y, z)$ is expressed in Cartesian coordinates. Following thermodynamics method of applying conservation of energy, we first define an infinitely small (differential) control volume, $\mathrm{d}x\mathrm{d}y\mathrm{d}z$ as shown in Fig. 8-2. Choosing to formulate the first law at an instant of time, the second step is to consider the energy processes that are relevant to this control volume. If there are temperature gradients, conduction heat transfer will occur across each of the control surfaces. The conduction heat rates perpendicular to each of the control surfaces at the x, y, and z coordinate locations are indicated by the terms Φ_x, Φ_y and Φ_z respectively. The conduction heat rates at the opposite surfaces can then be expressed as a Taylor series expansion where, neglecting higher-order terms:

$$\Phi_{x+\mathrm{d}x} = \Phi_x + \frac{\partial \Phi_x}{\partial x}\mathrm{d}x \tag{8-7a}$$

$$\Phi_{y+\mathrm{d}y} = \Phi_y + \frac{\partial \Phi_y}{\partial y}\mathrm{d}y \tag{8-7b}$$

$$\Phi_{z+\mathrm{d}z} = \Phi_z + \frac{\partial \Phi_z}{\partial z}\mathrm{d}z \tag{8-7c}$$

In words, Eq. (8-7a) simply states that the x component of the heat transfer rate at $x + \mathrm{d}x$ is equal to the value of this component at x plus the amount by which it changes with respect to x multiply $\mathrm{d}x$.

According to the general form of the conservation of energy:

$$\dot{E}_{in} + \dot{E}_g - \dot{E}_{out} = \dot{E}_{st} \tag{8-8}$$

Within the medium there may also be an energy source term associated with the rate of thermal energy generation. This term is represented as:

$$\dot{E}_g = \dot{q}\,dx\,dy\,dz \tag{8-9}$$

Where $\dot{q}$ is the rate at which energy is generated per unit volume of the medium (W/m^3). In addition, changes may occur in the amount of the internal thermal energy stored by the material in the control volume. For solids and incompressible fluids, the difference between the specific heat at constant pressure c_p and the specific heat at constant volume c_v is very small, then $c_p = c_v = c$. If the material is not experiencing a change in phase, latent energy effects are not pertinent, and the energy storage term may be expressed as:

$$\dot{E}_{st} = \rho c \frac{\partial t}{\partial \tau} dx\,dy\,dz \tag{8-10}$$

Where $\rho c \dfrac{\partial t}{\partial \tau}$ is the time rate of change of the sensible (thermal) energy of the medium per unit volume.

Once again it is important to note that the terms $\dot{E}_g$ and $\dot{E}_{st}$ represent different physical processes. The energy generation $\dot{E}_g$ term is a manifestation of some energy conversion process involving thermal energy on one hand and some other form of energy, such as chemical, electrical, or nuclear, on the other. The term is positive (a source) if thermal energy is being generated in the material at the expense of some other energy form; it is negative (a sink) if thermal energy is being consumed. In contrast, the energy storage term $\dot{E}_{st}$ refers to the rate of change of thermal energy stored by the matter.

Hence, recognizing that the conduction rates constitute the energy inflow $\dot{E}_{in}$ and outflow $\dot{E}_{out}$, and substituting Eq.(8-9) and Eq.(8-10), we obtain.

$$\Phi_x + \Phi_y + \Phi_z + \dot{q}\,dx\,dy\,dz - \Phi_{x+dx} - \Phi_{y+dy} - \Phi_{z+dz} = \rho c \frac{\partial t}{\partial \tau} dx\,dy\,dz \tag{8-11}$$

Substituting from Eq.(8-7), it follows that:

$$-\frac{\partial \Phi_x}{\partial x}dx - \frac{\partial \Phi_y}{\partial y}dy - \frac{\partial \Phi_z}{\partial z}dz + \dot{q}\,dx\,dy\,dz = \rho c \frac{\partial t}{\partial \tau} dx\,dy\,dz \tag{8-12}$$

The conduction heat rates in an isotropic material may be evaluated from Fourier's law:

$$\begin{cases} \Phi_x = -\lambda\,dy\,dz \dfrac{\partial t}{\partial x} \\ \Phi_y = -\lambda\,dx\,dz \dfrac{\partial t}{\partial y} \\ \Phi_z = -\lambda\,dx\,dy \dfrac{\partial t}{\partial z} \end{cases} \tag{8-13}$$

Substituting Eq.(8-13) into Eq.(8-12) and dividing out the dimensions of the control volume $dxdydz$, we obtain:

$$\frac{\partial}{\partial x}\left(\lambda \frac{\partial t}{\partial x}\right)+\frac{\partial}{\partial y}\left(\lambda \frac{\partial t}{\partial y}\right)+\frac{\partial}{\partial z}\left(\lambda \frac{\partial t}{\partial z}\right)+\dot{q}=\rho c \frac{\partial t}{\partial \tau} \tag{8-14}$$

Eq.(8-14) is the general form, in Cartesian coordinates, of the heat diffusion equation. This equation, often referred to as the heat equation, provides the basic tool for heat conduction analysis.

When the thermal conductivity λ is constant, the differential equation of heat conductivity can be simplified as:

$$\lambda\left(\frac{\partial^2 t}{\partial x^2}+\frac{\partial^2 t}{\partial y^2}+\frac{\partial^2 t}{\partial z^2}\right)+\dot{q}=\rho c \frac{\partial t}{\partial \tau}$$

$$\frac{\lambda}{\rho c}\left(\frac{\partial^2 t}{\partial x^2}+\frac{\partial^2 t}{\partial y^2}+\frac{\partial^2 t}{\partial z^2}\right)+\frac{\dot{q}}{\rho c}=\frac{\partial t}{\partial \tau} \tag{8-15}$$

$a=\dfrac{\lambda}{\rho c}$ is called thermal diffusivity, also known as temperature conductivity coefficient, and the unit is m^2/s. It is a thermal physical parameter that has an important influence on the process of unsteady heat conduction. Its value reflects the speed of temperature change in the body when the body is heated or cooled transiently, that is, the rate of heat diffusion in the body. The larger the a, the faster the heat transfer, the faster the internal temperature of the body tends to equilibrium, and the greater the ability of the temperature distribution of the body tends to be uniform.

According to the definition of thermal diffusivity $a=\lambda/\rho c$, we can know that: ① The molecule λ is the thermal conductivity of the object, the larger the λ, the more heat can be conducted under the same temperature gradient; ② The denominator ρc is a unit volume of object temperature rise 1 ℃ required heat (heat capacity or thermal inertia), ρc smaller, the temperature rise 1 ℃ absorbed less heat, can be left more heat to continue to transfer to the internal body, the body temperature of each point faster with the interface temperature rise and rise. The greater the thermal conductivity λ, and the smaller the heat capacity per unit volume c (itself heat storage or heat release energy is smaller) of the material, the greater the ability to diffuse heat, the greater the thermal diffusivity. In the process of unsteady heat conduction, the temperature of the material with large thermal diffusivity changes quickly, or the temperature of the whole material is relatively uniform, so the thermal diffusivity is also known as the temperature conductivity coefficient. For example, the thermal diffusivity of general wood is about $a=1.5\times10^{-7}$ m^2/s, and the thermal diffusivity of red copper (pure copper) is about $a=5.33\times10^{-5}$ m^2/s, 355 times that of wood. If the two hands hold the same length and thickness of the rod and the copper rod respectively, and put the other end into the hot stove, when the hand holding the copper rod is very hot, the hand holding the stick is not hot. This shows that the temperature changes much more rapidly in the copper rod than in the wood rod.

The heat equation may also be expressed in cylindrical and spherical coordinates. The expression of type as follows:

Cylindrical coordinates:

$$\frac{1}{r}\frac{\partial}{\partial r}\left(\lambda r\frac{\partial t}{\partial r}\right)+\frac{1}{r^2}\frac{\partial}{\partial \varphi}\left(\lambda\frac{\partial t}{\partial \varphi}\right)+\frac{\partial}{\partial z}\left(\lambda\frac{\partial t}{\partial z}\right)+\dot{q}=\rho c\frac{\partial t}{\partial \tau} \tag{8-16}$$

Spherical coordinates:

$$\frac{1}{r^2}\frac{\partial}{\partial r}\left(\lambda r^2\frac{\partial t}{\partial r}\right)+\frac{1}{r^2\sin^2\theta}\frac{\partial}{\partial \varphi}\left(\lambda\frac{\partial t}{\partial \varphi}\right)+\frac{1}{r^2\sin\theta}\frac{\partial}{\partial \theta}\left(\lambda\sin\theta\frac{\partial t}{\partial \theta}\right)+\dot{q}=\rho c\frac{\partial t}{\partial \tau} \tag{8-17}$$

8.2.2 Thermal Differential Equation Definite Conditions

Thermal differential equation is a mathematical expression describing common thermal process. In essence, solving the heat conduction problem is to solve the above-mentioned thermal differential equations. In order to obtain a particular solution of heat conduction problems, additional conditions of characterize this particular issue must be given, these conditions are called definite conditions attachments. For non-steady-state problem, definite conditions include two aspects, namely, the initial time given the initial conditions of temperature distribution and temperature distribution or heat transfer boundary conditions on the thermal conductivity of the object, that the boundary conditions. For steady-state problems, definite conditions require only boundary conditions, no initial conditions. Common boundary conditions of heat conduction problem can be summarized into the following three categories:

(1) First boundary condition: specifies the temperature distribution on the boundary, in the non-steady-state heat conduction problems, should give its variation with time. When $\tau>0$, $t_w=f(x,y,z,\tau)$. The most simple example of boundary conditions is the border of the surface temperature is held constant, i. e., $t_w=$ const, referred to as a constant wall temperature boundary condition.

(2) The second kind of boundary condition: the heat flux distribution on the boundary of the given object is changing with time, That is when $\tau>0$, $q_w=-\lambda\frac{\partial t}{\partial n}\big|_w=f(x,y,z,\tau)$. This kind of boundary condition is the simplest example of constant heat flux boundary on keeping. i. e., $q_w=$ const, referred to as the constant heat flux boundary condition. Among them, $q_w=0$, referred to as the adiabatic boundary condition.

(3) The third kind of boundary condition: giving the heat exchanging state on the object boundary, namely the convective heat transfer surface heat transfer coefficient h and the temperature t_f of the surrounding fluid. Generally written as $-\lambda\frac{\partial t}{\partial n}\big|_w=h(t_w-t_f)$.

8.3 Steady-State Conduction

8.3.1 Conduction of the Plane Wall

(1) Single plane wall

For one-dimensional Steady-State conduction in a plane wall without heat generation, as shown in Fig. 8-3, the thickness of the plane wall is δ, the temperature of the two surface respectively are t_1 and t_2. The mathematical description of the heat conduction is:

$$\frac{d^2 t}{dx^2} = 0 \tag{8-18}$$

Definite condition is:

$$x = 0, t = t_1; x = \delta, t = t_2 \tag{a}$$

Integrating twice with Eq. (8-18), its general solution is:

$$t = c_1 x + c_2 \tag{b}$$

Where c_1 and c_2 are integration constant, substituted to definite condition Eq.(b) we can get:

$$c_1 = \frac{t_2 - t_1}{\delta}, c_2 = t_1 \tag{c}$$

Plug Eq.(c) into Eq.(b) to get the temperature distribution:

$$t = \frac{t_2 - t_1}{\delta} x + t_1 \tag{8-19}$$

The heat flux through the wall is:

$$q = -\lambda \frac{t_2 - t_1}{\delta} = \lambda \frac{t_1 - t_2}{\delta} = \frac{\lambda}{\delta} \Delta t \tag{8-20}$$

The heat flow through the wall is:

$$\Phi = A \frac{\lambda}{\delta} \Delta t \tag{8-21}$$

It can be seen by the Eq.(8-18), In terms of the one-dimensional Steady-State conduction in a plane wall without heat generation, the temperature distribution is linear distribution. Adapt the Eq.(8-19) we can get the following form:

$$\Phi = \frac{\Delta t}{\delta/(A\lambda)} = \frac{\Delta t}{R} \tag{8-22}$$

Where the heat flow Φ is the transfer of the heat conduction process; the temperature difference Δt is the power in the process of transfer; denominator $\delta/(A\lambda)$ is the resistance of the heat transfer, setting $R = \delta/(A\lambda)$, referred to as the thermal resistance of the conduction. For the plane wall per unit area, the thermal resistance is $R_A = \delta/\lambda$.

The concept of thermal resistance is similar to the Ohm's law of the circuit; the establishment of the concept of thermal resistance has brought great convenience on the analysis of the complex heat transfer process. For example, we can borrow more familiar with the series and parallel resistance calculation formula in the circuit to calculate the total thermal resistance in the process of the complex heat transfer like the multi-layer flat walls, or combined heat conduction, convection, radiation and so on.

(2) Multilayer plane wall

The so-called multilayer flat wall is made of several layers of different materials stacked together to form a composite board. For example, the use of firebricks, insulation brick layer and common brick layer together formed boiler furnace wall is a multi-layer flat wall. The following three flat wall as an example to discuss. As shown in Fig. 8-4, the known thicknesses of the layers are $\delta_1, \delta_2, \delta_3$, the thermal conductivity of the layers are λ_1, λ_2, λ_3, the temperature of the left and right are t_1, t_4, respectively. Need to be determined heat flux of multi-level inter-wall and temperature of each layer.

Thermal conductivity through three flat walls can use concepts of thermal resistance. Application thermal resistance of each layer expression is as follows:

$$\frac{t_1 - t_2}{q} = \frac{\delta_1}{\lambda_1}, \frac{t_2 - t_3}{q} = \frac{\delta_2}{\lambda_2}, \frac{t_3 - t_4}{q} = \frac{\delta_3}{\lambda_3} \tag{8-23}$$

Application of the total thermal resistance is equal to the sum of a series of sub-processes of thermal resistance; the so-called principle of thermal resistance in series, i. e., the thermal resistance in the respective layers can be obtained in the total thermal resistance of the multilayer flat wall.

$$\frac{t_1 - t_4}{q} = \frac{\delta_1}{\lambda_1} + \frac{\delta_2}{\lambda_2} + \frac{\delta_3}{\lambda_3} \tag{8-24}$$

Therefore, heat flux density can be obtained which pass through a multilayer flat wall is:

$$q = \frac{t_1 - t_4}{\dfrac{\delta_1}{\lambda_1} + \dfrac{\delta_2}{\lambda_2} + \dfrac{\delta_3}{\lambda_3}} \tag{8-25}$$

After the calculated heat flux density, the interface between layers of the unknown temperature t_2, t_3 can be calculated using Eq. (8-21). For example:

$$t_2 = t_1 - q\frac{\delta_1}{\lambda_1} \tag{8-26}$$

8.3.2 Conduction through a Cylindrical Wall

(1) Single cylindrical wall

Suppose there is a length l, cylindrical wall which the inner, outer diameter of each is d_1 and d_2, and $l \gg d_2$, the inner, outer wall temperature of each is t_{w_1}, t_{w_2}, and $t_{w_1} > t_{w_2}$, as shown in Fig. 8-5. If use a cylindrical coordinate system, the problem becomes the problem of one-dimensional heat conduction along the radius direction, that is, the temperature field is one-dimensional, so the isothermal surface are concentric cylindrical surface. In order to facilitate the analysis, the coefficient of thermal conductivity of the material is assumed as a constant value λ.

Single cylinder wall heat conduction differential equation and boundary conditions as:

$$\frac{d}{dr}\left(r\frac{dt}{dr}\right) = 0 \tag{8-27}$$

$$r = r_1, t = t_1 \tag{d}$$

$$r = r_2, t = t_2 \tag{e}$$

Eq. (8-27) for two consecutive points, obtaining the general solution for:

$$t = c_1 \ln r + c_2 \tag{f}$$

The boundary condition of formula Eq. (d), and formula Eq. (e) is substituted into the

Eq. (f), was solved:

$$c_1 = \frac{t_2 - t_1}{\ln(r_2/r_1)}, c_2 = t_1 - \ln r_1 \frac{t_2 - t_1}{\ln(r_2/r_1)} \tag{g}$$

The Eq. (g) is substituted into Eq. (f), to obtain a temperature profile of:

$$t = t_1 + \frac{t_2 - t_1}{\ln(r_2/r_1)} \ln(r/r_1) \tag{8-28}$$

Thus, with the flat wall of the linear temperature profile is different from the cylindrical wall of the logarithmic curve of the temperature distribution. Derivative of Eq. (8-28) to give:

$$\frac{dt}{dr} = \frac{1}{r} \frac{t_2 - t_1}{\ln(r_2/r_1)}$$

Substituting into the Fourier heat conduction law:

$$q = -\lambda \frac{dt}{dr} = \frac{\lambda}{r} \frac{t_1 - t_2}{\ln(r_2/r_1)} \tag{8-29}$$

It can be seen, in the steady-state heat conduction cylindrical wall, the heat flux density is inversely proportional to the radius at different radius. However, the heat flow of the whole cylindrical wall is constant and does not change with the radius. On both sides of the Eq. (8-29) multiply $2\pi rl$ at the same time, it can be obtained that the heat rate is:

$$\Phi = 2\pi rlq = \frac{2\pi\lambda l(t_1 - t_2)}{\ln(r_2/r_1)} \tag{8-30}$$

We can obtain the thermal resistance through the cylindrical wall according to the definition of the thermal resistance.

$$R = \frac{\Delta t}{\Phi} = \frac{\ln(r_2/r_1)}{2\pi\lambda l} \tag{8-31}$$

(2) Multilayer cylindrical wall

Multi-wall thermal conductivity calculated round is like with multi-level wall, you can use the principles of thermal resistance in series, then can get multi-storey circular wall thermal heat flow:

$$\Phi = 2\pi rl = \frac{2\pi\lambda l(t_1 - t_2)}{\ln(r_2/r_1)} \tag{8-32}$$

8.3.3 The Thermal Conductivity of the Fins

In order to enhance heat transfer or meet other requirements, various shapes of ribs can be added in the surface of heat exchanger, to increase the convective heat transfer area, reduce the convection heat transfer coefficient, play a role in enhancing the heat transfer. Fig. 8-6 shows the fins of different forms of. The extended surface is connected with the flat walls are straight fin, it can be uniform, can also be a variable cross-section straight fin, the cross-sectional area decreases with distance from the flat wall changes. And the cylindrical in circumference contact for the ring rib, its section with the radial distance from the cylinder center line and change. Pin fin (also known as the column fin) is different, it is the cross section is extended surface circular. Pin fin can be uniform or be a variable cross section.

The main task of finned heat conduction analysis is to determine the temperature distribution and the quantity of fins along the height direction of the fins. Here discuss the simplest case, namely rectangular fins or needle rib section, as shown in Fig. 8-7. Each rib is connected with the surface which temperature is $t(0) = t_0$, and extends to the fluid whose temperature is t_∞.

Fins generally made of metal, the thermal conductivity is large. Any part of it can be considered a temperature distribution almost uniform surface can be assumed that only the x-direction (i. e. the direction of fin height) has a significant change in temperature. Meanwhile, the surface of the fins to the surrounding heat exchange fluid absorbs heat as the cold source. Thus, the temperature field inside the fin can be seen as a one-dimensional steady-state temperature field with a negative internal heat source. For a given fin, A_c is constant, $A_s = Px$, A_s is surface area from where the fin at the base to the x, P is the perimeter of the fin. Such as straight fins temperature field differential equations can be written as:

$$\frac{d^2 t}{dx^2} - \frac{hP}{\lambda A_c}(t - t_\infty) = 0 \tag{8-33}$$

To simplify the above formula, the definition of excess temperature θ instead of the dependent variable:

$$\theta(x) = t(x) - t_\infty \tag{8-34}$$

Due to the above formula t_∞ is constant, so $d\theta/dx = dt/dx$, The Eq.(8-34) substituting Eq.(8-33) to obtain:

$$\frac{d^2\theta}{dx^2} - m^2\theta = 0 \tag{8-35}$$

Where:

$$m^2 = \frac{hP}{\lambda A_c} \tag{8-36}$$

Eq. (8-35) is a linear, homogeneous, second-order differential equations with constant coefficients, in the form of its general solution is:

$$\theta(x) = C_1 e^{mx} + C_2 e^{-mx} \tag{8-37}$$

It substituting Eq. (8-35), it is easy to proof that it really is a solution Eq. (8-35).

In order to determine the Eq. (8-37) the constants C_1 and C_2, in particular to be given a suitable boundary conditions. Temperature fin group ($x=0$) can be used as one of the boundary condition:

$$\theta(0) = t_0 - t_\infty = \theta_0 \tag{8-38}$$

The second boundary condition at the end of the ribs can be used ($x=L$) conditions, can be considered $\frac{d\theta}{dx}\big|_{x=L} = 0$, that end portion of the fin does not heat. This two boundary conditions into Eq. (8-37) and can be obtained by finishing the fin temperature distribution expression for:

$$\theta = \theta_0 \frac{e^{-mx} + e^{-2mL} e^{mx}}{1 + e^{-2mL}} = \theta_0 \frac{\mathrm{ch}[m(L-x)]}{\mathrm{ch}(mL)} \tag{8-39}$$

In the steady state, without considering the end part of the radiation, from the surface of the fins of heat transmitted to the surrounding fluid by the ribs should be equal to the heat introduced into the substrate:

$$\Phi = \Phi_0 = \frac{hP}{m}\theta_0 \mathrm{th}(mL) \tag{8-40}$$

8.3.4 Two-Dimensional Steady-State Heat Conduction

The previously discussed contents are heat conduction problem only the temperature gradient in one coordinate direction, but in many cases, treated with a dimension, will cause some problems for is too simplified and can′t reflect the actual situation, so we must consider the multidimensional effect. For solving the multidimensional problem is mainly with the help of computer numerical simulation, the basic idea is to calculate the regional grid, the discretization, thus the continuous differential equation into algebraic equation set of discrete, and then using the iterative method to solve. In multi dimension, the simplest problem is the two-dimensional steady-state heat conduction problems.

Fig. 8-8 is a long cylinder of a two-dimensional thermal conductivity. About two of the wall is adiabatic surface, two upper and lower surfaces to keep the different temperatures t_1, t_2, with $t_1 > t_2$. Fig. 8-9 is the results of numerical simulation, showing the temperature distribution and heat flux in the two-dimensional steady-state heat conduction.

As can be seen from the figure, in the y direction, the temperature gradient in the object increases with y, Solid in a local heat flux density is a vector everywhere perpendicular to the isotherm. And the direction indicated by the arrow in the direction of heat flow in Fig. 8-9. It is worth noting that, all the isotherms at the top of the upper two corners are brought together physically unreasonable. This is mainly due to the four vertices given by the boundary conditions have two temperatures.

8.4 Transient Conduction

8.4.1 The Characteristics of Unsteady State Conduction

The temperature field which temperature of each point on the temperature field changes along with the time is called the unsteady temperature field. The process of heat conduction in the unsteady temperature field is unsteady heat conduction. It is caused by heating or cooling the object. It is frequently encountered in engineering, such as the start, stop and varying load of heating equipment, the heat treatment of parts and components and so on, all of they belong to unsteady heat conduction.

To illustrate the characteristics of the unsteady heat conduction process, taking the flat wall for example, as shown in Fig. 8-10. Assume the flat wall is surrounded by air and the temperature of the entire flat wall is t_0 at the initial time ($\tau = \tau_0$), when $\tau > \tau_0$, the temperature of the left side of the flat wall rises to t_1, and at the moment, the temperature of the right side of the flat wall which exposed to the air is still keeping at t_0. The temperature of the place closes to the left of the flat plate increases in turns with the passage of time, which shows the change process in Fig. 8-10(a). Finally, the temperature of each layer in the wall does not change with time, and the temperature distribution is linear when the flat plate reaches steady state. Fig. 8-10(b) show the changes of the heat transfer on unsteady state. Φ_1 represents the heat which enter in the system from the left surface of the flat plate. And Φ_2 represents the heat dissipation from the right surface of the flat plate. It is easily observed in the picture, when $\tau < \tau_4$, Φ_1 decreases in turns with the passage of time, because the temperature on the right side of the wall is always equal to t_0, there is no heat transfer to the air, so $\Phi_2 = 0$, when $\tau > \tau_4$, Φ_1 still lower and Φ_2 ratchets up. When $\tau = \tau_5$, heat gain is equal to heat dissipation ($\Phi_1 = \Phi_2$), the inner of the flat wall eventually established steady-state heat transfer condition at this moment.

From the above analysis we can see that there are significantly differences between unsteady heat conduction process and steady heat conduction state: ① In the process of unsteady heat conduction, the temperature of each layer in the wall gradually increases with the heat transfer, less than a certain amount of time, the temperature of a layer inside the wall would not rise. ② In the unsteady heat conduction process, the heat flow (Φ) of the cross section which perpendicular to the direction of heat flow would not be equal everywhere, the reduce of the heat flow is used to raise the temperature of the material in somewhere. ③ In the process of unsteady heat conduction, the instantaneous heat flow Φ_1 and Φ_2 are not constant, they are a function of time (τ), the shadow area of the Fig. 8-10(b) is expressed the material′s thermodynamic energy which accumulated in the process of unsteady heat conduction.

8.4.2 The Lumped Capacitance Method

(1) Analysis solution of the temperature field of lumped parameter method

A simple, yet common, transient conduction problem is one for which a solid experiences a sudden change in its thermal environment. Consider a hot metal forging that is initially at a uniform temperature t_i and is quenched by immersing it in a liquid of lower temperature $t_\infty < t_0$ (Fig. 8-11). If the quenching is said to begin at time $\tau = 0$, the temperature of the solid will decrease for time $\tau > 0$, until it eventually reaches t_∞. This reduction is due to convection heat transfer at the solid-liquid interface. The essence of the lumped capacitance method is the assumption that the temperature of the solid is spatially uniform at any instant during the transient process. This assumption implies that temperature gradients within the solid are negligible. From Fourier′s law, heat conduction in the absence of a temperature gradient implies the existence of infinite thermal conductivity. Such a condition is clearly impossible. However, the condition is closely approximated if the resistance to conduction within the solid is small compared with the resistance to heat transfer between the solid and its surroundings.

In neglecting temperature gradients within the solid, we can no longer consider the problem from within the framework of the heat equation, since the heat equation is a differential equation governing the spatial temperature distribution within the solid. Instead, the transient temperature response is determined by formulating an overall energy balance on the entire solid. This balance must relate the rate of heat loss at the surface to the rate of change of the internal energy. Applying thermal balance to the control volume of Fig. 8-16, this requirement takes the form:

$$-\dot{E}_{out} = \dot{E}_{st} \tag{8-41}$$

or

$$-hA(t - t_{\infty}) = \rho Vc \frac{dt}{d\tau} \tag{8-42}$$

Introducing the temperature difference:

$$\theta = t - t_{\infty} \tag{8-43}$$

And recognizing that $(d\theta/d\tau) = (dt/d\tau)$ if t_{∞} is constant, it follows that:

$$\frac{\rho Vc}{hA} \times \frac{d\theta}{d\tau} = -\theta$$

Separating variables and integrating from the initial condition, for which $\tau = 0, t(0) = t_0$, we then obtain:

$$\frac{\rho Vc}{hA} \int_{\theta_0}^{\theta} \frac{d\theta}{\theta} = -\int_0^{\tau} d\tau$$

Where $\theta_0 = t_0 - t_{\infty}$.

Evaluating the integrals, it follows that:

$$\frac{\rho Vc}{hA} \ln \frac{\theta_0}{\theta} = \tau \tag{8-44}$$

or

$$\frac{\theta}{\theta_0} = \frac{t - t_{\infty}}{t_0 - t_{\infty}} = \exp\left[-\left(\frac{hA}{\rho Vc}\right)\tau\right] \tag{8-45}$$

Eq. (8-44) may be used to determine the time required for the solid to reach some temperature t, Eq. (8-45) may be used to compute the temperature reached by the solid at some time τ. The foregoing results indicate that the difference between the solid and fluid temperatures must decay exponentially to zero as τ approaches infinity. The total quantity of heat transfer in the conduction process:

$$Q = \rho Vc\theta_0\left[1 - \exp\left(-\frac{hA}{pcV}\tau\right)\right] \tag{8-46}$$

Make the right exponential term of Eq. (8-45) changes to:

$$\frac{hA}{\rho Vc}\tau = \frac{hl}{\lambda} \frac{\lambda}{\rho c} \frac{\tau}{l^2} = \frac{hl}{\lambda} \frac{\alpha\tau}{l^2} = Bi \cdot Fo \tag{8-47}$$

$$Bi = \frac{hl}{\lambda} = \frac{l/\lambda}{1/h}, Fo = \frac{\alpha\tau}{l^2} \tag{8-48}$$

Where Bi is Biot number for the characteristic length l, Fo is the Fourier number for the characteristic length l, the Eq. (8-45) can be written as:

$$\frac{\theta}{\theta_0} = \exp(-Bi \cdot Fo) \tag{8-49}$$

(2) Applicable scope of the lumped parameter method

Eq. (8-49) show that the effect of the non steady state temperature distribution by *Fo* and *Bi* parameters. Among them, the Fourier number represent of unsteady-state heat conduction process in-depth dimensionless time, the numerical value is greater, the thermal disturbance is more deeply spread to inside the object, so the object of each point temperature is close to the temperature of the surrounding media. The *Bi* number is the ratio of internal to external objects thermal resistance of convection thermal resistance, thermal resistance is dimensionless. λ in the *Bi* is the number of the thermal conductivity of the solid material. The size of *Bi* determines the internal temperature distribution characteristics of the object. When $Bi \to \infty$, means the external thermal resistance tends to zero, the internal thermal resistance plays a dominant role in the heat transfer process, temperature drop occurs mainly in the object; when $Bi \to 0$, means that the internal thermal resistance tends to zero, the uniform object internal temperature, external thermal resistance plays a dominant role. Theoretical calculation shows that, when $Bi < 0.1$, if using the lumped parameter method to simplify calculation, the error is less than 5%. Therefore, the criterion of the lumped parameter method at the project application is:

$$Bi = \frac{hV/A}{\lambda} < 0.1M \tag{8-50}$$

In the formula, M is a number related to the geometry and with one dimensionless, as for infinite plate, infinite cylinder and sphere, respectively $M = 1$, $M = 1/2$, $M = 1/3$.

第9章 对流换热

本章导学

对流换热是流动的流体与固体壁面直接接触，当两者温度不同时相互间所发生的热传递过程，它既具有流体分子间微观导热作用，又具有流体宏观位移的热对流作用，所以对流换热过程必然受到导热规律和流体流动规律的双重支配。

本章介绍了牛顿冷却定律、对流换热微分方程、对流换热的影响因素，以及边界层理论，并根据质量守恒、能量守恒和动量守恒得到对流换热的微分方程组及简化的边界层微分方程组；阐述了相似原理和量纲分析的理论以及运用理论得到准则关联式；针对典型管内流动、管外流动及自然对流，应用实验准则关联式对其进行求解，以获得表面传热系数。

学习目标

熟练掌握并应用牛顿冷却公式，了解影响对流换热的因素。理解流动边界层和热边界层的概念，以及它们之间的相互关系，并能理解对流换热微分方程中各物理量的意义。理解特征数的物理内涵和特征数方程的意义。了解各种对流换热的换热特征，并能正确选用对流换热的特征数实验准则关联式进行对流换热的定量计算。

重点与难点

(1) 对流换热机理与条件。

(2) 边界层理论与相似原理。

(3) 对流换热问题的求解方法。

9.1 对流换热概述

9.1.1 对流换热机理

对流换热模式由两种机制组成。除了由随机的分子运动(扩散)导致的能量传输外，流体的整体或者说宏观运动也传输能量。这种流体运动在任何时刻都有聚集在一起的大量分子在运动。当存在温度梯度时，这种运动就会对传热起作用。由于聚集的大量分子保持着随机运动，所以总的传热是分子随机运动与流体整体运动所导致的能量传输的叠加。在工程上，最感兴趣的是运动的流体与固体界面处于不同温度时它们之间发生的传热。在流体与固体表面的交界面处($y=0$)流体的速度为零,分子的随机运动(扩散)起主要作用,即通过导热进行传热。随后,由固体表面贴壁处的流体通过整体流动传热给其附近的(通常称为边界层内)流体,随着边界层厚度的增加,最终传给边界层外的流体。

9.1.2 牛顿冷却定律

对流换热的基本公式为牛顿冷却定律,对单位面积有:

$$q = h\Delta t \tag{9-1}$$

对面积为 A 的接触面有:

$$\Phi = hA\Delta t_{\mathrm{m}} \tag{9-2}$$

式中,q 为热流密度;Φ 为热流量;h 为对流换热表面传热系数,是表征对流换热强度的量,数值上等于流体和壁面之间的温度差为 1 ℃时,每单位时间单位壁面的对流换热量;Δt 为流体和固体的温差;Δt_{m} 为换热面上流体与固体表面的平均温差。

将式(9-2)改写成:

$$\Phi = \frac{\Delta t_{\mathrm{m}}}{1/(hA)}$$

将上式与欧姆定律对比可知,式中,$1/(hA)$ 为对流换热的热阻。

牛顿冷却定律只是给出了对流换热表面传热系数 h 的一个定义式,它没有揭示出表面传热系数与影响它的有关物理量之间的内在联系。揭示这种内在联系正是研究对流换热的主要任务。

9.1.3 对流换热微分方程

由于流体流过壁面时,贴壁面的流体速度为零,热量传递完全依靠热传导,所以沿壁面横向任意位置 x 处的局部热流密度可由傅立叶定律得到:

$$q_{\mathrm{x}} = -\lambda\left(\frac{\partial t}{\partial y}\right)_{y=0} \tag{9-3}$$

联合牛顿冷却公式 $q_{\mathrm{x}} = h\Delta t$,可以得到局部对流换热表面传热系数:

$$h_{\mathrm{x}} = -\frac{\lambda}{\Delta t}\left(\frac{\partial t}{\partial y}\right)_{y=0} \tag{9-4}$$

式(9-4)把表面传热系数与流体的温度梯度联系起来,称为对流换热微分方程。而温度梯度的求解需要联合连续性方程、动量微分方程以及能量微分方程。对于一个具体的对流换热问题来说,我们更加感兴趣的是它的平均表面传热系数,平均表面传热系数定义为:

$$\bar{h} = \frac{1}{L}\int_0^L h_{\mathrm{x}}\mathrm{d}x \tag{9-5}$$

目前,表面传热系数的求取主要采用实验方法来获得其特征数的关联式。而实验方法的基础是相似性原理和量纲分析。

9.1.4 对流换热的影响因素

影响对流换热的因素很多,这是因为对流换热和流体的流动总是联系在一起的,同时又与固体壁面的形状、大小等有关,主要有以下几个方面。

(1) 流动产生的原因:按照引起流动的原因,可将对流换热区分为强制对流换热和自然对流换热两大类。前者是泵、风机或其他压差作用所造成的,而后者通常是流体内部的密度差所引起的。由于两者流动的推动力不同,它们的换热规律也不同。

（2）流体的流动状态：不论流动的起因如何，流体的流动存在两种不同的流态——层流和湍流。层流时流体微团沿着主流方向做有规则的分层流动，而湍流时流体各部分之间发生剧烈的混合，因而在其他条件相同时湍流换热的强度自然要较层流大。

（3）流体有无相变：无相变时对流中传热的热量交换是流体的显热变化而引起的；而在有相变的换热过程中（如沸腾或凝结），流体的相变潜热的释放或吸收常占据主导作用，故其传热规律与无相变时不同。

（4）流体的物理性质：影响对流换热过程的流体物性主要是导热系数、比热容、密度及黏度等。导热系数大的流体，贴壁处的导热热阻小，换热强。比热容和密度大的流体体积比热容大，即载热能力大，增强了流体与壁面之间的热交换。此外，黏度大的流体流动边界层增厚，对换热不利。在分析物性参数对换热的影响时，应注意物性的综合效果，不宜孤立分析每一个物性参数对换热的影响，否则会导致错误的结论。

（5）换热面的几何尺寸、形状和位置：在对流换热时，流体沿着壁面流动，所以壁面的几何尺寸、相对位置对流体的流动有很大影响，从而也影响对流换热的强弱。图 9-1(a)和图 9-1(b)示出了流体管内强制对流流动和流体横向绕过圆管强制对流流动的情况。显然，这是两种不同的流动情况，换热规律必定不同。图 9-1(c)和图 9-1(d)示出了水平壁面向空气自由流动换热的两种布置：一种热面朝上、一种热面朝下。前者气流受热上升容易展开流动；后者由于热面朝下抑制流动，h 减小。

图 9-1 换热面几何因素的影响

综上所述，影响表面传热系数的因素很多，强制对流换热和自由对流换热，层流和湍流，有相变和无相变，各种性质不同的流体以及换热面的形状、大小和相对位置等组成了多种不同规律的换热过程。因此，要了解表面传热系数的变化规律，只有对上述各种情况分门别类地进行分析和实验，才能获得反映各种情况的 h 的计算公式。一般地，表面传热系数可表示为如下函数形式：

$$h = f(u, l, \rho, \mu, \lambda, c_p) \tag{9-6}$$

9.1.5 对流换热的主要研究方法

研究对流换热的主要目的之一就是确定不同换热条件下表面传热系数的具体表达式，主要方法有 4 种。

（1）分析法

所谓分析法，就是用数学分析的方法求解描写对流换热的微分方程或积分方程及其定解条件。因此，根据对流换热的物理模型建立描写对流换热的数学模型是分析法的必要条件。尽管目前分析法只能求解简单的对流换热问题，但因其求解过程所依据的数学分析基础比较严谨，物理概念和逻辑推理比较清晰，求解结果以函数的形式表示，能清楚地显示各种因素对对流换热的影响，所以仍然是对流换热的基础内容。本书对于对流换热的数学描述和简单对流换热问题的分析结果加以简要介绍。

(2) 数值法

与导热问题的数值解法相比,对流换热控制方程的复杂性使其数值解法的难度和复杂性也随之加大。有关对流换热问题的数值解法本书不作介绍,感兴趣的读者可以参阅相关文献。

(3) 实验法

由于分析法的局限以及数值法的可靠性所限,相似理论指导下的实验研究仍然是解决复杂对流换热问题的主要方法,并且随着现代测量技术的进步,以前无法进行的对流换热微细结构和现象的观测现在得以实现,对尚未解决的对流换热(如湍流换热、沸腾换热等)机理的探索发挥着关键性的作用。由于测量精确度的提高,实验结果也常用来检验其他方法的准确性。本书将用一定篇幅介绍相似原理及其指导下的实验研究方法。

(4) 比拟法

所谓比拟法,就是利用热量传递与动量传递在机理上的共性建立起表面传热系数与摩擦系数之间的比拟关系式,由比较容易进行的流体流动实验获得摩擦系数的数据,再由比拟关系式求出表面传热系数。比拟法曾广泛用于求解湍流对流换热问题,但近些年来由于实验法和数值法的发展而很少被应用。

目前,理论分析、数值计算和实验研究相结合是科技工作者广泛采用的解决复杂对流换热问题的主要研究方式。

9.2 对流边界层

由对流机理可知,清晰认识边界层现象对理解对流换热非常重要。边界层概念是理解流体流过固体表面时对流换热的核心,边界层包括速度边界层和热边界层。

9.2.1 速度边界层

为引入边界层的概念,考虑图9-2所示的平板上的流动。当流体质点与表面接触时,它们的速度减小,在流体与固体的交界面处可认为速度为零。这些质点会阻碍邻近流体层中质点的运动,而后者又阻碍上一层质点的运动,依此类推,直到离开表面的距离 $y=\delta$ 时,才可以忽略这种影响。流体运动的受阻与作用在平行于流体速度的平面上的切应力 τ 有关(图9-2)。随着离开表面的距离 y 的增加,流体的 x 方向速度分量 u 必定增加,直到它接近自由流的值 u_∞(下标∞用于表示边界层外自由流的条件)。

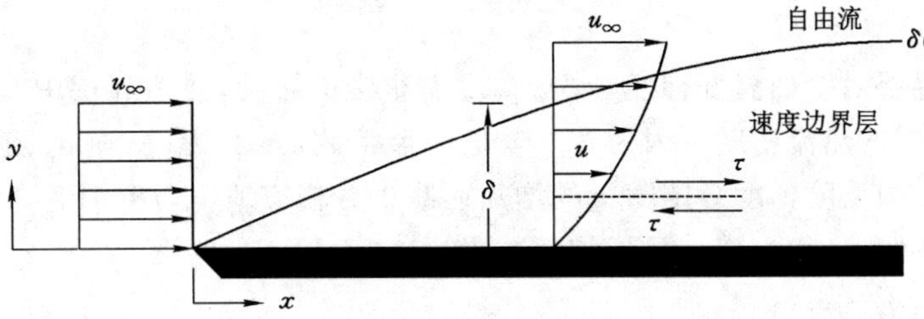

图9-2 平板上建立的速度边界层

δ 为边界层厚度,通常定义为 $u=0.99u_\infty$ 的 y 值。边界层速度分布是指边界层内 u 随

y 的变化方式。因此,流体的流动可分成两个不同的区域来描述:一个是很薄的流体层(边界层),其中速度梯度和切应力很大;另一个是边界层以外的区域,此处速度梯度和切应力可以忽略。随着离前缘的距离增加,黏性的影响逐步渗透进自由流,边界层也相应增厚(δ 随 x 增大)。由于与流体的速度有关,所以上述边界层可以更为明确地称为速度边界层。只要有流体流过表面,就会产生这种边界层,它在涉及对流换热的问题中非常重要。

9.2.2 热边界层

正如流体流过表面时产生速度边界层那样,如果流体的自由流和表面的温度不同,就必定形成热边界层。考虑等温平板上的流动(图 9-3),在前缘处,温度分布是均匀的,有 $t(y)=t_\infty$。然而,接触平板的流体质点达到热平衡,处于平板的表面温度下。依次地,这些质点和邻近流体层中的质点交换能量,并在流体中产生温度梯度,这个存在温度梯度的流体区域就是热边界层,通常其厚度 δ_t 定义为对应温度比 $(t_s-t)/(t_s-t_\infty)=0.99$ 的 y 值。随着离开前缘的距离增加,传热的影响逐步渗透进自由流,相应地,热边界层会增厚。

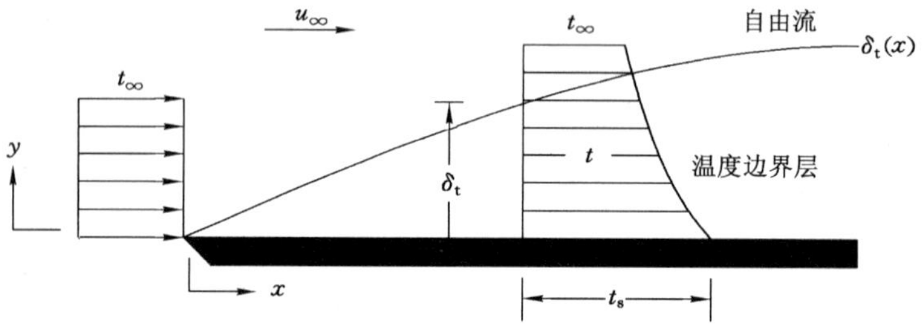

图 9-3 等温平板上建立的热边界层

很容易说明这个边界层中的状态与对流传热系数之间的关系。在离开前缘任意距离 x 处,局部表面热流密度可以通过对 $y=0$ 处的流体应用傅立叶定律得到,即

$$q=-\lambda_f \frac{\partial t}{\partial y}\bigg|_{y=0} \tag{9-7}$$

采用这个表达式是恰当的,因为在表面上不存在流体运动,能量的传递只能通过传导进行。根据牛顿冷却定律,有:

$$q=h(t_s-t_\infty) \tag{9-8}$$

将式(9-8)与式(9-7)联立,可得:

$$h=\frac{-\lambda_f(\partial t/\partial y|_{y=0})}{t_s-t_\infty} \tag{9-9}$$

因此,热边界层对壁面温度梯度 $\partial t/\partial y|_{y=0}$ 有很强的影响,热边界层中流体的温度分布决定着穿过边界层的热流密度。因为 t_s-t_∞ 是一个常数,与 x 无关,而 δ_t 随 x 的增加而增大,所以边界层中的温度梯度必定随 x 的增加而减小。相应地,$\partial t/\partial y|_{y=0}$ 的值也随 x 的增加而减小,由此可知,q 和 h 随 x 的增加而减小。

9.2.3 边界层的重要意义

对流过任意表面的流动,速度边界层总是存在的,因而存在表面摩擦和剪切力。同样地,如果表面和自由流的温度不同,就会存在热边界层,从而存在对流传热。速度边界层的范围是 $\delta(x)$,其特征是存在速度梯度和切应力。热边界层的范围是 $\delta_t(x)$,其特征是存在温度梯度和传热。工程上很多的实际问题既存在速度边界层,也存在热边界层。流动中的速度变化以及几乎全部的传热基本都发生在边界层内。因而,理解边界层的概念对求解流动传热问题非常重要。

9.3 对流换热微分方程组

对流换热微分方程组包括质量守恒、动量守恒及能量守恒这三大定律。质量守恒定律如下:

$$\frac{\partial \rho}{\partial \tau}=\frac{\partial(\rho u)}{\partial x}+\frac{\partial(\rho v)}{\partial y}=0 \tag{9-10}$$

在流体密度不变情况下,式(9-10)可以简化为:

$$\frac{\partial u}{\partial x}+\frac{\partial v}{\partial y}=0 \tag{9-11}$$

式(9-11)也称为连续性方程。

能量守恒方程推导如下:

假设流体系统为不可压缩、常物性、无内热源的二维系统。流体微元体的热量传递包括导热与对流两种形式,其中,对流部分能量的变换如图9-4所示。

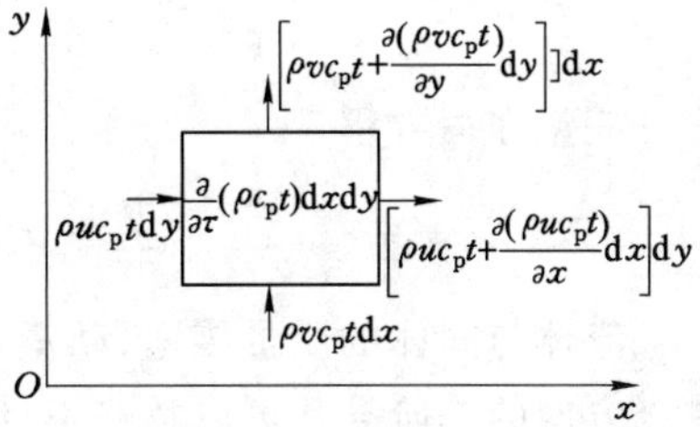

图9-4 微元体内能量的平衡

无内热源的能量守恒公式如式(9-12)所示:

$$\dot{E}_{in}-\dot{E}_{out}=\dot{E}_{st} \tag{9-12}$$

由于质量流入而进入微元体的热流量为:

$$\Phi_{in}=\rho u c_p t dy+\rho v c_p t dx \tag{9-13}$$

由于质量流出而离开微元体的热流量为:

$$\Phi_{out}=\left[\rho u c_p t+\frac{\partial(\rho u c_p t)}{\partial x}dx\right]dy+\left[\rho v c_p t+\frac{\partial(\rho v c_p t)}{\partial y}dy\right]dx \tag{9-14}$$

式(9-13)减去式(9-14)得到由于对流进入微元体的净热流量:

$$\begin{aligned}\Phi_f=\Phi_{in}-\Phi_{out}&=-\left[\frac{\partial(\rho u c_p t)}{\partial x}+\frac{\partial(\rho v c_p t)}{\partial y}\right]dxdy\\&=-\rho c_p\left(u\frac{\partial t}{\partial x}+v\frac{\partial t}{\partial y}+t\frac{\partial u}{\partial x}+t\frac{\partial v}{\partial y}\right)dxdy\end{aligned} \tag{9-15}$$

代入连续性方程消掉后两项得:

$$\Phi_f=-\rho c_p\left(u\frac{\partial t}{\partial x}+v\frac{\partial t}{\partial y}\right)dxdy \tag{9-16}$$

又由第8章相关知识得到由于导热而进入系统的净热流量:

$$\Phi_s=\lambda\left(\frac{\partial^2 t}{\partial x^2}+\frac{\partial^2 t}{\partial y^2}\right)dxdy \tag{9-17}$$

则由于导热部分和对流部分进入微元体的净能量为:

$$\dot{E}_{in}-\dot{E}_{out}=\Phi_f+\Phi_s=-\rho c_p\left(u\frac{\partial t}{\partial x}+v\frac{\partial t}{\partial y}\right)dxdy+\lambda\left(\frac{\partial^2 t}{\partial x^2}+\frac{\partial^2 t}{\partial y^2}\right)dxdy \tag{9-18}$$

系统内部的能量变化率为:

$$\dot{E}_{st}=\frac{\partial(\rho c_p t)}{\partial\tau}\mathrm{d}x\mathrm{d}y=\rho c_p\frac{\partial t}{\partial\tau}\mathrm{d}x\mathrm{d}y \tag{9-19}$$

将式(9-17)和式(9-18)代入式(9-12),并对公式化简可以得到:

$$\frac{\partial t}{\partial\tau}+u\frac{\partial t}{\partial x}+v\frac{\partial t}{\partial y}=\frac{\lambda}{\rho c_p}\left(\frac{\partial^2 t}{\partial x^2}+\frac{\partial^2 t}{\partial y^2}\right) \tag{9-20}$$

式(9-20)即为二维、不可压缩、常物性、无内热源流体的能量守恒方程。

动量守恒方程可以参考流体力学的推导过程,这里仅列出其公式。对三大方程进行汇总,可以得到二维、不可压缩、常物性、无内热源流体的微分方程组如下所示。

质量守恒方程:

$$\frac{\partial u}{\partial x}+\frac{\partial v}{\partial y}=0 \tag{9-21}$$

动量守恒方程:

$$\rho\left(\frac{\partial u}{\partial\tau}+u\frac{\partial u}{\partial x}+v\frac{\partial u}{\partial y}\right)=F_x-\frac{\partial p}{\partial x}+\mu\left(\frac{\partial^2 u}{\partial x^2}+\frac{\partial^2 v}{\partial y}\right) \tag{9-22}$$

$$\rho\left(\frac{\partial v}{\partial\tau}+u\frac{\partial v}{\partial x}+v\frac{\partial v}{\partial y}\right)=F_y-\frac{\partial p}{\partial y}+\mu\left(\frac{\partial^2 v}{\partial x^2}+\frac{\partial^2 v}{\partial y}\right) \tag{9-23}$$

能量守恒方程:

$$\frac{\partial t}{\partial\tau}+u\frac{\partial t}{\partial x}+v\frac{\partial t}{\partial y}=\frac{\lambda}{\rho c_p}\left(\frac{\partial^2 t}{\partial x^2}+\frac{\partial^2 t}{\partial y^2}\right) \tag{9-24}$$

流体与固体的质量传递、动量传递以及传热基本都发生在边界层内。一般地,边界层内主流方向的二阶导数可以忽略不计,将上述三大方程进行简化,可以得到二维、稳态、无内热源边界层内的控制方程组。

质量守恒方程:

$$\frac{\partial u}{\partial x}+\frac{\partial v}{\partial y}=0 \tag{9-25}$$

动量守恒方程:

$$u\frac{\partial u}{\partial x}+v\frac{\partial u}{\partial y}=-\frac{1}{\rho}\frac{\partial p}{\partial x}+\nu\frac{\partial^2 u}{\partial x^2} \tag{9-26}$$

能量守恒方程:

$$\frac{\partial t}{\partial\tau}+u\frac{\partial t}{\partial x}+v\frac{\partial t}{\partial y}=\alpha\left(\frac{\partial^2 t}{\partial x^2}+\frac{\partial^2 t}{\partial y^2}\right) \tag{9-27}$$

式中,ν 为运动黏度,$\nu=\mu/\rho$;α 为热扩散系数,$\alpha=\lambda/(\rho c_p)$。

上述微分方程组虽然形式上简单了一些,但一般仍然很难求得解析解,常用的方法是将偏微分方程组通过网格划分转化为代数方程组,在已知边界条件的情况下进行数值求解。通过求解得到边界层内的温度分布和温度梯度,进而可以求得对流换热表面传热系数。

9.4 相似原理及量纲分析

对流换热问题的目标之一是求得表面传热系数,而根据第 1 节所述,表面传热系数受多个因素的影响,如式(9-6)所示,受 6 个因素影响的话,按照常规方法:每个变量各变化 10 次,其他 5 个参数保持不变,共需要进行一百万次实验。如此庞大的实验数量让人难以接受;况且直接实验方法往往只能得出个别量之间的规律性关系,难以抓住现象的本质。此外,还有许多现象不宜进行直接实验,只能进行模型实验,模型实验结果必须和实际结果等同。相似原理和量纲分析即由此应运而生,一是如何减少实验次数,二是获得具有通用性的规律。本节以相似原理的基本内容为框架,采用量纲分析导出相似特征数之间的特征方程,从而得到具有较好通用性的普遍规律。

9.4.1 相似原理

相似的概念来自几何学,如果两个图形的对应边成比例,对应角相等,则两个图形几何相似。对于两个同类的物理现象,如果在对应的时刻及相应的地点与现象有关的物理量一一对应成比例,则称两现象彼此相似。相似原理研究相似物理现象之间的关系。需要注意的是,所谓同类现象,是指那些用相同形式并具有相同内容的微分方程所描写的现象。描述电场与导热问题的微分方程形式相仿但不相似,描述流动问题的温度场与速度场的微分方程形式相同,但由于它们各自研究的内容不同,因此不是同类问题,电场与导热温度场之间、流体中的温度场与速度场之间,只有“类比”或者“比拟”,不存在相似。

相似原理的基本内容为:两个现象相似,它们对应的同名已定特征数(准则数)相等,单值性条件相似。在对流传热中,常用的准则数有:

努塞尔数 $Nu=\dfrac{hl}{\lambda}$:表示流体的导热热阻与对流放热热阻之比,Nu 越大,则对流换热越强。

雷诺数 $Re=\dfrac{ul}{\nu}$:表示流体的惯性力与黏性力之比,Re 越大,惯性力也越大,流体运动就越剧烈,因此它的大小往往决定了流动是处于层流状态还是湍流状态。

普朗特数 $Pr=\dfrac{\nu}{\alpha}$:表示运动黏度与热扩散系数的比值,反映了动量扩散与热量扩散的相对大小,当 $Pr=1$ 时,表示动量扩散等于热量扩散。

格拉晓夫数 $Gr=\dfrac{g\beta\Delta tl^3}{\nu^2}$:表示流体所受的浮升力与黏性力的比值,$Gr$ 越大,表明浮升力作用越强。

对于对流传热问题,单值条件包括:

(1) 几何条件:指换热表面的几何形状、尺寸、位置及表面粗糙度等。

(2) 物理条件:指系统内流体的种类及物性,如密度、黏性等。

(3) 边界条件:指所研究系统的边界上(如进口、出口及壁面处等)的流速、温度等物理量的分布。

(4) 初始条件:稳态问题中不需要此条件,非稳态问题中是指物理量怎样随时间变化。

9.4.2 量纲分析

量纲分析是获得特征数的一种常用方法,其理论依据是 π 定律:一个表示 n 个物理量间关系的量纲一致的方程式,一定可以转换成包含 $n-r$ 个独立的无量纲物理量之间的关系式。其中,r 指 n 个物理量中所涉及的基本量纲的数目。这里我们不对其数学证明进行讨论,而只着眼于该定律的具体应用。

以管内强制对流换热为例,影响其对流传热系数的因素有:流体速度 u、特征长度 l 和流体的热物理性质包括黏性系数 μ、导热系数 λ、比热容 c_p 和密度 ρ。因此管内流动的对流传热系数可以写成上述参数的函数：

$$h = f(u, l, \mu, \lambda, c_p, \rho) \tag{9-28}$$

量纲分析的一般步骤：

(1) 找出与问题相关的各物理量量纲中的基本量的量纲。由式(9-28)可以发现，管内强制对流问题有 7 个物理量，它们的量纲均由 4 个基本量的量纲组成，即时间量纲 T，长度量纲 L，质量量纲 M 和温度量纲 Θ。因此 $n = 7, r = 4$，故可以组成的无量纲量为 $n - r = 3$ 个。从这 7 个物理量中选取 4 个作为基本物理量，该基本物理量必须包括上述基本量的量纲。本例中取 u, l, μ, λ 作为基本物理量。

(2) 把基本量逐一与其他量组成无量纲数。无量纲量总是采用幂指数形式，其指数待定。3 个无量纲数如下所示：

$$\pi_1 = hu^{a_1} l^{b_1} \lambda^{c_1} \mu^{d_1}$$

$$\pi_2 = \rho u^{a_2} l^{b_2} \lambda^{c_2} \mu^{d_2}$$

$$\pi_3 = c_p u^{a_3} l^{b_3} \lambda^{c_3} \mu^{d_3}$$

(3) 应用量纲和谐原理确定上述待定指数 a_1, a_2, a_3 等。以第一个无量纲量 π_1 为例：

$$\dim h = \mathrm{M}\Theta^{-1}\mathrm{T}^{-3}, \dim l = L, \dim \lambda = \mathrm{ML}\Theta^{-1}\mathrm{T}^{-3},$$

$$\dim \mu = \mathrm{ML}^{-1}\mathrm{T}^{-1}, \dim u = \mathrm{LT}^{-1}$$

将上述结果代入 π_1 公式，并把量纲相同的项归并在一起，可得：

$$\pi_1 = \mathrm{L}^{a_1+b_1+c_1-d_1} \mathrm{M}^{c_1+d_1+1} \Theta^{-1-c_1} \mathrm{T}^{-a_1-d_1-3c_1-3}$$

由于 π_1 是无量纲的，所以等号右边各量纲的指数为零，得到关于 a_1, b_1, c_1, d_1 的线性方程组，解该线性方程组得：

$$a_1 = 0, b_1 = 1, c_1 = -1, d_1 = 0$$

可以看出无量纲量 $\pi_1 = \dfrac{hl}{\lambda}$ 即前面所述的努塞尔数 Nu，其特征长度为 l。

类似地，可以得到 $\pi_2 = \dfrac{\rho ul}{\mu}$ 即雷诺数 $Re = \dfrac{ul}{\nu}$。$\pi_3 = \dfrac{c_p\mu}{\lambda} = \dfrac{\mu/\rho}{\lambda/(\rho c_p)} = \dfrac{\nu}{\alpha}$ 为普朗特数。由于对流换热表面传热系数 h 是待求的量，即 h 是个因变量，则包括 h 的准则数就是待定准则数，其他的准则数称为自变准则数或决定准则数。因此，式(9-28)所示的有因次方程转化为式(9-29)所示的无因次方程：

$$Nu = f(Re, Pr) \tag{9-29}$$

在自然对流中，对流换热还常与 Gr 有关，因此，对于对流换热过程，一般可以整理成如下的特征数之间的关系：

$$Nu = f(Re, Gr, Pr) \tag{9-30}$$

上述方程也称为特征数关联式。以特征数关联式为基础，通过实验确定各准则数之间的依变关系，从而确定其函数关系，得到 Nu，再通过转换得到表面传热系数 h。

9.5 管内流动

内部流动与外部流动的区别主要在于流动边界层与流道壁面之间的相对关系不同：在

外部流动中，换热壁面上的流动边界层可以自由发展，不会受到流道壁面的阻碍或限制；而在内部流动中，换热壁面上边界层的发展受到流道壁面的限制，其换热规律与外部流动有明显的区别。本节介绍内部流动中常见的管内流动。

9.5.1 管内强制对流流动与换热的特点

(1) 入口段与充分发展段

流体进入管内后，流动边界层有一个从零开始增长直到汇合于管子中心线的过程。类似地，当流体与管壁之间有热交换时，管子壁面上的热边界层也有一个从零开始增长直至汇合于管子中心线的过程。流动边界层及热边界层汇合于管子中心线后的管段称为充分发展段，此后换热强度将保持不变。而从进口到充分发展段之间的管段称为入口段。入口段的热边界层较薄，局部表面传热系数高于充分发展段，且沿主流方向逐渐降低[图 9-5(a)]。如果边界层内出现湍流，如图 9-5(b)所示，则因湍流的扰动作用，表面传热系数会有所提高，之后再趋向一个定值。实验研究表明，层流入口段长度由式(9-31)确定：

$$\frac{l}{d} \approx 0.05RePr \tag{9-31}$$

而湍流时，只要 $l/d > 60$，则平均表面传热系数就不受入口段的影响。工程上常利用入口段换热效果好这一特点来进行强化传热。

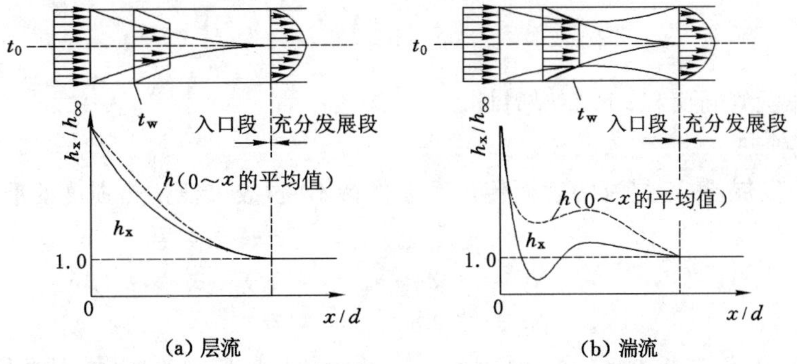

图 9-5 管内流动的特点及表面传热系数的沿程变化

(2) 均匀热流与均匀壁温下流体平均温度及流体与壁面的平均温差

实际的工程传热情况是多种多样的，为便于研究与应用，从复杂情况中抽象出两类典型的边界条件：① 均匀热流，如采用均匀缠绕的电热丝来加热壁面时，可认为壁面为均匀热流密度边界；② 均匀壁温，如采用蒸汽凝结来加热时或者液体沸腾来冷却时，壁面温度可以认为是均匀的。

图 9-6 给出了上述两种边界条件下沿主流方向流体截面平均温度 $t_f(x)$ 及壁面温度 $t_w(x)$ 的变化情况。均匀热流时，入口段流体与壁面的温差沿程增大，而到充分发展段后则保持不变。均匀壁温时，入口段流体与壁面的温差沿程减小，而到充分发展段后也基本保持不变。

9.5.2 管内强制对流换热的准则关联式

(1) 湍流换热

由第 4 节量纲分析得到 $Nu = f(Re, Pr)$，一般地，此函数是指数形式，可写成 $Nu =$

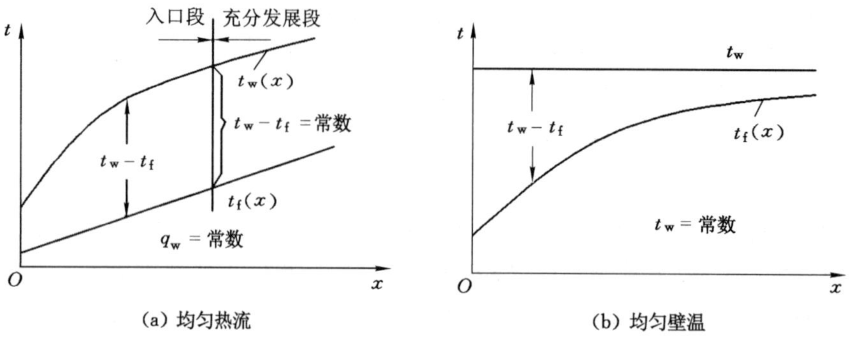

图 9-6 均匀热流和均匀壁温下流体截面平均温度与壁面温度的沿程变化

$CRe^n Pr^m$。湍流使用最广泛的关联式为迪图斯-贝尔特(Dittus-Boelter)公式:

$$Nu = 0.023 Re_f^{0.8} Pr_f^{0.4} (t_w > t_f) \tag{9-32}$$

$$Nu = 0.023 Re_f^{0.8} Pr_f^{0.3} (t_w < t_f) \tag{9-33}$$

此公式使用条件:对空气,温度差 < 50 ℃;对液体,温度差 < 20 ℃左右。使用参数范围为:$l/d \geqslant 60$,$Re_f > 10^4$,$Pr_f = 0.7 \sim 160$。特征温度为全管长流体平均温度 t_f,特征尺度为管内径 d。对于非圆管,采用当量直径作为特征尺度,当量直径的定义为:

$$d_e = \frac{4A_e}{P} \tag{9-34}$$

式中,A_e 为非圆管断面积;P 为湿周长。

(2) 层流换热

达到充分发展段后,层流的表面传热系数保持不变,故表征换热强度的准则数 Nu 也保持不变。

$$Nu = 4.36 \ (q_w = 常数) \tag{9-35}$$

$$Nu = 3.66 \ (t_w = 常数) \tag{9-36}$$

【例 9-1】 水流过长 $l = 4$ m,内径 $d = 50$ mm 的均匀壁温直管时,温度从 $t_{in} = 30$ ℃被加热至 $t_{out} = 70$ ℃,水在管内的流速 $u = 2$ m/s。求水在管内流动时的表面传热系数。已知水在 50 ℃时的物性参数为:$\nu = 0.556 \times 10^{-6}$ m²/s,$Pr = 3.54$,$\lambda = 0.648$ W/(m·K)。

【解】 水的平均温度为:

$$t_f = \frac{t_{in} + t_{out}}{2} = \frac{30 + 70}{2} = 50\ (℃)$$

以此为特征温度,管内水流的雷诺数为:

$$Re = \frac{ud}{\nu} = \frac{2 \times 0.05}{0.556 \times 10^{-6}} = 1.80 \times 10^5$$

又 $$l/d = 4/0.05 = 80 > 60$$

故流动处于充分发展段的旺盛湍流区。

采用式(9-32)得:

$$Nu_f = 0.023 Re_f^{0.8} Pr_f^{0.4} = 0.023 \times (1.8 \times 10^5)^{0.8} \times 3.54^{0.4} = 610.3$$

表面传热系数 h 为:

$$h = \frac{\lambda_f}{d} Nu_f = \frac{0.648}{0.05} \times 610.3 = 7.91 \times 10^3\ [\mathrm{W/(m^2 \cdot K)}]$$

9.6 管外流动

管外流动,由于换热壁面上的流动边界层与热边界层能自由发展,不会受到邻近流道壁面的影响。在边界层以外的区域,无论是速度梯度还是温度梯度都可以忽略,本节按照横掠单管和横掠管束分别介绍管外流动特点及对流换热准则关联式。

9.6.1 流体横掠单管

(1) 流体横掠单管的特点——边界层分离

所谓横掠单管,是指流体沿着垂直于管子轴线的方向流过管子表面。流体横掠单管流动除了具有边界层特征外,还要发生绕流脱体,并会产生回流、旋涡和涡束。如图 9-7 所示,当流体流过圆管时,由于流动截面的缩小,流速增加,压力降低,此后流体沿着圆管表面流动,边界层厚度逐渐增加,近壁面的流体层由于动量不大,在克服上升的压力时显得越来越困难,最终出现壁面处速度梯度为零,即 $\partial u/\partial y|_{y=0}=0$。随后产生与原来流动方向相反的回流,如图 9-7(a)所示。这一转折点称为绕流脱体起点(或称分离点)。从此起点起边界层内缘脱离壁面,如图 9-7(b)所示,故称流体脱体。脱体点的起点位置取决于 Re 。$Re<10$ 不会出现脱体。$10<Re<1.5\times10^5$ 时边界层为层流,脱体发生在 $\varphi=80°\sim85°$处。而当 $Re>1.5\times10^5$ 时,边界层在脱体前已经变为湍流,脱体的发生推后到 $\varphi=140°$处。

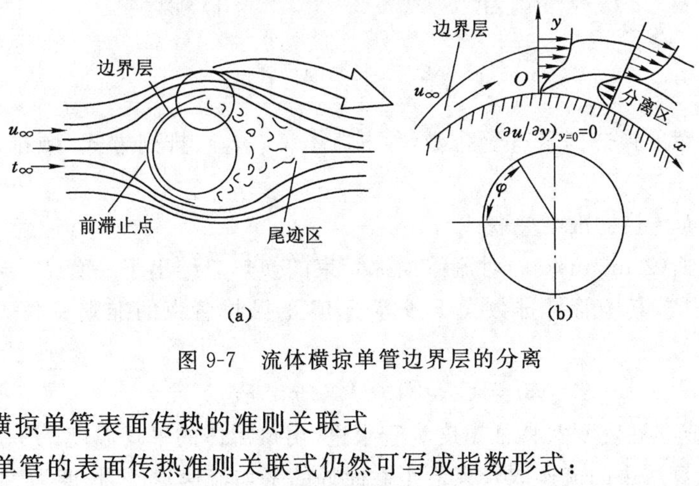

图 9-7 流体横掠单管边界层的分离

(2) 流体横掠单管表面传热的准则关联式

流体横掠单管的表面传热准则关联式仍然可写成指数形式:

$$Nu=CRe^nPr^{1/3} \tag{9-37}$$

式中,C 及 n 的值参见表 9-1;特征温度为 $(t_w+t_\infty)/2$;特征长度为管外径;特征速度为来流速度 u_∞。式(9-37)对空气的实验温度验证范围为 $t_\infty=15.5\sim980$ ℃,$t_w=21\sim1\ 046$ ℃。

表 9-1 式(9-37)中 C 和 n 值

Re	C	n
0.4～4	0.989	0.330
4～40	0.911	0.385
40～4 000	0.683	0.466
4 000～40 000	0.193	0.618
40 000～400 000	0.026 6	0.805

对于非圆形截面的管道，流体横掠换热的准则关联式也可以采用式(9-37)的形式，只是 C 和 n 的取值不同。

9.6.2 流体横掠管束

(1) 管束的排列方式及其对流动和传热的影响

横掠管束的换热在各种换热设备中最为常见。通常管子有叉排和顺排两者排列方式，如图 9-8 所示。叉排时流体在管间交替收缩和扩张的弯曲通道中流动，比顺排时在管间走廊通道的流动扰动剧烈，因而一般而言叉排时的换热比顺排时强。然而，也应注意到叉排管束的阻力损失大于顺排，对于需要冲刷清洗的管束而言，顺排有易于清洗的优点，故叉排和顺排的选择需要全面权衡。

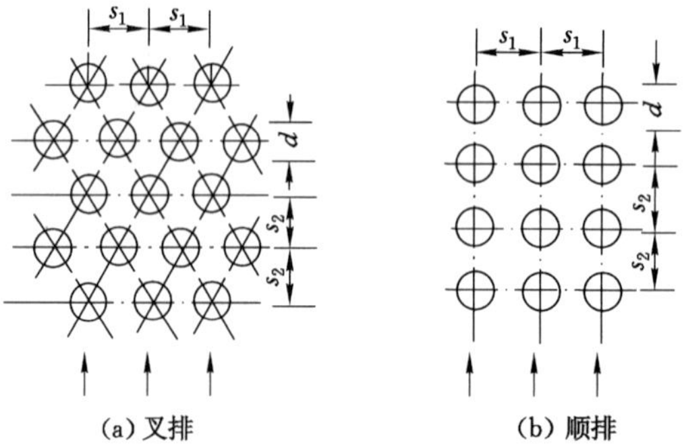

图 9-8 叉排与顺排管束

影响横掠管束的换热因素有管排数、管排定位(s_1，s_2)、排列方式(顺排、叉排)、流体物性及流体速度等。

(2) 管束表面传热的准则关联式

茹卡乌斯卡斯(Zhukauskas)对流体横掠管束的换热总结出了一套 $Pr = 0.6 \sim 500$ 范围内的准则关联式，其中管排排数大于或等于 16 的顺排管束的准则关联式写成如下指数形式：

$$Nu = C Re^n Pr_f^{0.36} (Pr_f / Pr_w)^{0.25} \tag{9-38}$$

式中，C 及 n 的值参见表 9-2；特征温度为管束进、出口流体的平均温度；Pr_w 按管束的平均壁温确定；Re 计算式中的流速取管束最小截面处的平均流速；特征长度为管子外径。

表 9-2 式(9-38)中 C 和 n 值

Re	C	n
1～100	0.90	0.40
100～1 000	0.52	0.50
1 000～20 000	0.27	0.63
20 000～200 000	0.033	0.80

对于管排排数小于 16 的管束需要进行另外的修正。

【例 9-2】 在一锅炉中，烟气横掠 4 排管组成的顺排管束。已知管外径 $d = 60$ mm，

$s_1/d = 2, s_2/d = 2$，烟气平均温度 $t_f = 600$ ℃，$t_w = 120$ ℃。烟气通道最窄处平均流速 $u = 8$ m/s。试求管束平均表面传热系数。已知 $\nu = 9.361 \times 10^{-5}$ m²/s，$Pr_f = 0.62$，$Pr_w = 0.686$，$\lambda_f = 7.42 \times 10^{-2}$ W/(m·K)，管排的修正系数 $\varepsilon = 0.91$。

【解】 管内烟气的雷诺数为：

$$Re = \frac{ud}{\nu} = \frac{8 \times 0.06}{93.61 \times 10^{-6}} = 5\ 128$$

查表 9-2 可得：　　　　　$C = 0.27, n = 0.63$

采用式(9-38)得：

$$Nu_f = 0.27Re^{0.63}Pr_f^{0.36}(Pr_f/Pr_w)^{0.25}$$
$$= 0.27 \times (5\ 128)^{0.63} \times 0.62^{0.36} \times (0.62/0.686)^{0.25} = 48.2$$

表面传热系数 h 为：

$$h = \frac{\lambda_f}{d}Nu_f = \frac{7.42 \times 10^{-2}}{0.06} \times 48.2 = 59.6\ [\mathrm{W/(m^2 \cdot K)}]$$

又已知管排的修正系数 $\varepsilon = 0.91$，故平均表面传热系数为：

$$h' = \varepsilon h = 0.91 \times 59.6 = 54.2\ [\mathrm{W/(m^2 \cdot K)}]$$

9.7　自然对流

不依靠泵或风机等外力推动，由流体自身温度场的不均匀所引起的流动称为自然对流。例如，暖气管道的散热以及不用风扇强制冷却的电器元件的散热等。不均匀的温度场会造成不均匀的密度场，由此产生的浮升力成为运动的动力。一般地，自然对流的传热热流密度低，仅 $10 \sim 10^2$ W/m² 的量级，但由于这种传热方式固有的特点——安全、经济、无噪声，仍然被广泛应用于多种工业技术中。功率密度较低的电子器件的冷却，如一般家用冰箱冷冻室、冷藏室中的气流流动等都是自然对流。按照周围空间大小的不同，自然对流可以分为大空间自然对流和有限空间自然对流。

9.7.1　大空间自然对流

(1) 竖直热壁自然对流换热的特点

现以大空间内一块竖直的温度分布均匀的热壁面附近形成的自然对流来分析。如图 9-9(a)所示，薄层内的速度分布呈现两头小中间大的特点。贴壁处，由于黏性作用速度为零，在薄壁外层由于温度不均匀作用消失，速度也等于零，故在接近热壁的中间处速度有一个峰值。热壁面附近自然对流引起的温度分布一般仅发生在靠近换热壁面的薄层内。在贴壁处，流体温度等于壁面温度 t_w，在离开壁面的方向上逐步降低，直至降到周围环境温度 t_∞，如图 9-9(b)所示。

自然对流同样有层流和湍流之分。竖直热壁的自然对流只要壁面足够高，在壁面下部流动刚开始形成时为层流，当达到某一高度时，上部流动会转变为湍流。在壁面下部往上，薄层厚度增加，热阻增大，局部表面传热系数会随高度增加而减小；进入湍流时，表面传热系数会有所增加。但是研究表明，剧烈湍流时局部表面传热系数几乎是常量，保持不变。

(2) 均匀壁温边界条件下大空间自然对流的准则关联式

自然对流的一个重要准则数为格拉晓夫数 $Gr = g\beta\Delta t l^3/\nu^2$，它在自然对流中的作用与

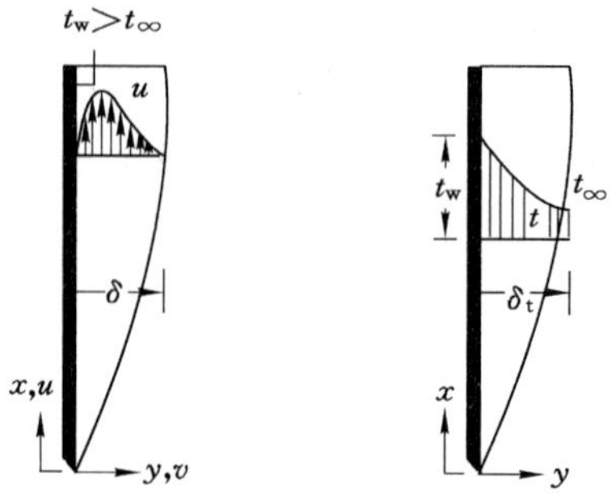

(a) 速度分布　　(b) 温度分布

图 9-9　热竖壁附近流体中的速度与温度分布示意图

雷诺数在强迫对流中的作用相当。Gr 为流体所受的浮升力与黏性力的比值，Gr 越大，表明浮升力作用越强。一般地，自然对流传热的准则关联式可以写成：

$$Nu = C(GrPr)^n \tag{9-39}$$

式中，C 及 n 的值参见表 9-3；特征温度采用边界层的算术平均温度 $t_m = (t_w + t_\infty)/2$；Gr 中的 Δt 为 t_w 与 t_∞ 之差，体积膨胀系数 $\beta = 1/T$。

表 9-3　式(9-39)中 C 和 n 值

加热表面形状	流态	Gr	C	n
竖平板及竖圆柱	层流	$1.43\times10^4 \sim 3\times10^9$	0.59	0.25
	过渡流	$3\times10^9 \sim 2\times10^{10}$	0.029 2	0.39
	湍流	$>2\times10^{10}$	0.11	0.33
横圆柱	层流	$1.43\times10^4 \sim 5.76\times10^8$	0.48	0.25
	过渡流	$5.76\times10^8 \sim 4.65\times10^9$	0.016 5	0.42
	湍流	$>4.65\times10^9$	0.11	0.33

9.7.2　有限空间自然对流

当自然对流发生在有限空间时，流体运动受到腔体的限制，流体的加热和冷却在腔体内同时进行，腔体的壁面存在高温与低温两部分，设温度分别为 t_h 和 t_c，如图 9-10 所示。此时 Gr 中的 Δt 为 $t_h - t_c$，特征温度为 $(t_h + t_c)/2$；特征尺度取冷、热两个表面的距离 δ。寒冷地区双层玻璃窗可认为是竖直夹层，水平放置的太阳能集热器则为水平夹层。对于竖直夹层，当 $Gr \leqslant 2\ 860$ 时，对于水平夹层，当 $Gr \leqslant 2\ 430$ 时夹层内的传热主要依靠导热。当 Gr 超过上述数值时，夹层内开始形成自然对流，并且随着 Gr 的增大，自然对流越来越强烈，当 Gr 达到一定数值时会出现从层流向湍流的过渡。

空气在夹层内的自然对流的准则关联式如下。

竖直夹层：

$$Nu = 0.197\,(GrPr)^{1/4}\,(H/\delta)^{-1/9},\ 8.6\times10^3 \leqslant Gr \leqslant 2.9\times10^5 \tag{9-40}$$

$$Nu = 0.073\,(GrPr)^{1/3}\,(H/\delta)^{-1/9},\ 2.9\times10^5 < Gr \leqslant 1.6\times10^7 \tag{9-41}$$

式(9-41)的实验范围为 $11 \leqslant \dfrac{H}{\delta} \leqslant 42$。

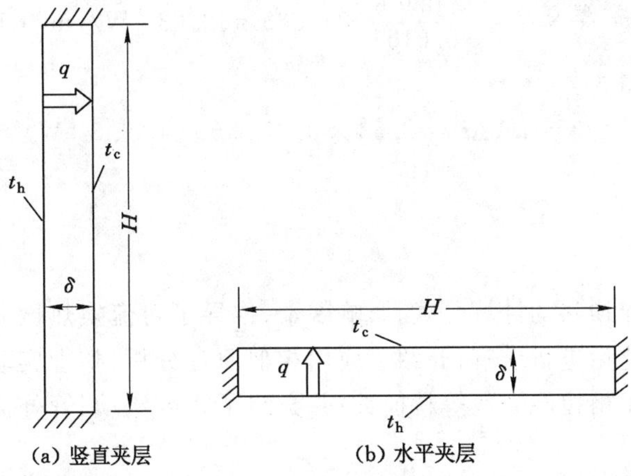

图 9-10　封闭空腔

水平夹层(底面向上散热)：

$$Nu = 0.212\,(GrPr)^{1/4},1.0 \times 10^4 \leqslant Gr \leqslant 4.6 \times 10^5 \tag{9-42}$$

$$Nu = 0.061\,(GrPr)^{1/3},Gr > 4.6 \times 10^5 \tag{9-43}$$

9.7.3　混合对流简介

在对流换热中，有时需要既考虑强制对流又考虑自然对流，这种对流称为混合对流。混合对流中强制对流受 Re 控制，自然对流受 Gr 控制。一般地，采用 Gr/Re^2 来判断自然对流的影响程度。当 $Gr/Re^2 \geqslant 0.01$ 时自然对流的影响不能忽略，而当 $Gr/Re^2 \geqslant 10$ 时强制对流的影响相对自然对流可以忽略不计。当 $0.01 < Gr/Re^2 < 10$ 时为混合对流，此时两种对流传热的作用都应加以考虑。

【例 9-3】　一个竖直封闭空腔夹层，两壁是边长 $H = 0.5$ m 的方形壁，两壁间距 $\delta = 15$ mm。两壁温度分别为 100 ℃和 40 ℃。试计算通过此空腔夹层的自然对流传热量。已知 70 ℃时空气的物性参数为：$\nu = 2.002 \times 10^{-5}$ m²/s，$\rho = 1.029$ kg/m³，$\lambda_f = 0.029$ W/(m·K)，$Pr = 0.694$。

【解】　特征温度为两壁面的平均温度：

$$t_m = \frac{t_{w1} + t_{w2}}{2} = \frac{100 + 40}{2} = 70\ (℃)$$

体积膨胀系数：

$$\beta = \frac{1}{t_m} = \frac{1}{343} = 2.915 \times 10^{-3}\,(\mathrm{K}^{-1})$$

计算 Gr：

$$Gr = \frac{g\beta\Delta t\delta^3}{\nu^2} = \frac{9.8 \times 2.915 \times 10^{-3} \times 60 \times 0.015^3}{(2.002 \times 10^{-5})^2} = 1.444 \times 10^4$$

又 $H/\delta = 0.5/0.015 = 33.3$，故可按式(9-40)进行计算，即

$$\begin{aligned} Nu_f &= 0.197(GrPr)^{1/4}(H/\delta)^{-1/9} \\ &= 0.197 \times (1.444 \times 10^4 \times 0.694)^{1/4} \times (33.3)^{-1/9} = 1.335 \end{aligned}$$

表面传热系数 h 为：

$$h = \frac{\lambda_f}{\delta} Nu_f = \frac{0.029\,6}{0.015} \times 1.335 = 2.63\ [\mathrm{W/(m^2 \cdot K)}]$$

自由对流传热量为：

$$\Phi = hA\Delta t = 2.63 \times 0.25 \times 60 = 39.5\ (\mathrm{W})$$

本 章 小 结

本章讨论了对流换热的计算公式、影响因素；推导了对流换热微分方程组，并应用边界层理论对方程组简化后进行求解；介绍了应用相似原理分析、整理实验数据的方法；应用无量纲分析方法推导了格拉晓夫数等特征数；并介绍了管内流动、管外流动、自然对流换热计算方法。

思考与练习题

（1）何谓对流换热表面传热系数？写出其定义式并说明其物理意义。

（2）用实例简要说明对流换热的主要影响因素。

（3）对流换热问题完整的数学描述应包括什么内容？既然大多数实际对流换热问题尚无法求得精确解，那么建立对流换热问题的数学描述有什么意义？

（4）速度边界层和热边界层有何区别和联系？

（5）简述边界层理论的基本内容。边界层理论对求解对流换热问题有何意义？

（6）什么叫作两个同类的物理现象相似？相似的物理现象有什么共性？怎么才能做到两个物理现象相似？

（7）简述相似理论的主要内容。相似理论对解决对流换热问题有何指导意义？

（8）分别写出努塞尔数 Nu、雷诺数 Re、普朗特数 Pr、格拉晓夫数 Gr 的表达式，并说明它们的物理意义。努塞尔数 Nu 和毕渥数 Bi 的表达式的形式完全相同，二者有何区别？

（9）外掠单管与管内流动这两个流动现象在本质上有什么不同？

（10）说明充分发展管内对流传热这一概念的含义。

（11）说明大空间自然对流与有限空间自然对流的区别，这与强制对流的外部流动和内部流动有什么异同？

（12）一流体流过平壁位置 x 处的温度分布为 $t(y) = a_0 + a_1 y + a_2 y^2$，式中，$a_0$，$a_1$ 和 a_2 是常数。已知流体与壁面的温差为 Δt，试求局部表面传热系数 h_x 的表达式。

（13）用平均温度为 50 ℃的空气来模拟平均温度为 400 ℃的烟气管内流动的强制对流换热，烟气流动速度在 10～15 m/s 范围内变化，模型采用与实际情况相同的管径。试问模型中空气的流速应在多大范围内变化？已知 400 ℃烟气的运动黏度 $\nu_s = 6.038 \times 10^{-5}$ $\mathrm{m^2/s}$，50 ℃空气的运动黏度 $\nu_a = 1.795 \times 10^{-5}\ \mathrm{m^2/s}$。

（14）水流过一直圆管，从 25.3 ℃被加热到 34.7 ℃。已知管长度为 5 m，管内直径为 20 mm，管壁面温度恒定，水在管内的平均流速为 2 m/s。试求水与圆管壁面间的表面传热系数。已知水在 30 ℃时的物性参数为：$\nu = 8.05 \times 10^{-7}\ \mathrm{m^2/s}$，$Pr = 5.42$，$\lambda = 0.618\ \mathrm{W/(m \cdot K)}$。

（15）在一摩托车引擎的壳体上有一条高 2 cm、长 12 cm 的散热片（其长度方向与车前进方向平行)，其表面温度为 150 ℃。若摩托车在 20 ℃的环境中逆风前进，车速为 30 km/h，风速为 2 m/s，计算此时散热片的散热量？已知空气在 85 ℃时的物性参数为：$\nu = 2.16 \times 10^{-5}$ m^2/s，$Pr = 0.691$，$\lambda = 0.0309$ W/(m·K)。流体沿壁面流动时对流换热的实验关联式为：$Nu = 0.664Re^{1/2}Pr^{1/3}$($Re \leqslant 5 \times 10^5$ 时)，$Nu = (0.037Re^{4/5} - 871)Pr^{1/3}$($Re > 5 \times 10^5$ 时)。

(16) 对有限空间的自然对流换热，有人经过计算得出其 $Nu = 0.5$。试利用所学过的传热学知识判别这个结果的正确性。

Chapter 9 Convection Heat Transfer

The convection heat transfer is describe to energy transfer between a surface and a fluid moving over the surface when they have different temperature, it includes energy transfer by both the bulk fluid motion (advection) and the random motion of fluid molecules (conduction or diffusion), so it is controlled by both the conduction heat transfer law and the law of fluid flow.

This chapter introduces the Newton′s law of cooling, convective heat differential equations and factors of convective heat. Secondly, it introduces the boundary layer theory, and according to the conservation of mass, conservation of energy and momentum conservation to get a set of differential equations of convective heat and simplified set of differential equations boundary. Then describe the similarity principle and dimensional analysis of theory and application of theory to obtain guidelines correlation. Finally, a typical tube flow tube flow and natural convection, the application of the test criteria correlation be solved in order to obtain the surface convective heat transfer coefficient.

9.1 Convection Heat Transfer Overview

9.1.1 Convection Mechanism

Convection heat transfer model composed by two mechanisms. In addition to the energy transmitted by the random molecular motion (diffusion) of the outer leads, the overall or macroscopic motion of the fluid is also transferred energy. This fluid motion has gathered together a large number of molecules in motion at any moment. When there is a temperature gradient, this movement will have the function about heat transfer. Due to the aggregation of a large number of molecules has random motion, so the total heat transfer is superimposed on the random motion of molecules and energy transfer caused by fluid overall movement. In Engineering, the project is interested in heat transfer between the movement of fluid and solid interface at different temperatures. At the interface between the fluid and the solid surface of the fluid velocity is zero, the random motion of molecules (diffusion) play a major role, i. e. the heat transfer by conduction. Subsequently, the fluid of solid surface adherent heat transfer to the vicinity through the fluid (usually called the boundary layer), the boundary layer thickness increases, eventually pass outside the boundary layer of fluid.

9.1.2 Newton's Law of Cooling

The basic formula of convective heat transfer is Newton's law of cooling, for unit area:

$$q = h\Delta t \tag{9-1}$$

For the area of contact surface A:

$$\Phi = hA\Delta t_m \tag{9-2}$$

Where, q is the convective heat flux, Φ is the heat transfer rate, h is the convection heat transfer coefficient, Δt and Δt_m is the temperature difference or average temperature difference between solid surface and fluid surface, respectively.

$$\Phi = \frac{\Delta t_m}{1/(hA)}$$

It is known by correlation with the Ohm's law, $1/(hA)$ is the thermal resistance of heat convective transfer.

Newton's law of cooling is only present the heat transfer coefficient (h) of convection heat transfer surface. It has not presented the inner link between the surface heat transfer coefficient and the related physical quantities. And reveal the inner link is the main object of the convection heat transfer study.

9.1.3 The Differential Equation of Convection Heat Transfer

Due to fluid flow through the wall, and the fluid velocity attach to the wall is zero, the heat transfer entirely depends on the heat conduction, so the partial heat flux density of the any transverse location of the wall can be obtained by the Fourier law.

$$q_x = -\lambda\left(\frac{\partial t}{\partial y}\right)_{y=0} \tag{9-3}$$

We can get the local surface heat transfer coefficient in accordance with the Newton cooling formula $q_x = h\Delta t$:

$$h_x = -\frac{\lambda}{\Delta t}\left(\frac{\partial t}{\partial y}\right)_{y=0} \tag{9-4}$$

It has linked the surface heat transfer coefficient with the fluid's temperature gradient, and it calls the differential equation of heat transfer. The solution about the temperature gradient requires combining the continuity equation, momentum equation and energy equation. For a specific convection heat transfer problem, we are more interested in the average convective heat transfer coefficient on the surface, and the define of the average surface heat transfer coefficient is:

$$\bar{h} = \frac{1}{L}\int_0^L h_x \mathrm{d}x \tag{9-5}$$

At present, the surface heat transfer coefficient which is mainly calculated by experiment method to obtain the relevance type of characteristic number. And the experimental method is based on the similarity principle and dimensional analysis.

9.1.4 The Influence Factors of Convective Heat Transfer

There is many factors affecting the convection heat transfer, it is because the convection heat transfer and fluid flow are always together, at the same time it is linked with the solid wall shape, size and so on, several main aspects are as follows:

(1) Causes of flow generated: according to the causes of flow, the convection heat transfer includes the forced convection and the natural convection two categories. The former is caused by pump, blower or other pressure difference, while the latter is usually caused by fluid density difference. Since the different driving force of two flows, their heat transfer rules are also different.

(2) The state of fluid flow: no matter how the flow causes, there are two different flow state in fluid flow—the laminar and turbulent flow. In the laminar flow, fluid micelles flow as a regular stratified flow along the main direction. While in the turbulent flow, intense mixing occurs between the various parts of the fluid, so under the same conditions, heat transfer of laminar flow is weaker than the turbulent flow.

(3) Phase change of fluid: heat exchange of convection heat transfer is caused by the change of sensible heat without phase change, while the heat transfer process with phase change(such as boiling or condensation), the release or absorption of fluid phase change latent heat often plays a dominant role, so the their heat transfer rule is different.

(4) The physical properties of the fluid: fluid properties affecting convection heat transfer process are mainly the thermal conductivity coefficient, specific heat capacity, density and viscosity. When the thermal conductivity coefficient of fluid is high, the thermal resistance of wall stickers is low, and the heat intensity is high. The volumetric heat capacity of the fluid is high when specific heat capacity and density are high, the heat-carrying capacity is high, and it enhanced the heat exchange between the fluid and the wall surface. In addition, the fluid flow boundary layer thickens when viscosity is high, and it is unfavorable for heat transfer. When analyzing the impact of physical parameters on heat transfer, we should pay attention to the comprehensive effect of properties, and cannot analyze the influence of each physical parameters on heat transfer in isolation, otherwise it will lead to a wrong conclusion.

(5) The geometric size, shape and position of the heat transfer surface: in the convection heat transfer, fluid flow along the wall. So the relative position of the wall flow geometry has great influence on the fluid, and it influent the strength of convective heat transfer. Fig. 9-1(a), (b) shows the conditions of fluid tube forced convection and fluid transverse bypass tube forced convection. Obviously, these are two kinds of different flow pattern, and heat transfer is different. Fig. 9-1(c), (d) shows two arrangement levels wall for free flow of air heat exchange: an upward hot surface, a downward hot surface. The former air flow heat up, which makes flow easily; the latter has inhibition flow because of downward hot surface, which reduces the h.

To sum up, there are many factors influencing the surface heat transfer coefficient, including forced convection and free convection heat transfer, laminar and turbulent flow, phase change and no phase transition flow, Various properties of different fluids as well as shape, size and relative position of the heat transfer surface etc. And they compose the heat transfer process of different rules. Therefore, to understand the variation of surface heat transfer coefficient, we should only arrange for analysis and test of the above circumstances, and it can obtain the formula of h reflecting the various situations. In general, surface heat transfer coefficient can be expressed as a function of the following form.

$$h = f(u, l, \rho, \mu, \lambda, c_p) \tag{9-6}$$

9.2 The Convection Boundary Layers

By convection mechanism, it shows that a clear understanding of the boundary layer phenomena is very important on the understanding of the convection heat transfer. The concept of the boundary layer is the core of the understanding of convection heat transfer when fluid flows through a solid surface. The convection boundary layer includes velocity boundary layer and thermal boundary layer.

9.2.1 Velocity Boundary Layer

To introduce the concept of a boundary layer, consider flow over the flat plate of Fig. 9-2. When fluid particles make contact with the surface, their velocity is reduced significantly relative to the fluid velocity upstream of the plate, and for most situations it is valid to assume that the particle velocity is zero at the wall. These particles then act to retard the motion of particles in the adjoining fluid layer, which act to retard the motion of particles in the next layer, and so on, until at a distance $y = \delta$ from the surface, the effect becomes negligible. This retardation of fluid motion is associated with shear stresses τ acting in planes that are parallel to the fluid velocity (Fig. 9-2). With increasing distance y from the surface, the x velocity component of the fluid u must increase until it approaches the free stream value u_∞. The subscript ∞ is used to designate conditions in the free stream outside the boundary layer.

The quantity δ is termed the boundary layer thickness, and it is typically defined as the value of y for $u = 0.99u_\infty$. The boundary layer velocity profile refers to the manner in which u varies with y through the boundary layer. Accordingly, the fluid flow is characterized by two distinct regions, a thin fluid layer (the boundary layer) in which velocity gradients and shear stresses large and a region outside the boundary layer in which velocity gradients and shear stresses are negligible. With increasing distance from the leading edge, the effects of viscosity penetrate farther into the free stream and the boundary layer grows (δ increases with x). Because it pertains to the fluid velocity, the foregoing boundary layer may be referred to more specifically as the velocity boundary layer. It develops whenever there is fluid flow over a surface, and it is of fundamental importance to problems involving convection transport.

9.2.2 Thermal Boundary Layer

As it generates velocity boundary layer when fluid flow through the surface, if the free flow of fluid is different from temperature of the surface, it will surely form the thermal boundary layer. Consider the flow on the isothermal plate (Fig. 9-3). At the leading edge, the distribution of temperature is uniform, that is $t(y) = t_\infty$. However, the fluid particles of the contact plate reach thermal equilibrium, and it is in the surface temperature of the plate. In turn, these particles and the adjacent fluid layer particle exchange energy, and generate a temperature gradient in the fluid. The fluid region where there is a temperature gradient is the thermal boundary layer, and the thickness δ_t is defined as the value of y corresponding to the temperature ratio of $[(t_s - t)/(t_s - t_\infty)] = 0.99$. With the increasing distance away from the leading edge, effects of heat gradually penetrate into the free stream; accordingly, the thermal boundary layer will be thicker.

It is easy to illustrate the relationship between the state of the boundary and the convective heat transfer coefficient. In any distance x away from the leading edge, the local surface heat flux can be gotten by Fourier's law through the fluid of $y=0$, that is:

$$q = -\lambda_f \left.\frac{\partial t}{\partial y}\right|_{y=0} \tag{9-7}$$

This expression is appropriate, because there is no fluid motion on the surface, the transfer of energy can be only by conduction. According to Newton's law of cooling, there is:

$$q = h(t_s - t_\infty) \tag{9-8}$$

The simultaneous equation of the Eq. (9-7), can be obtained:

$$h = \frac{-\lambda_f \partial t/\partial y|_{y=0}}{t_s - t_\infty} \tag{9-9}$$

Therefore, the thermal boundary layer has a strong influence on the wall temperature gradient $\partial t/\partial y|_{y=0}$, which determines the heat flux. Since $(t_s - t_\infty)$ is a constant, independent of x, and δ_t increases with the x increase, so must the temperature gradient in the boundary layer decreases with the increase of x. Accordingly, $\partial t/\partial y|_{y=0}$ value also decreases with an increase in x. Thus, q and h decreases with increasing x.

9.2.3 Significance of the Boundary Layer

Convection through the flow of any surface, the velocity boundary layer is always present and thus the surface friction and shear forces are presence. Similarly, if the temperature of the surface and the free stream is different, the thermal boundary layer will exist, so that there is the convective heat. Range speed boundary layer is $\delta(x)$, which is characterized by the presence of the velocity gradient and the shear stress. The thermal boundary layer in the range of $\delta_t(x)$, characterized in that there is a temperature gradient and heat transfer. Many practical problems of engineering exist on both velocity boundary layer and thermal boundary layer. Changes in flow velocity and almost all of the heat transfer occur in the boundary layer, therefore understand the concept of solving the boundary layer flow and heat transfer problems is very important.

9.3 Convection Transfer Differential Equations

Convection transfer differential equations include three laws of mass conservation, momentum conservation and energy conservation. The law of mass conservation as follows:

$$\frac{\partial \rho}{\partial \tau}=\frac{\partial (\rho u)}{\partial x}+\frac{\partial (\rho v)}{\partial y}=0 \tag{9-10}$$

When the fluid density unchanged, the above equation can be simplified as:

$$\frac{\partial u}{\partial x}+\frac{\partial v}{\partial y}=0 \tag{9-11}$$

The formula is also called continuity equation.

Energy conservation equation is derived as follows: assuming that the fluid system is two-dimensional system of uncompress, constant property and without inner heat source. Heat transfer fluid micro unit includes two forms of heat conduction and convection, where the convection part of the energy conversion as shown in Fig. 9-4.

Energy conservation formula without inner heat source is shown as follows:

$$\dot{E}_{\text{in}}-\dot{E}_{\text{out}}=\dot{E}_{\text{st}} \tag{9-12}$$

Heat transfer rate enter the micro unit due to mass inflow is:

$$\Phi_{\text{in}}=\rho u c_{\text{p}} t \mathrm{d}y+\rho v c_{\text{p}} t \mathrm{d}x \tag{9-13}$$

Heat transfer rate leave the micro unit due to mass outflow is:

$$\Phi_{\text{out}}=\left[\rho u c_{\text{p}} t+\frac{\partial (\rho u c_{\text{p}} t)}{\partial x}\mathrm{d}x\right]\mathrm{d}y+\left[\rho v c_{\text{p}} t+\frac{\partial (\rho v c_{\text{p}} t)}{\partial y}\mathrm{d}y\right]\mathrm{d}x \tag{9-14}$$

The net convective heat transfer rate [Eq. (9-13) minus Eq. (9-14)] flow into the micro unit is:

$$\Phi_{\text{f}}=\Phi_{\text{in}}-\Phi_{\text{out}}=-\left[\frac{\partial (\rho u c_{\text{p}} t)}{\partial x}+\frac{\partial (\rho v c_{\text{p}} t)}{\partial y}\right]\mathrm{d}x\mathrm{d}y$$

$$=-\rho c_{\text{p}}\left(u\frac{\partial t}{\partial x}+v\frac{\partial t}{\partial y}+t\frac{\partial u}{\partial x}+t\frac{\partial v}{\partial y}\right)\mathrm{d}x\mathrm{d}y \tag{9-15}$$

Substituted into the continuity equation and with the after two items has disappeared, then get:

$$\Phi_{\text{f}}=-\rho c_{\text{p}}\left(u\frac{\partial t}{\partial x}+v\frac{\partial t}{\partial y}\right)\mathrm{d}x\mathrm{d}y \tag{9-16}$$

The net heat transfer rate enter the micro unit due to conduction according Chapter 8 is:

$$\Phi_{\text{s}}=\lambda\left(\frac{\partial^2 t}{\partial x^2}+\frac{\partial^2 t}{\partial y^2}\right)\mathrm{d}x\mathrm{d}y \tag{9-17}$$

The net heat transfer rate enter the micro unit due to convection and conduction is:

$$\dot{E}_{\text{in}}-\dot{E}_{\text{out}}=\Phi_{\text{f}}+\Phi_{\text{s}}=-\rho c_{\text{p}}\left(u\frac{\partial t}{\partial x}+v\frac{\partial t}{\partial y}\right)\mathrm{d}x\mathrm{d}y+\lambda\left(\frac{\partial^2 t}{\partial x^2}+\frac{\partial^2 t}{\partial y^2}\right)\mathrm{d}x\mathrm{d}y \tag{9-18}$$

The energy change rate in the micro unit is:

$$\dot{E}_{st} = \frac{\partial(\rho c_p t)}{\partial \tau} dx dy = \rho c_p \frac{\partial t}{\partial \tau} dx dy \tag{9-19}$$

Take Eq. (9-17) and Eq. (9-18) into Eq. (9-12) and make a simplification, it can be obtained:

$$\frac{\partial t}{\partial \tau} + u\frac{\partial t}{\partial x} + v\frac{\partial t}{\partial y} = \frac{\lambda}{\rho c_p}\left(\frac{\partial^2 t}{\partial x^2} + \frac{\partial^2 t}{\partial y^2}\right) \tag{9-20}$$

That is the energy conservation equation of two-dimensional, incompressible, constant property, without inner heat source fluid system.

The momentum conservation equation can be derived from referencing fluid mechanics, here only lists the formula. Summing up the three equations, we can get the differential equations of fluid system under the conditions two-dimensional, incompressible, constant property, without inner heat source energy:

The mass conservation equation:

$$\frac{\partial u}{\partial x} + \frac{\partial v}{\partial y} = 0 \tag{9-21}$$

The momentum conservation equation:

$$\rho\left(\frac{\partial u}{\partial \tau} + u\frac{\partial u}{\partial x} + v\frac{\partial u}{\partial y}\right) = F_x - \frac{\partial p}{\partial x} + \mu\left(\frac{\partial^2 u}{\partial x^2} + \frac{\partial^2 v}{\partial y}\right) \tag{9-22}$$

$$\rho\left(\frac{\partial v}{\partial \tau} + u\frac{\partial v}{\partial x} + v\frac{\partial v}{\partial y}\right) = F_y - \frac{\partial p}{\partial y} + \mu\left(\frac{\partial^2 v}{\partial x^2} + \frac{\partial^2 v}{\partial y}\right) \tag{9-23}$$

Energy conservation equation:

$$\frac{\partial t}{\partial \tau} + u\frac{\partial t}{\partial x} + v\frac{\partial t}{\partial y} = \frac{\lambda}{\rho c_p}\left(\frac{\partial^2 t}{\partial x^2} + \frac{\partial^2 t}{\partial y^2}\right) \tag{9-24}$$

The quality transfer momentum transfer and heat transfer of the solid and fluid occurs in the basic boundary layer. In general, the two order derivative of the mainstream direction in the boundary layer can be ignored. Simplifying the above three equations, we can get the differential equations of boundary lawyer system under the conditions two-dimensional, incompressible, constant property, without inner heat source energy:

The mass conservation equation:

$$\frac{\partial u}{\partial x} + \frac{\partial v}{\partial y} = 0 \tag{9-25}$$

The mass conservation equation:

$$u\frac{\partial u}{\partial x} + v\frac{\partial u}{\partial y} = -\frac{1}{\rho}\frac{\partial p}{\partial x} + \nu\frac{\partial^2 u}{\partial x^2} \tag{9-26}$$

Energy conservation equation:

$$\frac{\partial t}{\partial \tau} + u\frac{\partial t}{\partial x} + v\frac{\partial t}{\partial y} = \alpha\left(\frac{\partial^2 t}{\partial x^2} + \frac{\partial^2 t}{\partial y^2}\right) \tag{9-27}$$

In the formula ν is kinematic viscosity, $\nu = \mu/\rho$, α is thermal diffusion coefficient, $\alpha = \lambda/\rho c_p$.

Although the differential equations are simple on the form, in generally, it is still difficult to get the analytical solution. The commonly used method is that making the partial differential equations into algebraic equations by grid division, and then get the numerical result in the known boundary condition. To obtain the temperature distribution and temperature gradient in the boundary layer by solving the differential equations, then we can calculate the convective heat transfer coefficient.

9.4 Similar Principle and Dimensional Analysis

One goal of the convective heat transfer problem is obtained convective surface heat transfer coefficient. On the basis of Section 9.1 above, a number of factors affected the surface heat transfer coefficient, as shown in Eq. (9-6), if influenced by 6 factors, according to the conventional method: each variable changes 10 times, the other 5 parameters remain unchanged, then there will be one million tests of all. Such a large number of experiments are unacceptable; moreover, direct experimental method can only get regularity relationship between individual variables, so it is difficult to grasp the essence of the phenomenon. In addition, it is not suitable for direct experiments of many other phenomena but model experiment only, and model experiment results must be the same with the actual results. So the similarity theory and dimensional analysis come into being, the first is how to reduce the number of trials, the second is to obtain a universal law of nature. This section treat the basic content of similar principles as a frame work, the characteristic equation is derived by dimensional analysis between the numbers of similar characteristics, in order to get the general rule of good universality.

9.4.1 Similar Principle

The concept of similar comes from geometry, if the corresponding edges of the graphic are proportional, and the corresponding angles are equal, then the two geometric is similar. As for the two similar physical phenomena, if the corresponding physical quantity related to the phenomenon which in the corresponding time and in the corresponding location are proportional, then the two phenomena are similar. Similarity principle is used to study the relationship between the similar physical phenomena. Note: the so-called similar phenomenon refers to the phenomenon described by differential equations with the same form and the same contents. The differential equations describing the problem of electric and thermal conductivity is resemble but not similar, the form of differential equations describing flow problems of temperature field and velocity field is same, but because of their different research content, it is not the similar problem. Between electric and thermal temperature field, temperature field and velocity field in the fluid, they are "analogy" or "resemble", not similar.

The basic content of similarity principle: if two phenomena is similar, their corresponding characteristic numbers (criterion numbers) are equal, and their single-value conditions are similar. In the convective heat, the frequently-used criterion numbers are as follows:

Nusselt number $Nu = \frac{hl}{\lambda}$: expressing the ratio of fluid thermal resistance and convection thermal resistance. The larger Nu, the stronger the convection heat transfers.

Reynolds $Re = \frac{ul}{\nu}$: expressing the ratio of inertial force and viscous force, the larger Re, the lager the inertial force, the more intense the fluid motion. Therefore, it often determines the size of the flow is laminar or turbulent flow conditions.

Prandtl number $Pr = \frac{v}{\alpha}$: expressing the ratio of kinematic viscosity and thermal diffusivity, reflecting the relative size of momentum diffusion and heat diffusion. When $Pr = 1$, the momentum diffusion is equal to the heat diffusion.

Grashof number $Gr = \frac{g\beta\Delta tl^3}{\nu^2}$: expressing the ratio of buoyancy and viscous forces of fluid. The larger Gr, the stronger buoyancy is.

For the convective heat transfer problems, there are single-value conditions:

(1) Geometry condition: refers to the heat transfer geometry surface, size, position and surface roughness.

(2) Physical condition: the type of fluid within the system and physical properties, such as density, viscosity, etc.

(3) Boundary Conditions: refers to flow rate, temperature and other physical quantities on the distribution of the studied system boundaries (such as import, export and at the wall, etc.).

(4) Initial conditions: refers to physical changes over time in the unsteady state problem. It does not require under the steady state conditions.

9.4.2 Dimensional Analysis

Dimensional analysis is a common way to get the number of features, and its theory is based on π law: a dimensional equation with the same unit among n physical parameters will be able to convert to the relationship include $n-r$ independent dimensionless parameters. Where r refers to the number of the basic dimensionless involved in the n physical parameters. Here we do not discuss their mathematical proof, but only focus on the specific application of the law.

For instance, in the forced convection heat transfer tube, factors affecting the convective heat transfer coefficient have fluid velocity u, Characteristic length l and Thermo-physical properties of the fluid including viscosity coefficient μ, Thermal conductivity λ, specific heat capacity c_p and density ρ. Therefore, the flow of convective heat transfer coefficient inside the tube can be written as a function of these parameters:

$$h = f(u, l, \mu, \lambda, c_p, \rho) \tag{9-28}$$

General steps of dimensional analysis method:

(1) Find out the basic dimension in the each physical dimension which related to this problem. We can find from Eq. (9-28), forced convection problem has seven physical variable inner tube, their dimension is composed of four basic variable dimension, namely:

time dimension T, length dimension L, quality dimension M and temperature dimension Θ. Therefore $n = 7, r = 4$. So the dimensionless parameters can be composed is $n - r = 3$. Selected 4 parameters as a basic physical variable from the 7 physical parameters, the basic physical quantities must include the dimension above basic variable. In the case of this example, the basic parameters is u, l, μ, λ.

(2) The basic variables one by one with other compositions construct the dimensionless variables. Dimensionless variable always using exponential, its index is pending. Three dimensionless variables are as follows:

$$\pi_1 = h u^{a_1} l^{b_1} \lambda^{c_1} \mu^{d_1}$$

$$\pi_2 = \rho u^{a_2} l^{b_2} \lambda^{c_2} \mu^{d_2}$$

$$\pi_3 = c_p u^{a_3} l^{b_3} \lambda^{c_3} \mu^{d_3}$$

(3) Application dimensional homogeneity determines the above pending index a_1, a_2, a_3 and so on. With the first dimensionless variable as an example,

$$\dim h = \mathrm{M\Theta^{-1}T^{-3}}, \dim l = \mathrm{L}, \dim \lambda = \mathrm{ML\Theta^{-1}T^{-3}},$$

$$\dim \mu = \mathrm{ML^{-1}T^{-1}}, \dim u = \mathrm{LT^{-1}}$$

The above results are substituted into the formula π_1, and to merge with the same dimension items, it can be obtained:

$$\pi_1 = \mathrm{L}^{a_1+b_1+c_1-d_1} \mathrm{M}^{c_1+d_1+1} \Theta^{-1-c_1} \mathrm{T}^{-a_1-d_1-3c_1-3}$$

Since π_1 is a dimensionless, so the right hand side of each dimension of the index is zero, then get a linear equation group about a_1, b_1, c_1, d_1, solve the equation group can get:

$$a_1 = 0, b_1 = 1, c_1 = -1, d_1 = 0$$

It can be see the dimensionless variable $\pi_1 = \frac{hl}{\lambda}$ is Nu which was previously described, and its characteristic length is l.

Similarly, it can be obtained that $\pi_2 = \frac{\rho u l}{\mu}$ is Reynolds number $Re = \frac{ul}{\nu}$, and $\pi_3 = \frac{c_p \mu}{\lambda} = \frac{\mu/\rho}{\lambda/(\rho c_p)} = \frac{\nu}{\alpha}$ is Prandtl number. In summary, by dimensional analysis, Eq. (9-28) with dimensional equations are transformed into a dimensionless equation formula:

$$Nu = f(Re, Pr) \tag{9-29}$$

In natural convection, convective heat transfer is often related to Gr number, therefore, for the convective heat transfer process, the general relationship among the criterion numbers can be concluded as follows:

$$Nu = f(Re, Gr, Pr) \tag{9-30}$$

The above equation is also called feature correlation. Basing on the feature correlation and experiments, it can be determined the relationship among criterion numbers, and eventually determine function relationship, get Nu number, and the convective heat transfer coefficient h is obtained by conversion.

9.5 Internal Tube Flow

The difference between internal and external liquidity flows mainly in different relative relationship boundary layer flow and the flow channel wall: in the external flow, heat transfer fluid boundary layer on the wall can be free development and will not be hindered or restricted by fluid wall. In the internal flow, the development of the boundary layer on the heat exchanger wall is restricted by flow channel wall, there are significant differences between the heat transfer law and external flows have. This section describes the common internal flow in tube flow.

9.5.1 Characteristics of Internal Tube Forced Convection Flow and Heat Transfer

(1) Entrance and fully developed section

After fluid enters the tube, the flow boundary layer grows from a scratch until have a convergence in the tube center. Similarly, when the heat exchange between the fluid and the tube wall, the thermal boundary layer on the wall of the tube also grow from a scratch until have a convergence in the tube center. The tube section which boundary layer flow and thermal boundary layer have convergence in the tube centerline called fully developed section. And then the strength of the heat exchanger remains unchanged. The tube section from imported into fully developed called entrance section. Entrance of the thermal boundary layer is thinner, the surface heat transfer coefficient is higher than that in the fully developed, and it reduced gradually along the main direction [Fig. 9-5 (a)]. If turbulence occurs in the boundary layer, as shown in Fig. 9-5(b), due to the disturbance of turbulence, the surface heat transfer coefficient will be increased, after then tends to a constant value. Experimental studies have shown that the entry length of the laminar flow is determined as follows.

$$\frac{l}{d} \approx 0.05 Re Pr \qquad (9\text{-}31)$$

And when in the turbulence, as long as $l/d > 60$, the average surface heat transfer coefficient will be not affected by the entry segment. The feature that good heat transfer effect in entrance often be used to enhance heat transfer.

(2) The average temperature of fluid and the average temperature difference between the fluid and the wall surface under the uniform heat flux and wall temperature

Heat transfer conditions in practical engineering are varied, in order to research and apply conveniently, abstracted from two typical boundary conditions in the complex situation: ① Uniform heat flux, if the use of a uniform electric wire to heat the wall, the wall can be considered as a uniform heat flux boundary. ② Uniform wall temperature, if using the steam condenses to heat or boiling liquid to cool, the wall temperature can be considered uniform.

Fig. 9-6 shows the changes of the average temperature $t_f(x)$ along the main direction of the fluid cross section under the two boundary conditions and wall temperature $t_w(x)$. When at uniform heat flux, temperature difference between the fluid and the wall of the entrance section along the path is increased, and at fully developed section remain unchanged. When at uniform wall temperature, the temperature difference between the fluid and the wall surface of the entrance section of decreasing along the way, and to fully developed section also remained basically unchanged.

9.5.2 Correlation Equations of Forced Convection Heat Transfer for Internal Tube Flow

(1) Turbulent heat transfer

By dimensional analysis at the fourth quarter we can get $Nu = f(Re, Pr)$ generally, this function is exponential form, can be written as $Nu = CRe^n Pr^m$. Turbulence uses the most widely relational Dittus-Boelter formula:

$$Nu = 0.023 Re_f^{0.8} Pr_f^{0.4} (t_w \geqslant t_f) \tag{9-32}$$

$$Nu = 0.023 Re_f^{0.8} Pr_f^{0.3} (t_w < t_f) \tag{9-33}$$

The condition of this formula: for the air, temperature difference < 50 ℃; for liquid, temperature difference < 20 ℃. Parameter range is: $l/d > 60$, $Re_f > 10^4$, $Pr_f = 0.7 \sim 160$. Characteristic temperature is average temperature for the whole length of the tube fluid t_f. Characteristic scale is tube diameter d. For non-tube, use the equivalent diameter as the characteristic scale. Equivalent diameter is defined as:

$$d_e = \frac{4A_e}{P} \tag{9-34}$$

In the formula, A_e is the non-circular tube cross-section area, P is wet perimeter.

(2) Laminar flow heat exchanger

After reaching fully developed, laminar flow heat transfer coefficient remains unchanged, it is characterized by the intensity of the heat exchanger Standards Nu also remains unchanged.

$$Nu = 4.36 \ (q_w = \text{const}) \tag{9-35}$$

$$Nu = 3.66 \ (t_w = \text{const}) \tag{9-36}$$

9.6 External Tube Flow

External tube flow, because of freedom development of the flow boundary layer and thermal boundary layer on the wall, it will not be affected by adjacent channel wall. Thus in an area outside the boundary layer, both velocity and temperature gradient or gradients can be ignored. According to across single tube and across tube bundles, this section introduces outside the tube flow characteristics and convective heat transfer criteria correlation.

9.6.1 Fluid Across Single Tube

(1) Characteristics of single tube cross flow—boundary layer separation

Single-tube cross flow is that fluid flow through the tube surface in the direction of along the vertical axis of the tube. Across single tube fluid flow not only has the characteristics of boundary layer, but also has flow separation, and will produce backflow, whirlpool and vortex beams. As shown in Fig. 9-7, when the fluid flows through the tube, since the reduction of the flow cross-section, the flow rate increases, the pressure decreases. After that fluid flow along the pipe surface, gradually increasing the boundary layer thickness. Due to Fluid layer near the wall momentum is not strong, and it is more difficult in overcoming the upward pressure. Eventually the velocity gradient is zero at the wall, that is $\partial u/\partial y|_{y=0}=0$. Then generate backflow which is opposite to the original flow direction, as shown in Fig. 9-7(a). The turning point called flow separation starting point (or separation point). From this starting point within the boundary layer from the edge of the wall, as shown in Fig. 9-7(b), so that is fluid separation. Separation point of starting position depends on the Reynolds number. It will not appear separation when $Re<10$. Boundary layer is laminar flow and separation occurred in $\varphi=80°\sim85°$ when $10<Re<1.5\times10^5$. When $Re>1.5\times10^5$ boundary layer has become turbulence before the separation, separation occurred in $\varphi=140°$.

(2) Correlation equations of single tube cross flow heat transfer

Correlation equation of across the circular tube surface heat transfer can still be written in exponential form:

$$Nu = CRe^n Pr^{1/3} \tag{9-37}$$

Where, C and n values are listed in Table 9-1; characteristic temperature is $(t_w+t_\infty)/2$; characteristic length is tube diameter; characteristic velocity of flow velocity u_∞; validation range that the type on test temperature of air is $t_\infty=15.5\sim980$ ℃, $t_w=21\sim1\,046$ ℃.

9.6.2 Fluid Flow across Tube Bundles

(1) The arrangement mode of the tube bundles and its effect on the flow and heat transfer

Heat transfer outside the tube bundles in heat exchange equipment is common. Usually the tubes have staggered and shun row two arrangements, as shown in Fig. 9-8. When in staggered, fluid flows in the curved path of alternating contraction and expansion between tubes, and the flow disturbances are severer than pipe corridor in aligned arrangement, so generally, staggered heat transfer is stronger than that in aligned arrangement. However, it should also be noted that resistance loss of staggered tube bundle is bigger than that in aligned arrangement tube bundle. For the need of cleaning the tube bundle, aligned arrangement has the advantage of easy cleaning, so the selection of staggered and aligned arrangement requires a comprehensive balance.

Factors of effecting the heat exchanger for outside the pipe bundles have the tube row number, tube row location (s_1, s_2), arrangement mode (staggered arrangement, aligned arrangement), fluid properties and fluid velocity etc.

(2) Correlation equations of tube bundles surface heat transfer

Zhukauskas summed up a set of guidelines correlations in the range $Pr = 0.6 \sim 500$ about heat transfer of fluid sweep outside the pipe bundles, in which correlations of order bundle of the tube row number greater than or equal to 16 written the exponential form as follows:

$$Nu = CRe^n Pr_f^{0.36} (Pr_f / Pr_w)^{0.25} \tag{9-38}$$

In the formula: C and n values are listed in Table 9-2; The characteristic temperature is the average temperature of the bundle inlet and outlet fluid; Pr_w is determined by an average wall temperature of tube bundle; Fluid velocity in Re number is the average fluid velocity of the minimum cross section of tube bundle; The characteristic length is the external diameter of the pipe.

9.7 Free Convection

The flow caused by the non-uniform fluid temperature field and not relied on the pump or fan of external force is called free convection. For example, the cooling of the heating pipes and electrical components without cooling fan, etc. Non-uniform temperature field will cause uneven density field, then the buoyancy resulted from it become motive power. In general, heat flux density of free convection is low, only $10 \sim 10^2$ W/m^2. However the heat transfer has the inherent characteristics—security, economy, no noise, and still widely used in many industrial technologies. The cooling of the lower power density electronic device, like the airflow in household freezer, freezer, etc., they are all free convection. According to the different sizes of the surrounding space, natural convection can be divided into a large space and limited space natural convection.

9.7.1 Natural Convection in Large Space

(1) The characteristics of natural convection heat transfer in vertical hot wall

Now analyze with a natural convection formed in the vicinity in a hot vertical temperature distribution over a large space of the wall surface. As shown in Fig. 9-9(a), the velocity distribution within a thin layer presents the features of two sides' small and middle big. In adherent place, the speed is zero due to the viscous effects and in the outer of thin-walled, due to the uneven temperature effect is disappeared, so the speed is equal to zero. Therefore, in the middle near the hot wall has a peak speed. The temperature distribution caused by natural convection near the hot wall usually occurs only in the thin layer near the heat exchanger wall. In adherent place, the wall temperature equal to the temperature of the fluid t_w, gradually reduced in a direction away from the wall until the it drops to the ambient temperature t_∞, as shown in Fig. 9-9(b).

Natural convection also has the distinction of laminar and turbulent flow equally. If the vertical hot wall free convection is high enough, the lower flow at the beginning of a laminar flow under the wall, when it reaches a certain height, the upper part of the flow becomes a turbulent flow. In the lower part of the wall up, the layer thickness increases, the thermal resistance increases, the local surface heat transfer coefficient will decrease with height increase. When entering turbulence, heat transfer coefficient will be increased. But studies show that in strong turbulence the local surface heat transfer coefficient is almost constant, unchanged.

(2) Large space free convection criteria correlation in uniform wall temperature boundary conditions

Grashof number is an important criterion number in Natural convection: $Gr = g\beta\Delta tl^3/\nu^2$. Its role in the natural convection equals to the role of the Reynolds number in forced convection. Gr is the ratio Buoyancy and viscous forces. The large Gr, the stronger buoyancy effect will be. In general, natural convection heat related criteria can be written:

$$Nu = C\,(GrPr)^n \tag{9-39}$$

Where, Value C and n in Table 9-3; characteristic temperature using arithmetic average temperature of boundary layer $t_m = (t_w + t_\infty)/2$; Gr is the difference between t_w and t_∞; Volume expansion coefficient is $\beta = 1/T$.

9.7.2 Limited Space Natural Convection

When natural convection occurs in the limited space, fluid flow is restricted by the cavity, and the heating and cooling of fluid get on in the cavity in the same time. Cavity wall has high temperature and low temperature in two parts, and temperatures were set to t_h and t_c, as shown in Fig. 9-10. Now, in Gr number, $\Delta t = t_h - t_c$, characteristics of temperature is $(t_h + t_c)/2$, the characteristic scale take the distance δ between cold and hot surface. Cold area double glazing can be thought of as a vertical interlayer, horizontal solar heat collector is horizontal interlayer. For a vertical interlayer, $Gr \leqslant 2\ 860$ and for horizontal interlayer, $Gr \leqslant 2\ 430$, the heat transfer in the interlayer mainly depends on the thermal conductivity. When Gr exceeds the above value, the interlayer began to form natural convection, and with the increase of Gr number, the natural convection is more and more intense, the transition from laminar to turbulent will emerge when the Gr number reaches a certain value.

Air in the dissection natural convection correlation criteria are as follows:

A vertical interlayer:

$$Nu = 0.197(GrPr)^{1/4}\,(H/\delta)^{-1/9},\ 8.6\times10^3 \leqslant Gr \leqslant 2.9\times10^5 \tag{9-40}$$

$$Nu = 0.073(GrPr)^{1/3}\,(H/\delta)^{-1/9},\ 2.9\times10^5 < Gr \leqslant 1.6\times10^7 \tag{9-41}$$

The above formula test range is $11 \leqslant \dfrac{H}{\delta} \leqslant 42$.

Horizontal dissection(underside up the heat):

$$Nu = 0.212(GrPr)^{1/4},\ 1.0\times10^4 \leqslant Gr \leqslant 4.6\times10^5 \tag{9-42}$$

$$Nu = 0.061(GrPr)^{1/3},\ Gr > 4.6\times10^5 \tag{9-43}$$

9.7.3 Mixed Convection Introduction

In convective heat transfer, and sometimes need to consider both forced convection and natural convection, this convection called mixed convection. In mixed convection the forced convection was controlled by Re, natural convection was controlled by Gr. In General Gr/Re^2 was used to determine the degree of influence of natural convection. When $Gr/Re^2 \geqslant 0.01$, the effect of natural convection can' t be ignored, but when $Gr/Re^2 \geqslant 10$, influence of forced convection is negligible compared with the natural convection. When $0.01 < Gr/Re^2 < 10$, it is the mixed convection, where the two effects should be considered.

第10章 辐射换热

本章导学

热辐射是热量传递的基本方式之一，以热辐射方式进行的热量交换称为辐射换热。辐射换热在热能动力工程、核能工程、冶金、化工、航天、太阳能利用、干燥技术，以及日常生活中的加热、供暖等方面具有非常广泛的应用。本章主要从宏观角度介绍热辐射的基本概念、基本定律以及辐射换热的计算方法。

学习目标

正确理解热辐射的本质，掌握有关辐射的概念：黑体、灰体、吸收比、发射率、有效辐射等。掌握热辐射的主要基本定律：普朗克定律、斯特藩-玻耳兹曼定律和基尔霍夫定律。理解角系数的物理意义和特征，能用代数法和图线法计算工程上常见的角系数。掌握网络法计算两个以上灰体构成的封闭系统的辐射换热。掌握空间辐射热阻和表面辐射热阻的概念。

重点与难点

(1) 热辐射与辐射换热。

(2) 黑体与黑体辐射。

(3) 基尔霍夫定律与灰体。

(4) 灰体表面间的辐射换热。

10.1 热辐射的基本概念

10.1.1 辐射的本质与特点

物质是由分子、原子、电子等粒子组成的，当原子内部的电子受激和振动时，产生交替变化的电场和磁场，发出电磁波向空间传播，物体通过电磁波来传递能量的方式称为辐射。发射辐射能是各类物质的固有特性。物体会因各种原因发出辐射能，其中因自身温度或热运动的原因而激发的电磁波传播的现象称为热辐射。只要物体的温度高于“绝对零度”(即0 K)，物体总是不断地把热能变为辐射能，向外发出热辐射。同时，物体亦不断地吸收周围物体投射到它上面的热辐射，并把吸收的辐射能重新转变成热能。辐射与吸收过程的综合结果就造成了以辐射方式进行的物体间的热量传递——辐射传热，也常称为辐射换热。辐射换热就是指物体之间相互辐射和吸收的总效果。当物体与环境处于热平衡时，辐射传热量等于零，但这是动态平衡，辐射与吸收过程仍在不停地进行。

不同频率的电磁波是产生方式不同造成的。各种热辐射也属于电磁波，具有一般辐射现象的共性，例如，各种电磁波都以光速在空间传播，这是电磁波辐射的共性，热辐射亦不例外。电磁波的性质取决于波长或频率，在热辐射分析中，通常用波长来描述电磁波。从理论上说，物体热辐射的电磁波波长可以包括整个波谱，即波长从零到无穷大。然而，在工业上所遇到的温度范围内，即 2 000 K 以下，有实际意义的热辐射波长位于 0.8～100 μm 之间，且大部分能量位于红外线区段的 0.76～20 μm 范围内。红外线又有近红外和远红外之分，大体上以 25 μm 为界限，波长在 25 μm 以下的红外线称为近红外线，25 μm 以上的红外线称为远红外线。20 世纪 70 年代初期发展起来的远红外加热技术，就是利用远红外辐射元件发射出的以远红外线为主的电磁波对物料进行加热的。微波炉就是利用远红外线来加热物体的。远红外线可以穿过塑料、玻璃及陶瓷制品，但却会被像水那样具有极性分子的物体吸收，在物体内部产生内热源，从而使物体比较均匀地得到加热。各类食品中的主要成分是水，因而远红外加热是一种比较理想的加热手段。如果把太阳辐射包括在内，热辐射的波长区段可放宽为 0.1～100 μm。本章下面所讨论的内容专指由于热的原因而产生的波长主要位于 0.1～100 μm 间的电磁辐射，常称为热射线，因为这一波长区段内的电磁波最容易被物体吸收并转化为热能。

与导热、对流相比，热辐射这种传递能量方式有两个特点：① 热辐射的能量传递不需要其他介质，而且在真空中传递的效率最高；② 在物体发射与吸收辐射能量的过程中不仅产生能量的转移，而且还伴随着能量形式的转换，即发射时从热能转换为辐射能，而吸收时又从辐射能转换为热能。这两个特点都是由辐射是电磁波的传递这个基本事实所决定的。

10.1.2　吸收、反射与透射

与可见光的情况一样，当热辐射能投射到实际物体表面上时，将有一部分被物体表面反射，有一部分被物体吸收，其余部分透过物体，如图 10-1 所示。

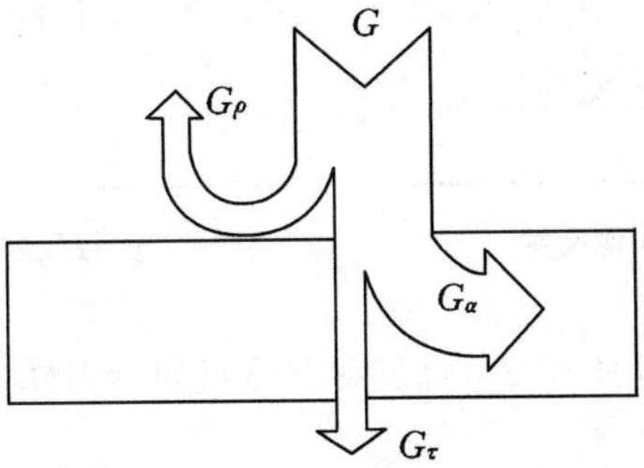

图 10-1　物体对热辐射的吸收、反射与透射示意图

单位时间内投射到单位面积物体表面上的全波长范围内的辐射能称为投入辐射，用 G 表示，单位为 W/m^2。其中，被物体吸收、反射和透射的部分分别为 G_α，G_ρ 和 G_τ，则 G_α，G_ρ 和 G_τ 在投入辐射 G 中所占的份额分别为：

$$\alpha = \frac{G_\alpha}{G},\rho = \frac{G_\rho}{G},\tau = \frac{G_\tau}{G}$$

式中，α、ρ 和 τ 分别称为物体对投射辐射能的吸收比、反射比与透射比。根据能量守恒，$G_\alpha + G_\rho + G_\tau = G$，于是有：

$$\alpha + \rho + \tau = 1 \tag{10-1}$$

如果投入辐射是某一波长 λ 的辐射能 G_λ，其中被物体吸收、反射和透射的部分分别为 $G_{\lambda\alpha}$、$G_{\lambda\rho}$ 和 $G_{\lambda\tau}$，则其所占的份额分别为：

$$\alpha_\lambda = \frac{G_{\lambda\alpha}}{G_\lambda}, \rho_\lambda = \frac{G_{\lambda\rho}}{G_\lambda}, \tau_\lambda = \frac{G_{\lambda\tau}}{G_\lambda}$$

式中，α_λ、ρ_λ 和 τ_λ 分别称为物体对该波长辐射能（即单色辐射）的光谱吸收比、光谱反射比和光谱透射比，或称为单色吸收率、单色反射率和单色透射率。

与式(10-1)类似，有：

$$\alpha_\lambda + \rho_\lambda + \tau_\lambda = 1 \tag{10-2}$$

α_λ、ρ_λ、τ_λ 属于物体的辐射特性，取决于物体的种类、温度和表面状况，一般是波长 λ 的函数。但全波长特性 α、ρ 和 τ 不仅取决于物体的性质，还与投入辐射的波长分布有关。

实际上，当热辐射投射到固体或液体表面时，一部分被反射，其余部分在很薄的表面层内就被完全吸收了。对于金属，这一表面层的厚度只有 1 μm 的量级；对于绝大多数非金属材料，这一表面层的厚度也小于 1 mm。因此，对于固体和液体，可以认为对热辐射的透射比为零，式(10-1) 简化为：$\alpha + \rho = 1$。

物体表面对热辐射的反射有两种现象：镜反射与漫反射。镜反射的特点是反射角等于入射角，如图 10-2(a)所示。漫反射时，被反射的辐射能在物体表面上方空间的各个方向均匀分布，如图 10-2(b)所示。物体表面对热辐射的反射情况取决于物体表面的粗糙程度和投入辐射的波长。当物体表面粗糙尺度小于投入辐射的波长时，就会产生镜反射，如高度抛光的金属表面就会产生镜反射；当物体表面的粗糙尺度大于投入辐射的波长时，就会产生漫反射。对全波长范围的热辐射能完全镜反射或完全漫反射的实际物体是不存在的，绝大多数工程材料对热辐射的反射都近似于漫反射。

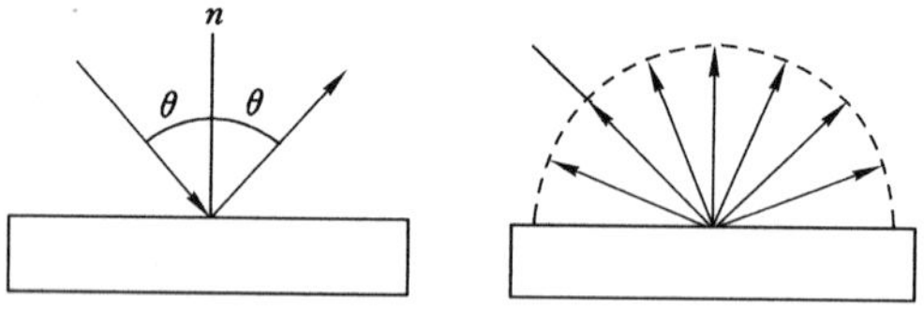

(a) 镜反射　　(b) 漫反射

图 10-2　镜反射与漫反射示意图

10.1.3　灰体与黑体

物体的光谱辐射特性随波长的变化给辐射换热分析带来很大的困难。为了工程上分析计算简便，引进灰体的概念。所谓灰体，是指光谱辐射特性不随波长而变化的假想物体，即 α_λ、ρ_λ 和 τ_λ 分别等于常数。对于灰体，由前面 α_λ 与 α、ρ_λ 与 ρ、τ_λ 与 τ 的关系式可得：

$$\alpha_\lambda = \alpha, \rho_\lambda = \rho, \tau_\lambda = \tau$$

即灰体的光谱吸收比、光谱反射比、光谱透射比分别等于吸收比、反射比、透射比，大小与波长无关，只取决于灰体本身的性质。在热辐射的波长范围内，绝大多数工程材料都可以近似地作为灰体处理。吸收比 $\alpha=1$ 的物体称为绝对黑体，简称黑体。黑体将所有投射在它上面的辐射能全部吸收，在所有物体之中，它吸收热辐射的能力最强。后面将证明，在温度相同的物体之中，黑体发射辐射能的能力也最强。黑体和灰体一样，是一种理想物体，在自然界是不存在的，但可以人工制造出接近黑体的模型。图 10-3 所示是一个人工黑体模型：一个内表面吸收比较高的空腔，空腔的壁面上有一个小孔。只要小孔的尺寸与空腔相比足够小，则从小孔进入空腔的辐射能经过空腔壁面的多次吸收和反射后，几乎全部被吸收，相当于小孔的吸收比接近1，即接近黑体。

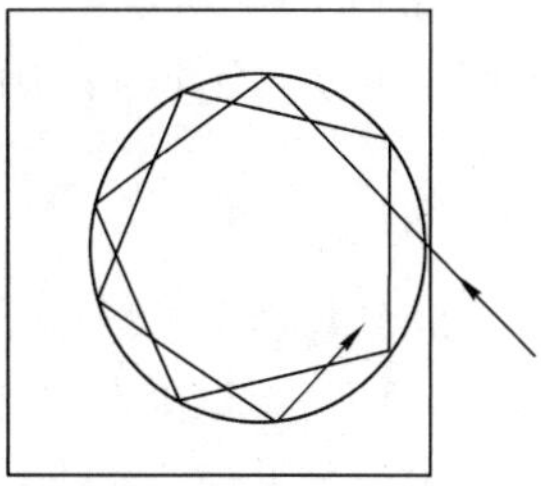

图 10-3　人工黑体示意图

黑体的引进对热辐射规律的研究具有重要意义：由于实际物体的热辐射特性和规律非常复杂，所以人们首先研究黑体辐射的性质和规律，把实际物体的辐射特性与之比较，找出与黑体辐射的区别，再将黑体辐射的规律进行修正后用于实际物体。

反射比 $\rho=1$ 的物体称为镜体（或白体）。透射比 $\tau=1$ 的物体称为绝对透明体。镜体、绝对透明体与灰体、黑体一样，都是一种理想物体，自然界中并不存在。

这里所说的黑体、白体与日常生活中所说的白色物体和黑色物体不同，颜色只是对可见光（可见光的波长范围为 0.38～0.76 μm）而言的，而可见光在热辐射的波长范围中只占很小部分，所以不能凭物体颜色的黑白来判断它对热辐射吸收比的大小。例如，白雪对可见光的吸收比很小，对热射线（如红外线）的吸收比高达 0.94，非常接近黑体；白布和黑布对可见光的吸收比差别很大，但对红外线的吸收比基本相同。又如，玻璃对可见光是透明的，但对一般温度下的热辐射则变成不透明的。因此，用玻璃做温室的透明盖板，就能产生让阳光进入，不让室内热辐射逸出的“温室效应”。决定物体吸收比和反射比的是物体的表面性质、本身所处的温度和发射体的性质、温度，而不是物体表面的颜色。

10.1.4　辐射强度

为了说明物体表面发射的辐射能在空间各个方向的分布规律，引进辐射强度的概念。因为涉及立体角的定义，下面先对立体角加以解释。

平面几何中，在一个半径为 r 的圆上，弧长 s 所对应的圆心角是平面角，大小为 $\theta=s/r$，单位是 rad（弧度）。而半径为 r 的球面上的面积 A 与球心所对应的是一个空间角度，用 Ω 表示，称为立体角，其大小定义为：

$$\Omega=\frac{A}{r^2}$$

立体角的单位叫球面度，用 sr 表示。由上式可以算出，半个球面所对应的立体角为 2πsr。

如图 10-4 所示，在微元面积 dA_1 上方半径为 r 的球面上，在(θ,φ)方向上有一个由经、纬线切割的微元面积 dA_2，面积为：

$$dA_2=rd\theta\cdot r\sin\theta d\varphi=r^2\sin\theta d\theta d\varphi$$

dA_2对球心所占的微元立体角为 $d\Omega$：

$$d\Omega=\frac{dA_2}{r^2}=\sin\theta d\theta d\varphi \tag{10-3}$$

如图 10-5 所示，单位时间内微元面积 dA_1 向 dA_2 所发射的辐射能为 $d\Phi$，dA_1 在 θ 方向的投影面积为 $dA_1\cos\theta$，则单位投影面积所发出的包含在单位立体角内的辐射能可表示为：

$$L(\theta,\varphi)=\frac{d\Phi}{dA_1\cos\theta d\Omega} \tag{10-4}$$

式中，$L(\theta,\varphi)$称为 dA_1 在(θ,φ)方向的辐射强度，或称为定向辐射强度，单位为 $W/(m^2\cdot sr)$。因为 $dA_1\cos\theta$ 就是从 θ 方向看过来可见面积 dA_1 的大小，所以辐射强度也可以说是单位时间内从单位可见面积上发出的包含在单位立体角内的辐射能。

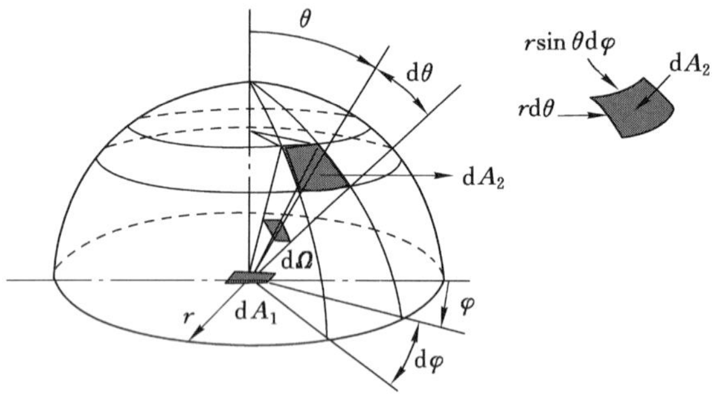

图 10-4　立体角定义示意图

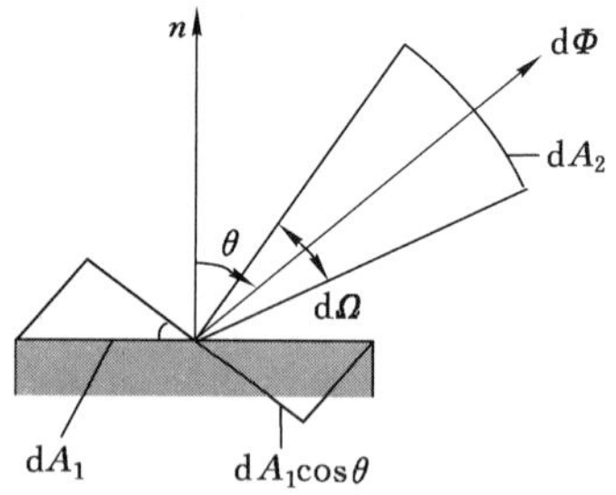

图 10-5　辐射强度定义示意图

辐射强度不仅取决于物体种类、表面性质、温度，还与方向有关，对于各向同性的物体表面，辐射强度与角 φ 无关，$L(\theta,\varphi)=L(\theta)$。以下的讨论仅限于各向同性物体表面。

只对某一波长辐射能而言的辐射强度称为光谱辐射强度或单色辐射强度。用符号 $L_\lambda(\theta)$ 表示。辐射强度与光谱辐射强度之间的关系可表示为：

$$L(\theta)=\int_0^\infty L_\lambda(\theta)d\lambda \tag{10-5}$$

如果波长的单位用 m，则光谱辐射强度的单位是 $W/(m^3\cdot sr)$；如果波长的单位用 μm，则光谱辐射强度的单位为 $W/(m^2\cdot\mu m\cdot sr)$。

10.1.5　辐射力

单位时间内，单位面积的物体表面向半球空间发射的全部波长的辐射能的总和称为该物体表面的辐射力，用符号 E 表示，单位为 W/m^2。

某一波长辐射能的辐射力称为光谱辐射力或单色辐射力，用符号 E_λ 表示，单位为 W/m^3。辐射力与光谱辐射力之间的关系可以表示为：

$$E=\int_0^\infty E_\lambda d\lambda \tag{10-6}$$

单位时间内，单位面积物体表面向某个方向发射的单位立体角内的辐射能，称为该物体表面在该方向上的定向辐射力，用符号 E_θ表示，单位为 $W/(m^2\cdot sr)$。

与辐射力、辐射强度的定义相对比，可知定向辐射力与辐射力之间的关系为：

$$E=\int_{\Omega=2\pi}E_\theta d\Omega \tag{10-7}$$

定向辐射力与辐射强度之间的关系为：

$$E_\theta=L(\theta)\cos\theta \tag{10-8}$$

于是，辐射力与辐射强度之间的关系可表示为：

$$E = \int_{\Omega=2\pi} L(\theta)\cos\theta \mathrm{d}\Omega \tag{10-9}$$

10.2 热辐射的基本定律

10.2.1 普朗克定律

1900 年，普朗克(M. Planck)在量子假设的基础上，从理论上确定了黑体辐射的光谱分布规律，给出了黑体的光谱辐射力 $E_{b\lambda}$ 与热力学温度 T、波长 λ 之间的函数关系，称为普朗克定律：

$$E_{b\lambda} = \frac{C_1\lambda^{-5}}{e^{C_2/(\lambda T)} - 1} \tag{10-10}$$

式中，λ 为波长，m；T 为热力学温度，K；C_1 为普朗克第一常数，3.743×10^{-16} W·m²；C_2 为普朗克第二常数，1.439×10^{-2} m·K。

不同温度下黑体的光谱辐射力随波长的变化如图 10-6 所示。可以看出，黑体的光谱辐射力随波长和温度的变化具有下述特点：① 温度越高，同一波长下的光谱辐射力越大；② 在一定的温度下，黑体的光谱辐射力随波长连续变化，并在某一波长下具有最大值；③ 随着温度的升高，光谱辐射力取得最大值的波长 λ_{max} 越来越小，即 λ 在坐标中的位置向短波方向移动。

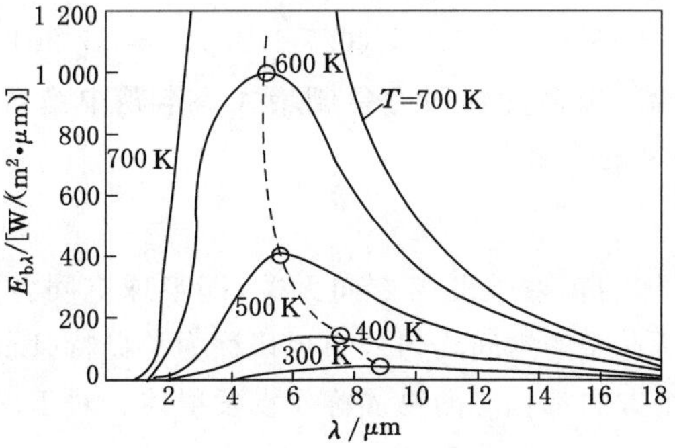

图 10-6 黑体的光谱辐射力[$E_{b\lambda}=f(\lambda,T)$]

在温度不变的情况下，由普朗克定律表达式(10-10)求极值，可以确定黑体的光谱辐射力取得最大值的波长 λ_{max} 与热力学温度 T 之间的关系：

$$\lambda_{max}T = 2.897\times10^{-3} \approx 2.9\times10^{-3}\,(\mathrm{m\cdot K}) \tag{10-11}$$

式(10-11)称为维恩(Wien)位移定律。

根据维恩位移定律，可以确定任一温度下黑体的光谱辐射力取得最大值的波长。例如，太阳可以近似为表面温度约为 5 800 K 的黑体，由式(10-11)可求得太阳光谱辐射力取得最大值的波长 $\lambda_{max}=0.5$ μm，位于可见光的范围内。所以，可见光的波长范围虽然很窄(0.38～0.76 μm)，但所占太阳辐射能的份额却很大(约为 44.6%)。再如，工业上常见的高温一般低于 2 000 K，由式(10-11)可以确定，2 000 K 温度下黑体的光谱辐射力取得最大值的波长 $\lambda_{max}=1.45$ μm，处于红外线范围内。加热炉中铁块升温过程中颜色的变化也能体现黑体辐射的特点：当铁块的温度低于 2 000 K 时，所发射的热辐射主要是红外线，人的眼睛感受不到，看起来还是暗黑色的。随着温度的升高，铁块的颜色逐渐变为暗红色、鲜红色、橘黄色、亮白色，这是由于随着温度的升高，铁块发射的热辐射中可见光的比例逐渐增大。

10.2.2 斯特藩-玻耳兹曼定律

斯特藩(J. Stefan)-玻耳兹曼(D. Boltzmann)定律确定了黑体的辐射力 E_b 与热力学温度 T 之间的关系，它首先由斯特藩于 1879 年从实验中得出，后来玻耳兹曼于 1884 年运用热力学理论进行了证明。其表达式为：

$$E_b = \sigma T^4 \tag{10-12}$$

式中，$\sigma = 5.67 \times 10^{-8}\ \mathrm{W/(m^2 \cdot K^4)}$，称为斯特藩-玻耳兹曼常数，又称为黑体辐射常数。

斯特藩-玻耳兹曼定律说明黑体的辐射力 E_b 与热力学温度 T 的四次方呈正比，故又称为四次方定律。斯特藩-玻耳兹曼定律表达式可以直接根据辐射力与光谱辐射力之间的关系式(10-6)由普朗克定律表达式导出：

$$E_b = \int_0^{\infty} E_\lambda \mathrm{d}\lambda = \int_0^{\infty} \frac{C_1 \lambda^{-5}}{e^{C_2/(\lambda T)} - 1} \mathrm{d}\lambda$$

【例 10-1】 一黑体置于室温 27 ℃的房间中，试求热平衡条件下，黑体表面的辐射力。若把黑体加热到 627 ℃，其辐射力又是多少？

【解】 热平衡时，黑体温度与室温相同，此时辐射力为：

$$E_{b1} = \sigma T^4 = 5.67 \times 10^{-8} \times (27 + 273)^4 = 459\ (\mathrm{W/m^2})$$

当温度为 627 ℃时，黑体辐射力为：

$$E_{b2} = \sigma T^4 = 5.67 \times 10^{-8} \times (627 + 273)^4 = 37\ 201\ (\mathrm{W/m^2})$$

可见，随着温度的升高，黑体的辐射力急剧增大。本题中绝对温度只是变为原来的 3 倍，而辐射力却增加到原来的 81 倍。

10.2.3 兰贝特定律

理论上可以证明，黑体的辐射强度与方向无关，即半球空间各方向上的辐射强度都相等。这种黑体辐射强度所遵循的空间均匀分布规律称为兰贝特(Lambert)定律。

辐射强度在空间各个方向都相等的物体称作漫发射体。对于漫发射体：

$$L(\theta) = L = 常数 \tag{10-13}$$

根据定向辐射力 E_θ 与辐射强度的关系式(10-8)，有：

$$E_\theta = L\cos\theta = E_n \cos\theta \tag{10-14}$$

式中，E_n 称为表面法线方向的定向辐射力。式(10-13)、式(10-14)称为兰贝特定律表达式。因为定向辐射力随方向角 θ 按余弦规律变化，所以兰贝特定律也称为余弦定律。

对于漫发射体，根据辐射力与辐射强度之间的关系式(10-9)与式(10-3)，有：

$$E = \int_0^{2\pi} \mathrm{d}\varphi \int_0^{\pi/2} L\sin\theta\cos\theta \mathrm{d}\theta = L\int_0^{2\pi} \mathrm{d}\varphi \int_0^{\pi/2} \sin\theta\cos\theta \mathrm{d}\theta = \pi L \tag{10-15}$$

即漫发射体的辐射力是辐射强度的 π 倍。

既漫发射又漫反射的物体称为漫射体。

10.2.4 实际物体的辐射特性与基尔霍夫定律

实际物体的辐射特性与黑体有很大的区别，下面分别介绍实际物体的发射特性和吸收特性以及二者之间的关系。

一切实际物体的辐射力都小于同温度下黑体的辐射力，为了说明实际物体的发射特性，引入发射率的概念：把任一实际物体的辐射力 E 与同温度下黑体的辐射力 E_b 之比称为该物体的发射率（习惯上称为黑度），用符号 ε 表示，即

$$\varepsilon=\frac{E}{E_b} \tag{10-16}$$

发射率的大小反映了物体发射辐射能能力的大小，表征物体辐射力接近黑体辐射力的程度。

实际物体的光谱辐射力与同温度下黑体的光谱辐射力之比称为该物体的光谱发射率（或称为光谱黑度或单色黑度），用符号 ε_λ 表示：

$$\varepsilon_\lambda=\frac{E_\lambda}{E_{b\lambda}} \tag{10-17}$$

发射率与光谱发射率之间的关系为：

$$\varepsilon=\frac{\int_0^\infty \varepsilon_\lambda E_{b\lambda}\mathrm{d}\lambda}{E_b} \tag{10-18}$$

对于灰体，光谱辐射特性不随波长而变化，ε_λ = 常数。由式(10-18)可得：

$$\varepsilon=\frac{\varepsilon_\lambda\int_0^\infty E_{b\lambda}\mathrm{d}\lambda}{E_b}=\varepsilon_\lambda$$

因此，灰体的光谱辐射力随波长的变化趋势与黑体相同。

实际物体光谱辐射力随波长的变化较大。图 10-7 是同温度下黑体、灰体和实际物体的光谱辐射力随波长变化的示意图，可以看出，实际物体的光谱辐射力随波长的变化规律完全不同于黑体和灰体。图 10-8 是黑体、灰体和实际物体的光谱发射率随波长变化的示意图。

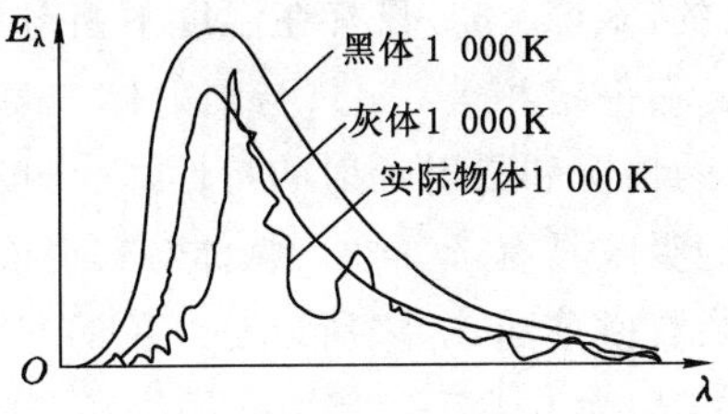

图 10-7 光谱辐射力随波长变化的示意图

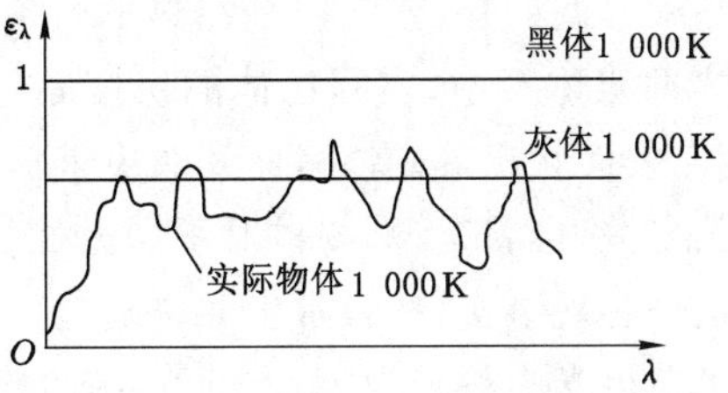

图 10-8 光谱发射率随波长变化的示意图

在工程计算中，实际物体的辐射力 E 可以根据发射率的定义式(10-16)由式(10-19)计算：

$$E=\varepsilon E_b=\varepsilon\sigma T^4 \tag{10-19}$$

应该指出，实际物体的辐射力并不严格与热力学温度的四次方呈正比，所存在的偏差由实验确定的发射率 ε 来修正，常见物体的黑度 ε 可查阅相关资料。

物体的吸收比 α 和发射率 ε 是关系物体之间辐射换热中能量收、支的两个指标，它们之间的关系由基尔霍夫定律确定。如图 10-9 所示，设有表面积为 F_b 的封闭空腔，其内表面为绝对黑体，壁温为 T_b，空腔内放置表面积为 F_1 的物体，其表面温度为 T_1，且 $T_1=T_b=$ 定值。

该系统是孤立的，并处于热力学平衡状态，即每一表面所辐射的能量必然等于其所

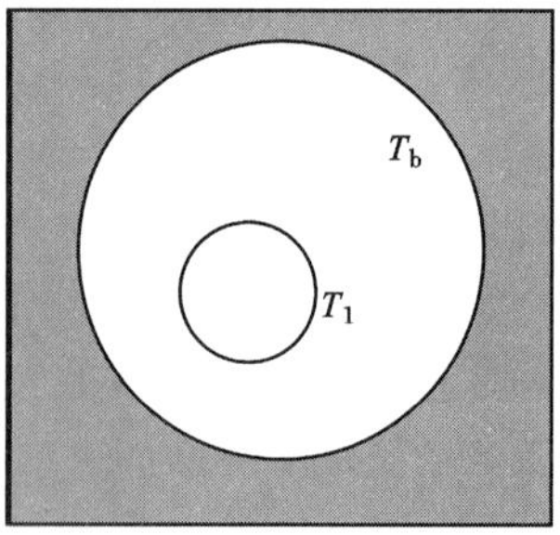

图 10-9　基尔霍夫定律推导

吸收的能量。物体 1 的本身辐射能 E 和它的反射辐射能 $(1-\alpha)E_b$，均被绝对黑体空腔表面所吸收；同时，空腔内表面也在辐射能量，其值为 E_b。因处于热平衡状态，则空腔本身所放出的能量等于其吸收的能量：

$$E_b = E + (1-\alpha)E_b$$

于是有：

$$E = \alpha E_b$$

或

$$\frac{E}{\alpha} = E_b \tag{10-20}$$

可想而知，若以另一物体替代物体 1，亦可得同样结果。

式(10-20)为基尔霍夫定律的数学表达式。

由基尔霍夫定律可得出如下结论：

(1) 物体的辐射力越大，则其吸收率也越大，即善于发射辐射能的物体必善于吸收辐射能。

(2) 因为实际物体的吸收比永远小于 1，所以同温度下，黑体的辐射力最大。

(3) 由式(10-20)可得 $\alpha = E/E_b$，把它与发射率定义式 $\varepsilon = E/E_b$ 相对照，则有：

$$\alpha(T) = \varepsilon(T) \tag{10-21}$$

即在与黑体处于热平衡的条件下，任何物体对来自黑体辐射的吸收比等于同温度下该物体的黑度(即发射率)。

(4) 对于光谱辐射：

$$\alpha_\lambda(T) = \varepsilon_\lambda(T) \tag{10-22}$$

必须强调的是，该定律是在热平衡条件下导出的，式(10-22)只有在温度平衡时才成立。在非热平衡体系中，实际物体的黑度 ε 与吸收率 α 不一定相等。ε 只与辐射体本身的情况(温度及表面状况等)有关，而 α 取决于辐射体自身的情况和投射辐射的特性，它与投射辐射的波长有关，也就是说，物体对不同波长的辐射能的吸收具有选择性。物体对辐射能选择性吸收这一客观事实，在日常生活中随处可见，如一朵红颜色的花，之所以是红色的，是因为它吸收了白色光中波长为 0.4～0.5 μm 的蓝色光和波长为 0.55～0.6 μm 的绿色光，仅仅反射了波长为 0.6～0.7 μm 的红色光。花本身没有色彩，光才是色彩的源泉，物体对光的选择性吸收才是它呈现颜色的主要原因。在工农业生产中利用物体的辐射能选择性吸收特性来达到一定的目的。例如，利用黑镍、黑铬、黑钴等金属及金属氧化物、金属硫化物和半导体等发色体粒子的电子跃迁能级与可见光谱区的光子能量较为匹配的材料，制成各种太阳能选择性涂层，使太阳能利用技术取得了变革性的进展。利用玻璃对波长小于 2.2 μm 的辐射能透射率很高，而对一般低温物体所能发出的波长在 3 μm 以上的辐射能透射率很小的特性，建造了植物生长玻璃温室。在检测领域，人们利用气体的选择性吸收特性，开发出了多种物料成分测试分析仪。在干燥领域，研究材料的选择性吸收特性，解决辐射干燥能量利用不匹配问题等。但是，实际物体的辐射能选择性吸收特性，使研究实际物体的热辐射问题变得很复杂，给工程计算造成了一定的麻烦。为简化问题，将实际物体理想化，而引入了“灰体”的概念。

如前面所述，灰体是指光谱辐射特性不随波长而变化的假想物体，即光谱吸收比不随波长发生变化，即 $\alpha_\lambda =$ 常数。这一定义表明，不管投射辐射的分布如何，吸收比 α 都是同一个常数，也就是说灰体的吸收比只取决于灰体本身的情况，与外界无关。不论投射辐射的波长和温度等如何变化，也不论投射辐射是否来自黑体，基尔霍夫定律都完全适用于灰体，即灰体的黑度恒等于它在同温度下的吸收比（$\varepsilon = \alpha$）。就辐射和吸收而言，灰体与黑体完全相似，只不过灰体在辐射和吸收的数量上小于黑体。

可见，灰体的辐射力遵守斯特藩-玻耳兹曼定律：

$$E = \varepsilon E_b = \varepsilon\sigma T^4 \tag{10-23}$$

自然界并不存在灰体，灰体是一种理想物体。工程上常见的温度范围 $T \leqslant 2\,000$ K，其热辐射主要位于红外线波长范围内（绝大部分能量位于 0.76～10 μm 之间）。在此范围内，可以将大多数工程材料近似地看作灰体。由于不论投射辐射是否来自黑体，也不论是否处于温度平衡条件，其吸收比恒等于同温度下的发射率。所以，已知发射率的数值就可以确定吸收比的数值，不会引起较大的误差，这一结论将大大简化辐射换热条件下吸收比的确定。

但在太阳能利用中研究物体表面对太阳能的吸收和本身的热辐射时，就不能简单地将物体当作灰体，而错误地认为对太阳能的吸收比等于自身辐射的发射率。这是因为，近50%的太阳辐射位于可见光的波长范围内，而物体自身热辐射位于红外波长范围内，由于实际物体的光谱吸收比对投入辐射的波长具有选择性，所以一般物体对太阳辐射的吸收比与自身辐射的发射率有较大的差别。例如，常温下各种颜色油漆的发射率约为 0.9，但白漆对可见光的吸收比只有 0.1～0.2。现在已开发出应用于太阳能集热器上的选择性表面涂层材料，其对太阳能的吸收比高达 0.9，而自身发射率只有 0.1 左右。这样既有利于太阳能的吸收，又减少了自身的辐射散热损失。

10.3　辐射换热计算

研究热辐射的目的之一，就是计算物体间的辐射换热量。温度不同的诸表面，当表面间被透明介质（如真空、空气等）或半透明介质（如烟气）所分隔时，此时表面间会有辐射换热，如空气夹层两壁面间的辐射换热，锅炉炉膛与水冷壁管间的辐射换热，采暖辐射板对整个车间的辐射换热，以及采暖辐射板对一个人或一个工作台的辐射换热等。物体间的辐射换热不仅与物体的温度、物体的黑度有关，还与物体换热表面的形状以及物体间的相对位置有很大关系。图 10-10 所示为两板间的辐射换热在不同位置的情况。在图 10-10(a)中，两板平行且相距很近，因此一个板表面发射的辐射能几乎全部落到另一板上，辐射换热量最大；在图 10-10(b)中，两板垂直，一个板表面发射的辐射能只有一部分落到另一板表面上，辐射换热量较小；在图 10-10(c)中，一个板表面发出的辐射能几乎都无法投射到另一板表面上去，辐射换热量最小，几乎为零。

10.3.1　角系数

为反映物体的几何形状、尺寸及相对位置等几何因素对辐射换热的影响，引入角系数概念。将表面 1 发出（自身发射与反射）的总辐射能中直接投射到表面 2 上的辐射能所占总辐射能的百分数称为表面 1 对表面 2 的角系数，用符号 φ_{12} 表示。同理，表面 2 对表面 1 的角

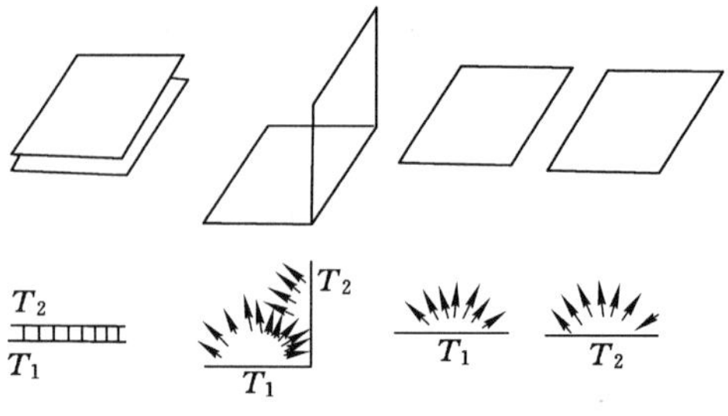

图 10-10　相对位置对辐射换热的影响

系数用 φ_{21} 表示。要计算任意两表面的辐射换热,必须先求出它们之间的角系数。

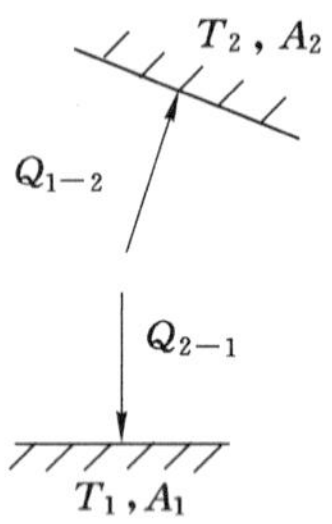

图 10-11　两黑体表面的辐射换热

角系数反映的是物体的几何形状、尺寸和相对位置等几何因素对物体间辐射换热的影响,与辐射体物性($\varepsilon,\alpha,\rho,\tau$ 及表面温度等)无关,是一个纯几何因子,可以用数学分析法、查曲线图法、投影法或几何图形法等方法来确定。为了研究方便,在讨论角系数时,我们把物体作为黑体。

设有任意放置的两个黑体表面,表面积分别为 A_1 和 A_2,并分别维持均匀温度 T_1 和 T_2,辐射力分别为 E_{b1} 与 E_{b2},如图 10-11 所示。

在单位时间内,由表面 A_1 到达表面 A_2 的辐射能为 $E_{b1}A_1\varphi_{12}$。与此同时,由表面 A_2 到达表面 A_1 的辐射能为 $E_{b2}A_2\varphi_{21}$。因为两个表面均为黑体,落到各表面上的能量全部被吸收。物体的辐射换热量是物体失去与获得的辐射能的差额,故两黑体表面之间的辐射换热量为:

$$Q_{12} = E_{b1}A_1\varphi_{12} - E_{b2}A_2\varphi_{21} \tag{10-24}$$

由式(10-24)可看出,由于黑体表面的温度和面积均已知,故求黑体表面之间的辐射换热问题就归结为求角系数的问题。

为确定角系数,先介绍角系数具有的两个特性。

(1) 互换性(相对性)

如图 10-11 所示的两黑体表面的温度如果相等($T_1 = T_2$),则它们处于热平衡状态,净辐射换热量 $Q_{12} = 0$,且 $E_{b1} = E_{b2}$,由式(10-24) 知:

$$A_1\varphi_{12} = A_2\varphi_{21} \tag{10-25}$$

式(10-25)表示了角系数的互换性,这种性质在辐射换热计算中十分有用。需要指出的是,尽管这一关系是在热平衡条件下得到的,但因角系数纯系几何因子,与表面性质、温度无关,所以,实际上在非平衡情况下或系统为非黑体表面时,此性质仍是存在的。由式(10-25)可知,只要知道其中一个角系数,就可以根据相对性求出另一个角系数。

(2) 完整性

从辐射换热的角度看,任何物体都处于其他物体(实际物体或假想物体,如太空背景)的包围之中。换句话说,任何物体都与其他所有参与辐射换热的物体构成一个封闭空腔。它所发出的辐射能百分之百地落在封闭空腔的各个表面之上,也就是说,它对构成封闭空腔的所有表面的角系数之和等于 1,即式(10-26)成立:

$$\sum_{j=1}^{n}\varphi_{i,j}=\varphi_{i,1}+\varphi_{i,2}+\cdots+\varphi_{i,i}+\cdots+\varphi_{i,n}=1 \tag{10-26}$$

式(10-26)称为角系数的完整性。对于非凹表面,表面 i 对自身的角系数 $\varphi_{i,i}=0$。

利用角系数特性可以方便地确定一些常见情况下的角系数,例如:

① 两块大平行平板,每一表面的辐射能可以认为全部落到另一表面上。按角系数定义,必然有 $\varphi_{12}=\varphi_{21}=1$。

② 一凸形表面(或平面)1为另一表面2所包围,因表面1发出的辐射能全部投到表面2上,则 $\varphi_{12}=1$。根据角系数的互换性,即按下式计算 φ_{21},可得:

$$\varphi_{21}=\frac{A_1}{A_2}$$

③ 由3个凸面(或平面)组成的一个封闭体(该封闭体在垂直于纸面方向上足够长,从两端开口逸出的辐射能可以略去不计算),设3个表面的面积分别为 A_1,A_2,A_3,根据角系数的完整性,有:

$$\varphi_{12}+\varphi_{13}=1,\varphi_{21}+\varphi_{23}=1,\varphi_{31}+\varphi_{32}=1$$

又根据角系数的互换性有:

$$A_1\varphi_{12}=A_2\varphi_{21},A_1\varphi_{13}=A_3\varphi_{31},A_2\varphi_{23}=A_3\varphi_{32}$$

联立求解上述六元一次方程组,即可分别求得未知的3个角系数:

$$\varphi_{12}=\frac{A_1+A_2-A_3}{2A_1}=\frac{l_1+l_2-l_3}{2l_1}$$

$$\varphi_{13}=\frac{A_1+A_3-A_2}{2A_1}=\frac{l_1+l_3-l_2}{2l_1}$$

$$\varphi_{23}=\frac{A_2+A_3-A_1}{2A_2}=\frac{l_2+l_3-l_1}{2l_2}$$

式中,l_1,l_2,l_3 分别为表面1,2,3的横断面线段长度。

工程上为了计算方便起见,将常见几何形状物体在某些相对位置时的角系数绘制成线算图,可从有关传热学书籍中查得。

10.3.2 黑体表面间的辐射换热

任意位置的两个黑体表面1和2,由式(10-24)结合角系数的互换性,可得两黑体表面间的辐射换热计算公式:

$$Q_{12}=(E_{b1}-E_{b2})A_1\varphi_{12}=(E_{b1}-E_{b2})A_2\varphi_{21} \tag{10-27}$$

10.3.3 灰体表面间的辐射换热

灰体表面间的辐射换热要比黑体表面复杂,这是因为投射到灰体表面的辐射能只有一部分被吸收,其余部分则被反射出去,结果形成辐射能在表面之间多次吸收和反射的现象。如果采用射线跟踪法,即跟踪一部分辐射能,累计它每次被吸收和反射的数量,计算就显得非常烦琐。对于漫灰表面,它自身发射和反射的辐射能都是漫分布的,所以在计算辐射换热时没有必要分别考虑,引进有效辐射的概念,可以使计算大为简化。

本身辐射 E:物体在单位时间内,单位表面积对外辐射的能量。

投入辐射 G:单位时间内,周围诸物体投射到某物体单位表面积上的能量之总和。

有效辐射 J:单位时间内离开单位面积表面的总辐射能。对于黑体,$J_b=E_b$;对于灰体,

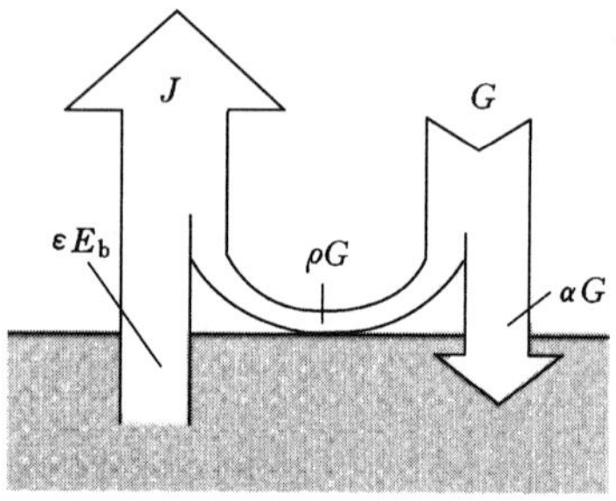

图 10-12 灰体表面的辐射能

有效辐射是物体本身辐射和反射辐射之和(图 10-12),即

$$J = E + \rho G = \varepsilon E_b + (1 - \alpha)G \tag{10-28}$$

根据表面的辐射平衡,单位面积的辐射换热量应该等于有效辐射与投入辐射之差,即

$$Q = (J - G)A$$

同时也等于自身辐射力与吸收的投入辐射之差,即

$$Q = (J - G)A = (\varepsilon E_b - \alpha G)A$$

考虑灰体表面特性 $\alpha = \varepsilon$,可得:

$$Q = \frac{A\varepsilon}{1 - \varepsilon}(E_b - J) \tag{10-29}$$

对于两个灰体表面 1,2 构成的封闭空腔中的辐射换热,并假设 $T_1 > T_2$,由式(10-29)知,表面 1 净损失的热量为:

$$Q_1 = \frac{A_1\varepsilon_1}{1 - \varepsilon_1}(E_{b1} - J_1)$$

表面 2 净获得的热量为:

$$Q_2 = \frac{A_2\varepsilon_2}{1 - \varepsilon_2}(J_2 - E_{b2})$$

根据有效辐射的定义,灰体表面 1,2 之间净辐射换热量为单位时间内由表面 1 到达表面 2 的辐射能 $J_1A_1\varphi_{12}$ 与同时由表面 2 到达表面 1 的辐射能 $J_2A_2\varphi_{21}$ 之差,结合角系数的互换性,可得:

$$Q_{12} = J_1A_1\varphi_{12} - J_2A_2\varphi_{21} = (J_1 - J_2)A_1\varphi_{12} = (J_1 - J_2)A_2\varphi_{21}$$

由于表面 1,2 构成一个封闭空腔,所以:

$$Q_1 = Q_2 = Q_{12}$$

联立上述公式可得构成封闭空腔的两灰体表面间的辐射换热的一般计算公式:

$$Q_{12} = \frac{E_{b1} - E_{b2}}{\dfrac{1 - \varepsilon_1}{A_1\varepsilon_1} + \dfrac{1}{A_1\varphi_{12}} + \dfrac{1 - \varepsilon_2}{A_2\varepsilon_2}} \tag{10-30}$$

10.3.4 辐射换热网络法

辐射换热网络法是根据热电现象的相似性,把辐射换热模拟成相应的电路,引入辐射热阻的概念,然后借助电路理论来求解,从而得到辐射换热问题解答的一种方法。利用网络法求解辐射换热问题,除使得换热过程直观、物理概念清楚外,还使得多个表面间的辐射换热问题的求解变得十分简便。辐射换热网络法以空腔法为基础,要求在求解过程中将物体表面间组成一个封闭空腔,如果表面间有开口,需设一假想面来加以封闭。

将黑体表面间的辐射换热计算式(10-27)写成电学中欧姆定律表达式的形式:

$$Q_{12} = \frac{E_{b1} - E_{b2}}{\dfrac{1}{A_1\varphi_{12}}} \tag{10-31}$$

式中,分子 $E_{b1} - E_{b2}$ 相当于电势差,是产生辐射换热的推动力;辐射换热量 Q_{12} 相当于电流;分母 $1/(A_1\varphi_{12})$ 相当于电阻,由于它仅与换热表面间的相对位置有关,所以称为空间辐射热阻,可以理解为由于两个表面的几何形状、大小及相对位置产生的它们之间辐射换热的阻力。

如果两个黑体表面构成封闭空腔，则式(10-31)计算的辐射换热量 Q_{12} 既是表面1净损失的热量，也是表面2净获得的热量。表面1,2之间的辐射换热可以用图10-13所示的辐射换热网络来表示(图中，由于黑体表面对任何的投射辐射都能全部吸收，所以网络图的另一侧相当于“接地”)，其中的 E_{b1}，E_{b2} 相当于直流电源。

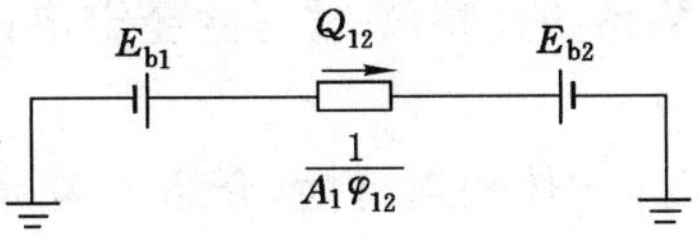

图10-13　两个黑体表面构成封闭空腔的辐射换热网络

如果由 n 个黑体表面构成封闭空腔，那么每个表面的净辐射换热量应该是该表面与封闭空腔的所有表面之间辐射换热量的代数和，即

$$Q_i = \sum_{j=1}^{n} Q_{i,j} = \sum_{j=1}^{n} A_i \varphi_{i,j} (E_{bi} - E_{bj})$$

对于构成封闭空腔的两灰体表面间的辐射换热，由式(10-30)可知，两个灰体表面之间的辐射换热热阻由3个串联的辐射热阻组成，可由图10-14所示的辐射换热网络表示，其中 $\frac{1-\varepsilon_1}{A_1\varepsilon_1}$ 与 $\frac{1-\varepsilon_2}{A_2\varepsilon_2}$ 是由于物体表面为非黑体而形成的热阻，称为表面辐射热阻，黑度越大，表面辐射热阻越小，若表面为黑体，此热阻为零；$\frac{1}{A_1\varphi_{12}}$ 为空间辐射热阻。

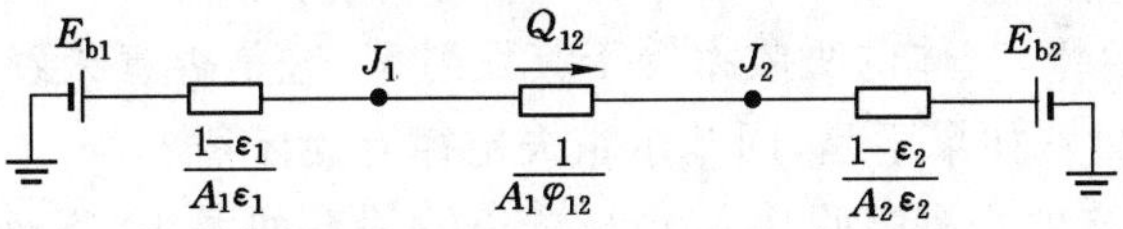

图10-14　两个灰体表面构成封闭空腔的辐射换热网络

对于两块平行灰体大壁面构成的封闭空腔，由于 $A_1 = A_2 = A$，$\varphi_{12} = \varphi_{21} = 1$，式(10-30)可简化为：

$$Q_{12} = \frac{A(E_{b1} - E_{b2})}{\frac{1}{\varepsilon_1} + \frac{1}{\varepsilon_2} - 1} = A\varepsilon_s (E_{b1} - E_{b2})$$

式中，$\varepsilon_s = \frac{1}{\frac{1}{\varepsilon_1} + \frac{1}{\varepsilon_2} - 1}$ 称为该辐射换热系统的系统黑度。若两物体为黑体，则 $\varepsilon_1 = \varepsilon_2 = 1$，此时 $\varepsilon_s = 1$，$Q_{12} = A(E_{b1} - E_{b2})$。由此可见，系统黑度是指在其他条件相同时，灰体间的换热量与黑体间的换热量之比，因为灰体 ε_1 和 ε_2 均小于1，故 $\varepsilon_s < 1$，即可视系统黑度是因实际表面为非黑体时而引入的一个修正系数。

利用有效辐射的概念，还可以计算多个灰体表面构成的封闭空腔内的辐射换热。无论是2个表面间还是3个或3个以上表面间的辐射换热，利用表面辐射热阻和空间辐射热阻计算的前提是它们必须构成封闭系统，不是封闭系统不可采用上述公式。如果计算表面不能构成封闭空腔，可虚构一些表面构成封闭空腔后再进行分析计算。

本章小结

本章讨论了物体热辐射的基本概念和辐射能的定量描述。

介绍了黑体的光谱分布规律:普朗克定律;由普朗克定律得出的维恩位移定律;确定黑体辐射力与热力学温度之间的关系的斯特藩-玻耳兹曼定律;说明黑体的辐射强度与方向无关的兰贝特定律。

引入发射率概念,讨论了实际物体辐射特性,介绍了揭示物体吸收辐射能的能力与发射辐射能的能力之间的关系的基尔霍夫定律。

讨论了角系数的定义和计算方法,介绍了经过假设简化的辐射换热的计算方法。

思考与练习题

(1) 辐射换热有何特点? 它与导热、对流换热有何区别?

(2) 何谓吸收比、反射比、穿透比? 三者之间有何关系?

(3) 什么叫黑体? 在热辐射理论中为什么要引入这一概念?

(4) 说明灰体的定义以及引入灰体的简化对工程辐射换热计算的意义。

(5) 什么叫光谱吸收比? 在不同光源的照耀下,物体常呈现不同的颜色,如何解释?

(6) 某物体的黑度 $\varepsilon=0.95$,求温度为 627 ℃时,此物体的辐射力。

(7) 对于一般物体,吸收比等于发射率在什么条件下成立?

(8) 按基尔霍夫定律,灰体的吸收率等于其发射率,这是否意味着灰体所吸收的能量总是等于它所辐射的能量? 如果不是,同基尔霍夫定律有无矛盾?

(9) 已知材料 A,B 的光谱吸收比 $\alpha(\lambda)$ 与波长的关系如图 10-15 所示,试估计这两种材料的发射率 ε 随温度变化的特性,并说明理由。

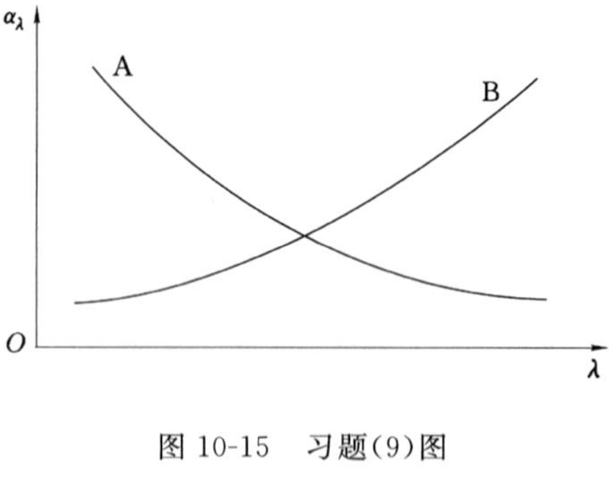

图 10-15 习题(9)图

(10) 何谓“大气温室效应”,为什么减少 CO_2 的排放就可以降低温室效应?

(11) 选择太阳能集热器的表面涂层时,该涂料表面光谱吸收比随波长的变化最佳曲线是什么? 有人认为取暖用的辐射采暖片也需要涂上这种材料,你认为合适吗?

(12) 有一个放置于室外的冷库,从减小冷库冷量损失的角度出发,冷库外壳颜色应涂成深色还是浅色?

(13) 人造地球卫星在返回地球表面时为何容易被烧毁?

(14) 北方深秋季节的清晨，树叶叶面上常常结霜。试问树叶上、下表面的哪一面结霜？为什么？

(15) 在波长 $\lambda<2\ \mu m$ 的短波范围内，木板的光谱吸收比小于铝板，而在长波($\lambda>2\ \mu m$)范围内则相反。在木板和铝板同时长时间放在太阳光下时，哪个温度高？为什么？

(16) 窗玻璃对红外线几乎不透明，但为什么隔着玻璃晒太阳却使人感到暖和？

(17) 角系数有哪些特性？这些特性的物理依据是什么？

(18) 什么是一个表面的本身辐射、投入辐射及有效辐射？有效辐射概念的引入对灰体表面系统辐射换热的计算有什么作用？

(19) 什么是表面辐射热阻？什么是空间辐射热阻？试分析由几个黑度不同的表面拼成的表面与外界辐射换热时的热阻。

Chapter 10 Radiation Heat Transfer

Radiation is the energy emitted by matter in the form of electromagnetic waves (or photons) as a result of the changes in the electronic configurations of the atoms or molecules. Unlike conduction and convection, the transfer of energy by radiation does not require the presence of an intervening medium. In fact, energy transfer by radiation is fastest (at the speed of light) and it suffers no attenuation in a vacuum. This is how the energy of the sun reaches the earth. In heat transfer studies we are interested in thermal radiation, which is the form of radiation emitted by bodies because of their temperature. It differs from other forms of electromagnetic radiation such as X-rays, gamma rays, microwaves, radio waves, and television waves that are not related to temperature. All bodies at a temperature above absolute zero emit thermal radiation. Radiation is a volumetric phenomenon, and all solids, liquids, and gases emit, absorb, or transmit radiation to varying degrees. However, radiation is usually considered to be a surface phenomenon for solids that are opaque to thermal radiation such as metals, wood, and rocks since the radiation emitted by the interior regions of such material can never reach the surface, and the radiation incident on such bodies is usually absorbed within a few microns from the surface.

10.1 Basic Concept of Thermal Radiation

10.1.1 The Essence of Radiation

Radiation propagates in the form of electromagnetic waves. The frequency v and wavelength λ of electromagnetic waves in a medium are related by $\lambda = c/v$, where c is the speed of propagation in that medium. All matter whose temperature is above absolute zero continuously emits thermal radiation as a result of vibrational and rotational motions of molecules, atoms, and electrons of a substance. Temperature is a measure of the strength of these activities at the microscopic level.

Radiation differs from the other two heat transfer mechanisms in that it does not require the presence of a material medium to take place. In fact, energy transfer by radiation is fastest (at the speed of light) and it suffers no attenuation in a vacuum. Also, radiation transfer occurs in solids as well as liquids and gases. In most practical applications, all three modes of heat transfer occur concurrently at varying degrees. But heat transfer through an evacuated space can occur only by radiation. For example, the energy of the sun reaches the earth by radiation. You will recall that heat transfer by conduction or convection takes place in the direction of decreasing temperature; that is, from a high-temperature medium to a lower-temperature one. It is interesting that radiation heat transfer can occur between two bodies separated by a medium colder than both bodies. For example, solar radiation reaches the surface of the earth after passing through cold air layers at high altitudes. Also, the radiation absorbing surfaces inside a greenhouse reach high temperatures even when its plastic or glass cover remains relatively cool.

The short-wavelength gamma rays and X-rays are primarily of concern to nuclear engineers, while the long-wavelength microwaves and radio waves are of concern to electrical engineers. The type of electromagnetic radiation that is pertinent to heat transfer is the thermal radiation emitted as a result of energy transitions of molecules, atoms, and electrons of a substance. Temperature is a measure of the strength of these activities at the microscopic level, and the rate of thermal radiation emission increases with increasing temperature. Thermal radiation is continuously emitted by all matter whose temperature is above absolute zero. That is, everything around us such as walls, furniture, and our friends constantly emits (and absorbs) radiation. Thermal radiation is also defined as the portion of the electromagnetic spectrum that extends from about 0.1 to 100 μm, since the radiation emitted by bodies due to their temperature falls almost entirely into this wavelength range. Thus, thermal radiation includes the entire visible and infrared (IR) radiation as well as a portion of the ultraviolet (UV) radiation.

In heat transfer studies, we are interested in the energy emitted by bodies because of their temperature only. Therefore, we will limit our consideration to thermal radiation, which we will simply call radiation. The relations developed below are restricted to thermal radiation only and may not be applicable to other forms of electromagnetic radiation.

10.1.2 Absorptivity, Reflectivity, and Transmissivity

Everything around us constantly emits radiation, and the emissivity represents the emission characteristics of those bodies. This means that every body, including our own, is constantly bombarded by radiation coming from all directions over a range of wavelengths. Recall that radiation flux incident on a surface is called irradiation and is denoted by G.

When radiation strikes a surface, part of it is absorbed, part of it is reflected, and the remaining part, if any, is transmitted, as illustrated in Fig. 10-1. The fraction of irradiation absorbed by the surface is called the absorptivity α, the fraction reflected by the surface is called the reflectivity ρ, and the fraction transmitted is called the transmissivity τ. That is:

$$\alpha = \frac{G_\alpha}{G}, \rho = \frac{G_\rho}{G}, \tau = \frac{G_\tau}{G}$$

Where G is the radiation energy incident on the surface, and G_α, G_ρ, and G_τ are the absorbed, reflected, and transmitted portions of it, respectively. The First Law of

Thermodynamics requires that the sum of the absorbed, reflected, and transmitted radiation energy be equal to the incident radiation. That is:

$$G_\alpha + G_\rho + G_\tau = G$$

Dividing each term of this relation by G yields:

$$\alpha + \rho + \tau = 1 \tag{10-1}$$

For opaque surfaces, $\tau = 0$, and thus $\alpha + \rho = 1$. This is an important property relation since it enables us to determine both the absorptivity and reflectivity of an opaque surface by measuring either of these properties.

These definitions are for total hemispherical properties, since G represents the radiation flux incident on the surface from all directions over the hemispherical space and over all wavelengths. Thus, α, ρ, and τ are the average properties of a medium for all directions and all wavelengths. However, these properties can also be defined for a specific wavelength. Likewise, the spectral absorptivity, spectral reflectivity of a surface, and the spectral transmissivity of a medium are defined as:

$$\alpha_\lambda = \frac{G_{\lambda\alpha}}{G_\lambda}, \rho = \frac{G_{\lambda\rho}}{G_\lambda}, \tau = \frac{G_{\lambda\tau}}{G_\lambda}$$

$$\alpha_\lambda + \rho_\lambda + \tau_\lambda = 1 \tag{10-2}$$

Where G_λ is the spectral irradiation [in $W/(m^2 \cdot \mu m)$] incident on the surface.

10.1.3 Gray Body and Blackbody

A gray body is defined such that the monochromatic emissivity ε_λ of the body is independent of wavelength. The monochromatic emissivity is defined as the ratio of the monochromatic emissive power of the body to the monochromatic emissive power of a blackbody at the same wavelength and temperature.

Radiation is a complex phenomenon as it is, and the consideration of the wavelength and direction dependence of properties, assuming sufficient data exist, makes it even more complicated. Therefore, the gray and diffuse approximations are often utilized in radiation calculations. A surface is said to be diffuse if its properties are independent of direction, and gray if its properties are independent of wavelength. Therefore, the emissivity of a gray, diffuse surface is simply the total hemispherical emissivity of that surface because of independence of direction and wavelength.

A body at a temperature above absolute zero emits radiation in all directions over a wide range of wavelengths. The amount of radiation energy emitted from a surface at a given wavelength depends on the material of the body and the condition of its surface as well as the surface temperature. Therefore, different bodies may emit different amounts of radiation per unit surface area, even when they are at the same temperature. Thus, it is natural to be curious about the maximum amount of radiation that can be emitted by a surface at a given temperature. Satisfying this curiosity requires the definition of an idealized body, called a blackbody, to serve as a standard against which the radiative properties of real surfaces may be compared.

A blackbody is defined as a perfect emitter and absorber of radiation. At a specified temperature and wavelength, no surface can emit more energy than a blackbody. A blackbody absorbs all incident radiation, regardless of wavelength and direction. Also, a blackbody emits radiation energy uniformly in all directions per unit area normal to direction of emission. That is, a blackbody is a diffuse emitter. The term diffuse means "independent of direction."

The concept of a blackbody is an idealization; that is, a perfect blackbody does not exist—all surfaces reflect radiation to some extent, however slight. A blackbody may be approximated very accurately, however, in the following way. A cavity is constructed, as shown in Fig. 10-3. So that it is very large compared with the size of the opening in the side. An incident ray of energy is reflected many times on the inside before finally escaping from the side opening. With each reflection there is a fraction of the energy absorbed corresponding to the absorptivity of the inside of the cavity. After the many absorptions, practically all the incident radiation at the side opening is absorbed. It should be noted that the cavity of Fig. 10-3 behaves approximately as a blackbody emitter as well as an absorber.

10.1.4 Radiation Intensity

Radiation is emitted by all parts of a plane surface in all directions into the hemisphere above the surface, and the directional distribution of emitted (or incident) radiation is usually not uniform. Therefore, we need a quantity that describes the magnitude of radiation emitted (or incident) in a specified direction in space. This quantity is radiation intensity, denoted by I. Before we can describe a directional quantity, we need to specify direction in space. The direction of radiation passing through a point is best described in spherical coordinates in terms of the zenith angle θ and the azimuth angle $\boldsymbol{\Phi}$, as shown in Fig. 10-5. Radiation intensity is used to describe how the emitted radiation varies with the zenith and azimuth angles.

Let us try to quantify the size of a slice of pizza. One way of doing that is to specify the arc length of the outer edge of the slice, and to form the slice by connecting the endpoints of the arc to the center. A more general approach is to specify the angle of the slice at the center, as shown in Fig. 10-4. An angle of 90° (or $\pi/2$ radians), for example, always represents a quarter pizza, no matter what the radius is. For a circle of unit radius, the length of an arc is equivalent in magnitude to the plane angle it subtends (both are 2π for a complete circle of radius $r=1$).

Now consider a watermelon, and let us attempt to quantify the size of a slice. Again we can do it by specifying the outer surface area of the slice (the green part), or by working with angles for generality. Connecting all points at the edges of the slice to the center in this case will form a three-dimensional body (like a cone whose tip is at the center), and thus the angle at the center in this case is properly called the solid angle. The solid angle is denoted by $\boldsymbol{\Omega}$, and its unit is the steradian (sr). In analogy to plane angle, we can say that the area of a surface on a sphere of unit radius is equivalent in magnitude to the solid angle it subtends (both are 4π for a sphere of radius $r=1$).

For a hemisphere, which is more relevant to radiation emitted or received by a surface, it is $\Omega=2\pi$sr.

The differential solid angle $d\Omega$ subtended by a differential area dA_2 on a sphere of radius r can be expressed as:

$$d\Omega = \frac{dA_2}{r^2} = \sin\theta d\theta d\varphi \tag{10-3}$$

The radiation intensity for emitted radiation $L(\theta,\varphi)$ is defined as the rate at which radiation energy $d\Phi$ is emitted in the (θ,φ) direction per unit area normal to this direction and per unit solid angle about this direction. That is:

$$L(\theta,\varphi) = \frac{d\Phi}{dA_1 \cos\theta d\Omega} \tag{10-4}$$

$$L(\theta) = \int_0^{\infty} L_\lambda(\theta) d\lambda \tag{10-5}$$

Where θ is the angle between the normal of the surface and the direction of viewing, and thus $dA_1 \cos\theta$ is the normal (or projected) area to the direction of viewing.

10.1.5 Emissive Power

If all surfaces emitted radiation uniformly in all directions, the emissive power would be sufficient to quantify radiation, and we would not need to deal with intensity. The radiation emitted by a blackbody per unit normal area is the same in all directions, and thus there is no directional dependence. But this is not the case for real surfaces. Before we define intensity, we need to quantify the size of an opening in space.

When the variations of spectral radiation fluxes E_λ with wavelength λ are known, the total radiation fluxes can be determined by integration over the entire wavelength spectrum as:

$$E = \int_0^{\infty} E_\lambda d\lambda \tag{10-6}$$

The radiation flux for emitted radiation is the emissive power E (the rate at which radiation energy is emitted per unit area of the emitting surface), which can be expressed in differential form as:

$$E = \int_{\Omega=2\pi} E_\theta d\Omega \tag{10-7}$$

$$E_\theta = L(\theta)\cos\theta \tag{10-8}$$

$$E = \int_{\Omega=2\pi} L(\theta)\cos\theta d\Omega \tag{10-9}$$

10.2 Basic Law of Thermal Radiation

10.2.1 Planck's Law

Sometimes we need to know the spectral blackbody emissive power, which is the amount of radiation energy emitted by a blackbody at an absolute temperature T per unit time, per unit surface area, and per unit wavelength about the wavelength λ. For example, we are more interested in the amount of radiation an incandescent light bulb emits in the visible wavelength spectrum than we are in the total amount emitted.

The relation for the spectral blackbody emissive power $E_{b\lambda}$ was developed by Max Planck in 1901 in conjunction with his famous quantum theory. This relation is known as Planck's law and is expressed as:

$$E_{b\lambda} = \frac{C_1 \lambda^{-5}}{e^{C_2/(\lambda T)} - 1} \tag{10-10}$$

Where $C_1 = 3.743 \times 10^{-16}\ \mathrm{W \cdot m^2}$; $C_2 = 1.439 \times 10^{-2}\ \mathrm{m \cdot K}$.

The variation of the spectral blackbody emissive power with wavelength is plotted in Fig. 10-6 for selected temperatures. Several observations can be made from this figure:

(1) The emitted radiation is a continuous function of wavelength. At any specified temperature, it increases with wavelength, reaches a peak, and then decreases with increasing wavelength.

(2) At any wavelength, the amount of emitted radiation increases with increasing temperature.

(3) As temperature increases, the curves shift to the left to the shorter-wavelength region. Consequently, a larger fraction of the radiation is emitted at shorter wavelengths at higher temperatures.

(4) The radiation emitted by the sun, which is considered to be a blackbody at 5 780 K (or roughly at 5 800 K), reaches its peak in the visible region of the spectrum. Therefore, the sun is in tune with our eyes. On the other hand, surfaces at $T \leqslant 800$ K emit almost entirely in the infrared region and thus are not visible to the eye unless they reflect light coming from other sources.

As the temperature increases, the peak of the curve in Fig. 10-6 shifts toward shorter wavelengths. The wavelength at which the peak occurs for a specified temperature is given by Wien's displacement law as:

$$\lambda_{max} T = 2.897 \times 10^{-3} \approx 2.9 \times 10^{-3}\ (\mathrm{m \cdot K}) \tag{10-11}$$

This relation was originally developed by Willy Wien in 1894 using classical thermodynamics, but it can also be obtained by differentiating Eq. (10-10) with respect to λ while holding T constant and setting the result equal to zero. A plot of Wien's displacement law, which is the locus of the peaks of the radiation emission curves, is also given in Fig. 10-6.

An electrical resistance heater starts radiating heat soon after it is plugged in, and we can feel the emitted radiation energy by holding our hands facing the heater. But this radiation is entirely in the infrared region and thus cannot be sensed by our eyes. The heater would appear dull red when its temperature reaches about 1 000 K, since it will start emitting a detectable amount [about 1 $\mathrm{W/(m^2 \cdot \mu m)}$] of visible red radiation at that temperature. As the temperature rises even more, the heater appears bright red and is said to be red hot. When the temperature reaches about 1 500 K, the heater emits enough radiation in the entire visible range of the spectrum to appear almost white to the eye, and it is called white hot.

Although it cannot be sensed directly by the human eye, infrared radiation can be detected by infrared cameras, which transmit the information to microprocessors to display visual images of objects at night. Rattlesnakes can sense the infrared radiation or the "body heat" coming off warm-blooded animals, and thus they can see at night without using any instruments. Similarly, honeybees are sensitive to ultraviolet radiation. A surface that reflects all of the light appears white, while a surface that absorbs all of the light incident on it appears black. (Then how do we see a black surface?)

It should be clear from this discussion that the color of an object is not due to emission, which is primarily in the infrared region, unless the surface temperature of the object exceeds about 1 000 K. Instead, the color of a surface depends on the absorption and reflection characteristics of the surface and is due to selective absorption and reflection of the incident visible radiation coming from a light source such as the sun or an incandescent lightbulb. A piece of clothing containing a pigment that reflects red while absorbing the remaining parts of the incident light appears "red" to the eye. Leaves appear "green" because their cells contain the pigment chlorophyll, which strongly reflects green while absorbing other colors.

10.2.2 Stefan-Boltzmann Law

The radiation energy emitted by a blackbody per unit time and per unit surface area was determined experimentally by Joseph Stefan in 1879 and expressed as:

$$E_b = \sigma T^4 \tag{10-12}$$

Where $\sigma = 5.67 \times 10^{-8}\ \mathrm{W/(m^2 \cdot K^4)}$ is the Stefan-Boltzmann constant and T is the absolute temperature of the surface in K. This relation was theoretically verified in 1884 by Ludwig Boltzmann. Eq. (10-12) is known as the Stefan-Boltzmann law and E_b is called the blackbody emissive power. Note that the emission of thermal radiation is proportional to the fourth power of the absolute temperature.

10.2.3 Lambert Law

The intensity of radiation emitted by a surface, in general, varies with direction (especially with the zenith angle). But many surfaces in practice can be approximated as being diffuse. For a diffusely emitting surface, the intensity of the emitted radiation is independent of direction and thus

$$L(\theta) = L = \text{const} \tag{10-13}$$

$$E_\theta = L\cos\theta = E_n \cos\theta \tag{10-14}$$

Noting that $\int_0^{2\pi} \mathrm{d}\varphi \int_0^{\pi/2} \sin\theta\cos\theta \mathrm{d}\theta = \pi$, the emissive power relation reduces in this case to:

$$E = \int_0^{2\pi} d\varphi \int_0^{\pi/2} L\sin\theta\cos\theta d\theta = L\int_0^{2\pi} d\varphi \int_0^{\pi/2} \sin\theta\cos\theta d\theta = \pi L \tag{10-15}$$

10.2.4 Radiation Properties of Real Materials and Kirchhoff's Law

The emissivity of a surface represents the ratio of the radiation emitted by the surface at a given temperature to the radiation emitted by a blackbody at the same temperature. The emissivity of a surface is denoted by ε, and it varies between zero and one, $0 \leqslant \varepsilon \leqslant 1$. Emissivity is a measure of how closely a surface approximates a blackbody, for which $\varepsilon = 1$.

The emissivity of a real surface is not a constant. Rather, it varies with the temperature of the surface as well as the wavelength and the direction of the emitted radiation. Therefore, different emissivities can be defined for a surface, depending on the effects considered. The most elemental emissivity of a surface at a given temperature is the spectral directional emissivity, which is defined as the ratio of the intensity of radiation emitted by the surface at a specified wavelength in a specified direction to the intensity of radiation emitted by a blackbody at the same temperature at the same wavelength.

The emissivity of a surface represents the ratio of the radiation emitted by the surface at a given temperature to the radiation emitted by a blackbody at the same temperature. Different emissivities are defined as:

$$\varepsilon = \frac{E}{E_b} \tag{10-16}$$

Therefore, the total hemispherical emissivity (or simply the "average emissivity") of a surface at a given temperature represents the ratio of the total radiation energy emitted by the surface to the radiation emitted by a blackbody of the same surface area at the same temperature.

In practice, it is usually more convenient to work with radiation properties averaged over all directions, called hemispherical properties. Noting that the integral of the rate of radiation energy emitted at a specified wavelength per unit surface area over the entire hemisphere is spectral emissive power, the spectral hemispherical emissivity can be expressed as:

$$\varepsilon_\lambda = \frac{E_\lambda}{E_{b\lambda}} \tag{10-17}$$

$$\varepsilon = \frac{\int_0^\infty \varepsilon_\lambda E_{b\lambda} d\lambda}{E_b} \tag{10-18}$$

Note that the emissivity of a surface at a given wavelength can be different at different temperatures since the spectral distribution of emitted radiation (and thus the amount of radiation emitted at a given wavelength) changes with temperature.

$$E = \varepsilon E_b = \varepsilon\sigma T^4 \tag{10-19}$$

$$\alpha(T) = \varepsilon(T) \tag{10-20}$$

The total hemispherical emissivity of a surface at temperature T is equal to its total hemispherical absorptivity for radiation coming from a blackbody at the same temperature. This relation, which greatly simplifies the radiation analysis, was first developed by Gustav Kirchhoff in 1860 and is now called Kirchhoff's law. Note that this relation is derived under the condition that the surface temperature is equal to the temperature of the source of irradiation, and the reader is cautioned against using it when considerable difference (more than a few hundred degrees) exists between the surface temperature and the temperature of the source of irradiation.

The derivation above can also be repeated for radiation at a specified wavelength to obtain the spectral form of Kirchhoff's law:

$$\alpha_\lambda(T) = \varepsilon_\lambda(T) \tag{10-21}$$

This relation is valid when the irradiation or the emitted radiation is independent of direction. The form of Kirchhoff's law that involves no restrictions is the spectral directional form expressed as $\varepsilon_{\lambda,\theta}(T)=\alpha_{\lambda,\theta}(T)$. That is, the emissivity of a surface at a specified wavelength, direction, and temperature is always equal to its absorptivity at the same wavelength, direction, and temperature.

It is very tempting to use Kirchhoff's law in radiation analysis since the relation $\varepsilon=\alpha$ together with $\rho=1-\alpha$ enables us to determine all three properties of an opaque surface from a knowledge of only one property. Although Eq. (10-20) gives acceptable results in most cases, in practice, care should be exercised when there is considerable difference between the surface temperature and the temperature of the source of incident radiation.

10.3 The Calculation of Radiation Heat Transfer

We have considered the fundamental aspects of radiation and the radiation properties of surfaces. We are now in a position to consider radiation exchange between two or more surfaces, which is the primary quantity of interest in most radiation problems.

10.3.1 The View Factor

Radiation heat transfer between surfaces depends on the orientation of the surfaces relative to each other as well as their radiation properties and temperatures, as illustrated in Fig. 10-10. For example, a camper will make the most use of a campfire on a cold night by standing as close to the fire as possible and by blocking as much of the radiation coming from the fire by turning her front to the fire instead of her side. Likewise, a person will maximize the amount of solar radiation incident on him and take a sunbath by lying down on his back instead of standing up on his feet.

To account for the effects of orientation on radiation heat transfer between two surfaces, we define a new parameter called the view factor, which is a purely geometric quantity and is independent of the surface properties and temperature. It is also called the shape factor, configuration factor, and angle factor. The view factor based on the assumption that the surfaces are diffuse emitters and diffuse reflectors is called the diffuse view factor, and the view factor based on the assumption that the surfaces are diffuse emitters but specular reflectors is called the specular view factor. In this book, we will consider radiation exchange between diffuse surfaces only, and thus the term view factor will simply mean diffuse view factor.

The view factor from a surfacel to a surface 2 is denoted by φ_{12}, and is defined as:

φ_{12} = the fraction of the radiation leaving surface 1 that strikes surface 2 directly

The notation φ_{12} is instructive for beginners, since it emphasizes that the view factor is for radiation that travels from surface 1 to surface 2.

The value of the view factor ranges between zero and one. The limiting case $\varphi_{12}=0$ indicates that the two surfaces do not have a direct view of each other, and thus radiation leaving surface 1 cannot strike surface 2 directly. The other limiting case $\varphi_{12}=1$ indicates that surface 2 completely surrounds surface 1, so that the entire radiation leaving surface 1 is intercepted by surface 2.

The view factor has proven to be very useful in radiation analysis because it allows us to express the fraction of radiation leaving a surface that strikes another surface in terms of the orientation of these two surfaces relative to each other. The underlying assumption in this process is that the radiation a surface receives from a source is directly proportional to the angle the surface subtends when viewed from the source. This would be the case only if the radiation coming off the source is uniform in all directions throughout its surface and the medium between the surfaces does not absorb, emit, or scatter radiation. That is, it will be the case when the surfaces are isothermal and diffuse emitters and reflectors and the surfaces are separated by a nonparticipating medium such as a vacuum or air.

Consider two black surfaces of arbitrary shape maintained at uniform temperatures T_1 and T_2. Recognizing that radiation leaves a black surface at a rate of E_b per unit surface area and that the view factor φ_{12} represents the fraction of radiation leaving surface 1 that strikes surface 2, the net rate of radiation heat transfer from surface 1 to surface 2 can be expressed as:

$$Q_{12} = E_{b1}A_1\varphi_{12} - E_{b2}A_2\varphi_{21} \tag{10-22}$$

(1) The reciprocity relation

$$A_1\varphi_{12} = A_2\varphi_{21} \tag{10-23}$$

This relation is referred to as the reciprocity relation or the reciprocity rule, and it enables us to determine the counterpart of a view factor from a knowledge of the view factor itself and the areas of the two surfaces. When determining the pair of view factors φ_{12} and φ_{21}, it makes sense to evaluate first the easier one directly and then the more difficult one by applying the reciprocity relation.

(2) The symmetry rule

The radiation analysis of a surface normally requires the consideration of the radiation coming in or going out in all directions. Therefore, most radiation problems encountered in practice involve enclosed spaces. When formulating a radiation problem, we usually form an enclosure consisting of the surfaces interacting radiatively. Even openings are treated as imaginary surfaces with radiation properties equivalent to those of the opening. The conservation of energy principle requires that the entire radiation leaving any surface i of an enclosure be intercepted by the surfaces of the enclosure. Therefore, the sum of the view factors from surface i of an enclosure to all surfaces of the enclosure, including to itself, must equal unity. This is known as the summation rule for an enclosure and is expressed as:

$$\sum_{j=1}^{n} \varphi_{i,j} = \varphi_{i,1} + \varphi_{i,2} + \cdots + \varphi_{i,i} + \cdots + \varphi_{i,n} = 1 \tag{10-24}$$

Where n is the number of surfaces of the enclosure.

10.3.2 Radiation Heat Transfer: Black Surfaces

From Eq. (10-22) and the reciprocity relation of the view factor:

$$Q_{12} = (E_{b1} - E_{b2})A_1\varphi_{12} = (E_{b1} - E_{b2})A_2\varphi_{21} \tag{10-25}$$

10.3.3 Radiation Heat Transfer: Gray Surfaces

The analysis of radiation transfer in enclosures consisting of black surfaces is relatively easy, as we have seen above, but most enclosures encountered in practice involve nonblack surfaces, which allow multiple reflections to occur. Radiation analysis of such enclosures becomes very complicated unless some simplifying assumptions are made.

To make a simple radiation analysis possible, it is common to assume the surfaces of an enclosure to be opaque, diffuse, and gray. That is, the surfaces are nontransparent, they are diffuse emitters and diffuse reflectors, and their radiation properties are independent of wavelength. Also, each surface of the enclosure is isothermal, and both the incoming and outgoing radiation are uniform over each surface. But first we review the concept of radiosity discussed.

Surfaces emit radiation as well as reflect it, and thus the radiation leaving a surface consists of emitted and reflected parts. The calculation of radiation heat transfer between surfaces involves the total radiation energy streaming away from a surface, with no regard for its origin. The total radiation energy leaving a surface per unit time and per unit area is the radiosity and is denoted by J (Fig. 10-12).

For a surface that is gray and opaque, the radiosity can be expressed as:

$$J = E + \rho G = \varepsilon E_b + (1 - \alpha)G \tag{10-26}$$

During a radiation interaction, a surface loses energy by emitting radiation and gains energy by absorbing radiation emitted by other surfaces. A surface experiences a net gain or a net loss of energy, depending on which quantity is larger. The net rate of radiation heat transfer from a surface of surface area A is denoted by Q and is expressed as:

$$Q = (J - G)A = (\varepsilon E_b - \alpha G)A$$

For gray surfaces $\alpha = \varepsilon$:

$$Q = \frac{A\varepsilon}{1 - \varepsilon}(E_b - J) \tag{10-27}$$

The radiation network of this two-surface enclosure consists of two surface resistances and one space resistance. In an electrical network, the electric current flowing through these resistances connected in series would be determined by dividing the potential difference between points A and B by the total resistance between the same two points. The net rate of radiation transfer is determined in the same manner and is expressed as:

$$Q_{12} = \frac{E_{b1} - E_{b2}}{\dfrac{1-\varepsilon_1}{A_1\varepsilon_1} + \dfrac{1}{A_1\varphi_{12}} + \dfrac{1-\varepsilon_2}{A_2\varepsilon_2}} \tag{10-28}$$

10.3.4 Network Methods of Solving Radiation Problems

In the radiation analysis of an enclosure, either the temperature or the net rate of heat transfer must be given for each of the surfaces to obtain a unique solution for the unknown surface temperatures and heat transfer rates. The network method, a method commonly used to solve radiation problems, is based on the electrical network analogy.

The network method was first introduced by A. K. Oppenheim in the 1950s and found widespread acceptance because of its simplicity and emphasis on the physics of the problem. The application of the method is straightforward: draw a surface resistance associated with each surface of an enclosure and connect them with space resistances. Then solve the radiation problem by treating it as an electrical network problem where the radiation heat transfer replaces the current and radiosity replaces the potential. The network method is not practical for enclosures with more than three or four surfaces, however, because of the increased complexity of the network. Next we apply the method to solve radiation problems in two- and three surface enclosures.

In an electrical analogy to Ohm's law, substituting into Eq. (10-25) yields:

$$Q_{12} = \frac{E_{b1} - E_{b2}}{\dfrac{1}{A_1\varphi_{12}}} \tag{10-29}$$

Where $1/(A_1\varphi_{12})$ is the space resistance to radiation. The quantity $E_{b1} - E_{b2}$ corresponds to a potential difference and the net rate of radiation heat transfer corresponds to current in the electrical analogy, as illustrated in Fig. 10-13.

In an n-surface enclosure, the conservation of energy principle requires that the net heat transfer from surface i be equal to the sum of the net heat transfers from surface i to each of the n surfaces of the enclosure. That is:

$$Q_i = \sum_{j=1}^{n} Q_{i,j} = \sum_{j=1}^{n} A_i\varphi_{i,j}(E_{bi} - E_{bj})$$

Which has the electrical analogy interpretation that the net radiation flow from a surface through its surface resistance is equal to the sum of the radiation flows from that surface to all other surfaces through the corresponding space resistances.

In an electrical analogy to Ohm's law, for Eq. (10-28), $\dfrac{1-\varepsilon_1}{A_1\varepsilon_1}$ and $\dfrac{1-\varepsilon_2}{A_2\varepsilon_2}$ are the surface resistance to radiation, and $1/(A_1\varphi_{12})$ is the space resistance. The quantity

$E_{b1} - E_{b2}$ corresponds to a potential difference and the net rate of radiation heat transfer corresponds to current in the electrical analogy, as illustrated in Fig. 10-14.

For two very large parallel plates, $A_1 = A_2 = A$, $\varphi_{12} = \varphi_{21} = 1$, the net rate of radiation heat transfer between the two surfaces per unit surface area of the plates is determined from Eq. (10-28) to be:

$$Q_{12} = \frac{A(E_{b1} - E_{b2})}{\dfrac{1}{\varepsilon_1} + \dfrac{1}{\varepsilon_2} - 1} = A\varepsilon_s(E_{b1} - E_{b2})$$

Where, $\varepsilon_s = \dfrac{1}{\dfrac{1}{\varepsilon_1} + \dfrac{1}{\varepsilon_2} - 1}$ is system emissivity.

第 11 章　传热过程及换热器

本章导学

在实际工程中,大量的热量传递都是在换热器中进行的,这种热量传递往往是导热、对流和辐射 3 种基本传热方式组合在一起的复杂传热过程。本章重点讨论了平壁和圆管壁的传热过程的分析计算方法,介绍了换热器的工作原理及结构,阐述了强化和削弱传热过程的方法。

学习目标

掌握传热过程的分析方法,熟悉通过平壁和圆管壁的传热过程的计算方法。了解换热器的基本概念、工作原理及分类。了解强化和削弱传热的途径。

重点与难点

(1) 传热系数的计算与分析。

(2) 应用传热学知识分析强化和削弱传热的途径。

11.1　传热过程

前面讨论了热量传递的三种基本方式。但实际上,大多数传热过程往往是几种传热方式同时起作用的结果。如热力设备和热力管道表面的散热,它既有热表面与空气之间的对流换热,又有与周围环境表面间的辐射换热。凡有几种热量传递方式同时起作用的换热过程统称为复合换热。

工程中经常出现这样一种换热现象:冷热两种流体被固体间隔壁隔开,热量由热流体通过间隔壁传给冷流体,通常称这种热量传递过程为传热过程。传热过程是工程中广泛应用的热量传递过程,如锅炉中的过热器及制冷循环中的蒸发器、冷凝器等设备的热量传递都是传热过程。在这些换热设备中,热量通过固体壁面时,是以单纯的热传导的方式进行的,而流体与固体壁面之间的热量传递则依靠对流换热或热辐射的方式进行,或者是两种方式联合作用进行的。一般来说,可将整个传热过程均看作由 3 个串联环节组成,分别为:① 热流体与壁面的复合传热过程;② 固体壁面的导热过程;③ 冷流体与壁面的复合传热过程。

在热工换热设备中,常见的换热壁面主要是平壁和圆管壁,下面就以这两种典型形状的固体壁面为例进行传热过程的分析与计算。

11.1.1 通过平壁的传热过程

如图 11-1 所示,冷、热流体被一无限大平壁隔开。已知平壁两侧流体温度分别为 t_{f1} 和 t_{f2},且 $t_{f1} > t_{f2}$;平壁厚度为 δ,平壁表面积为 A,导热系数为 λ;热、冷流体对壁面的对流换热系数分别为 α_1 和 α_2。试确定热流量 Q 及壁面两侧温度。

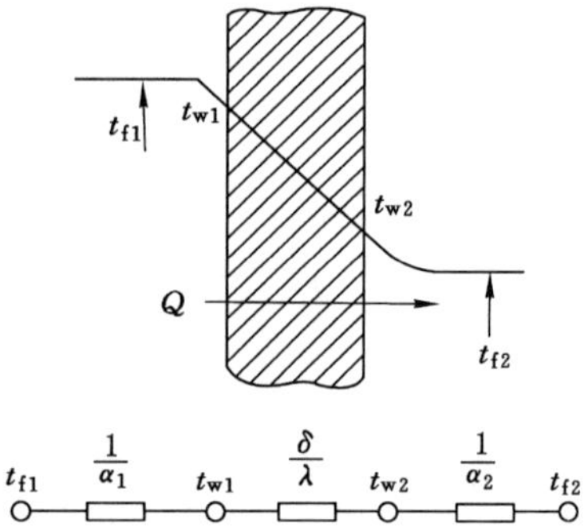

图 11-1 通过平壁的传热过程

热流体传给壁面一侧的对流换热量为:

$$Q_1 = \alpha_1(t_{f1} - t_{w1})A \tag{a}$$

平壁高温侧传给低温侧的热量为:

$$Q_2 = \frac{\lambda}{\delta}(t_{w1} - t_{w2})A \tag{b}$$

平壁另一侧传给冷流体的对流换热量为:

$$Q_3 = \alpha_2(t_{w2} - t_{f2})A \tag{c}$$

对于稳态传热,则有:

$$Q_1 = Q_2 = Q_3$$

整理式(a)、式(b)、式(c)可得:

$$Q = k(t_{f1} - t_{f2})A \tag{11-1}$$

式中,k 为传热系数,W/(m^2 · K),$k = \dfrac{1}{\dfrac{1}{\alpha_1} + \dfrac{\delta}{\lambda} + \dfrac{1}{\alpha_2}}$,表示当冷、热流体的温差为 1 ℃时,单位时间通过单位平壁面积所传递的热量。由于壁面两侧的面积相等,故传热系数不论对哪侧来说均相等。

根据式(a)至式(c),可容易算出壁面两侧的温度分别为:

$$t_{w1} = t_{f1} - \frac{Q}{\alpha_1 A}$$

$$t_{w2} = t_{f2} + \frac{Q}{\alpha_2 A}$$

如果将式(11-1)改写成 $Q = \dfrac{\Delta t}{1/(kA)}$,可知 $1/(kA)$ 为传热的热阻。可知:

$$\frac{1}{kA} = \frac{1}{\alpha_1 A} + \frac{\delta}{\lambda A} + \frac{1}{\alpha_2 A} \tag{11-2}$$

式(11-2)所表示的热阻为三个分热阻之和,即壁面的对流换热热阻、导热热阻、另一侧壁面对流换热热阻之和。热量依次经过三个热阻才从热流体传给冷流体,这与电流依次通过串联的三个电阻的情况类似,所以这时的总热阻为串联的热阻之和。这里,热阻是对整个平壁传热面积而言的。

按照上述热阻串联的概念,可知多层平壁的传热热阻为:

$$\frac{1}{kA} = \frac{1}{\alpha_1 A} + \sum_{i=1}^{n} \frac{\delta_i}{\lambda_i A} + \frac{1}{\alpha_2 A} \tag{11-3}$$

式中,n 为多层平壁的层数。

传热量为:

$$Q=\frac{t_{f1}-t_{f2}}{\frac{1}{\alpha_1 A}+\sum_{i=1}^{n}\frac{\delta_i}{\lambda_i A}+\frac{1}{\alpha_2 A}}$$

【例 11-1】 已知室内空气温度为 20 ℃，墙体厚度为 240 mm，室外环境温度为 −10 ℃，墙体的导热系数 $\lambda=0.95$ W/(m·℃)，室内空气与内墙面的对流换热系数 $\alpha_1=8$ W/(m²·℃)，室外空气与外墙面的对流换热系数 $\alpha_2=22$ W/(m²·℃)，试求此时墙体的热流密度和墙体内侧的温度。

【解】 对门、窗、墙角处的传热不予考虑（实际情况下需要考虑），将房间的墙看作无限大的平壁。根据平壁传热公式求得热流密度：

$$q=\frac{Q}{A}=\frac{t_{f1}-t_{f2}}{\frac{1}{\alpha_1}+\frac{\delta}{\lambda}+\frac{1}{\alpha_2}}=\frac{20-(-10)}{\frac{1}{8}+\frac{240\times10^{-3}}{0.95}+\frac{1}{22}}=\frac{30}{0.4231}=70.91\ (\mathrm{W/m^2})$$

$$t_{w1}=t_{f1}-\frac{q}{\alpha_1}=20-\frac{70.91}{8}=11.1\ (℃)$$

11.1.2　通过圆管壁的传热过程

有一根长为 l 的圆管，管内外直径分别为 d_1 和 d_2，内、外两侧的流体温度分别为 t_{f1} 和 t_{f2}（$t_{f1}>t_{f2}$），管内外两侧的对流换热系数分别为 α_1 和 α_2，管壁材料的导热系数为 λ，如图 11-2 所示。试确定热流量 Q。对于稳态传热，热流体对圆管壁的放热量等于通过圆管壁的导热量，也等于圆管壁对冷流体的放热量。即

$$Q_1=\alpha_1\pi d_1 l(t_{f1}-t_{w1})\qquad\text{(a)}$$

$$Q_2=\frac{\pi l(t_{w1}-t_{w2})}{\frac{1}{2\lambda}\ln\frac{d_2}{d_1}}\qquad\text{(b)}$$

$$Q_3=\alpha_2\pi d_2 l(t_{w2}-t_{f2})\qquad\text{(c)}$$

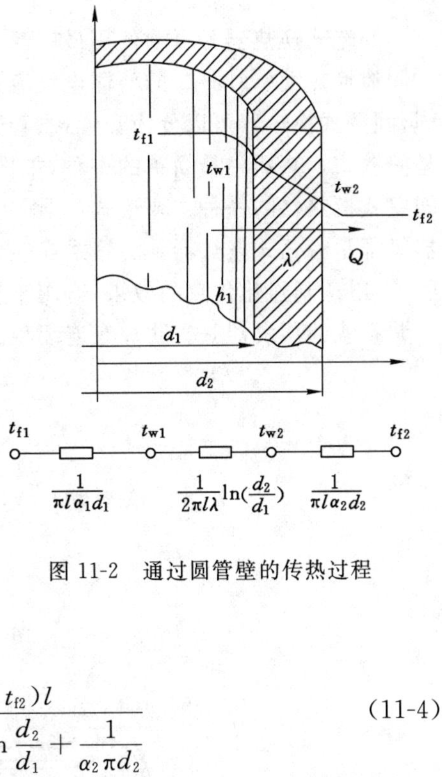

图 11-2　通过圆管壁的传热过程

由 $Q_1=Q_2=Q_3$ 整理后可得：

$$Q=\frac{t_{f1}-t_{f2}}{\frac{1}{\alpha_1\pi d_1 l}+\frac{1}{2\pi\lambda l}\ln\frac{d_2}{d_1}+\frac{1}{\alpha_2\pi d_2 l}}$$

由此可写出：

$$Q=k_1\Delta t l=\frac{(t_{f1}-t_{f2})l}{\frac{1}{\alpha_1\pi d_1}+\frac{1}{2\pi\lambda}\ln\frac{d_2}{d_1}+\frac{1}{\alpha_2\pi d_2}}\qquad(11\text{-}4)$$

式中，k_1 为单位管长的传热系数：

$$k_1=\frac{1}{\frac{1}{\alpha_1\pi d_1}+\frac{1}{2\pi\lambda}\ln\frac{d_2}{d_1}+\frac{1}{\alpha_2\pi d_2}}\qquad(11\text{-}5)$$

必须指出，k_1 是对单位管长而言的传热系数，平壁的传热系数是对单位面积而言的，二者不同。在传热问题中，壁面的几何形状不同，传热系数的定义不同，在应用计算公式时，需

注意传热系数的定义。

同理，对于多层圆管壁，其 k_1 为：

$$k_1 = \frac{1}{\dfrac{1}{\alpha_1 \pi d_1} + \sum_{i=1}^{n} \dfrac{\ln(d_{i+1}/d_i)}{2\pi\lambda_i} + \dfrac{1}{\alpha_2 \pi d_{n+1}}} \tag{11-6}$$

11.2 换　热　器

换热器也称热交换器，是实现热量从热流体传递给冷流体的装置。换热器种类繁多，应用甚广，是工程传热中的重要设备，广泛应用于石油、化工、动力、建筑、机械等工业部门。比如车用水散热器(俗称水箱)，空调系统中的蒸发器、冷凝器，热电循环中的凝汽器等。

11.2.1 换热器的种类

工程上常用的换热器种类很多，按其工作原理大致可分为 3 类：间壁式(或称表面式)、回热式(或称蓄热式)和混合式(或称接触式)。下面分别介绍各类换热器的原理及特点。

(1) 间壁式换热器

间壁式换热器是目前使用最广泛的一种换热器。在这类换热器中，冷热两种流体同时在固体换热面(即间壁)的两侧连续流过，热流体通过壁面把热量传给冷流体。根据结构不同，间壁式换热器可以分为套管式、管壳式、肋管式、板式、螺旋板式等类型。套管式换热器是间壁式换热器中最简单的一种，如图 11-3 所示。管壳式换热器又称列管式换热器，这是间壁式换热器的一种主要形式，如图 11-4 所示，化工厂、电厂及空调系统中的蒸发器、冷凝器等都是管壳式换热器。肋管式换热器是为了增大换热面积，减小管外热阻而在管外加有肋片，以达到强化传热的效果，适用于管内液体和管外气体之间的换热，如汽车水箱散热器、空调系统的蒸发器等。肋片有连续型的[图 11-5(a)]和非连续型的[图 11-5(b)]。

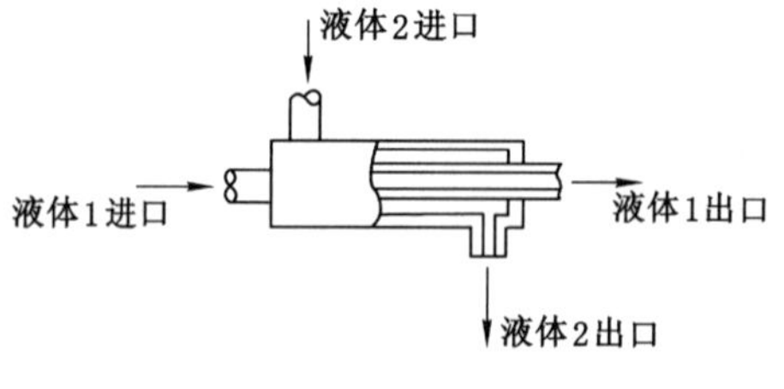

图 11-3　套管式换热器

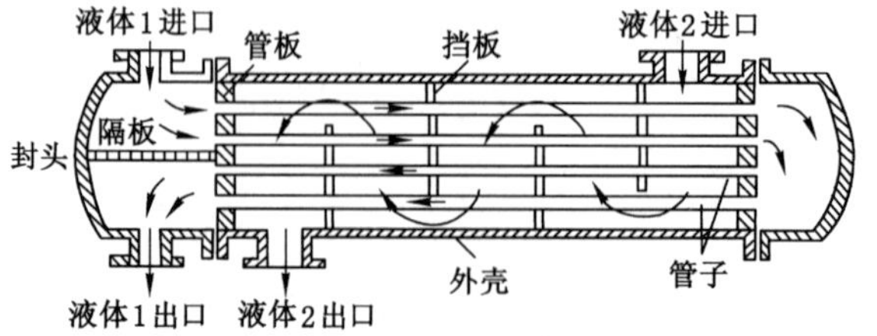

图 11-4　管壳式换热器

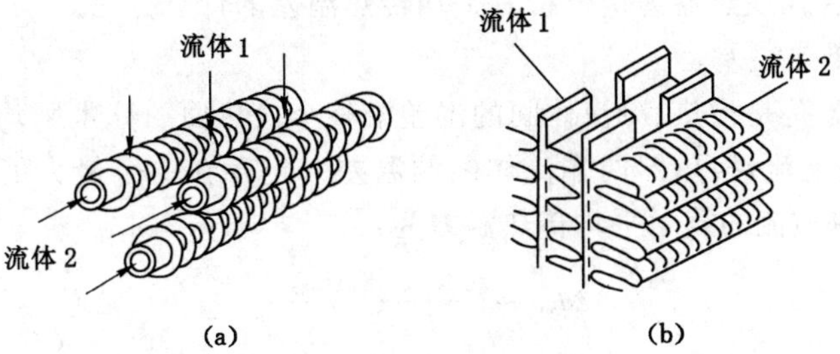

图 11-5　肋管式换热器

(2) 回热式换热器

回热式换热器的特点是冷、热流体交替地流过同一换热面。当热流体流过换热器时，换热面吸热，温度升高，储蓄热量。当冷流体流过换热器时，则储蓄的热量传给冷流体。如此反复进行，实现热流体将热量传给冷流体的目的。如钢厂热风炉、大型电厂锅炉的回热式空气预热器就是这类换热器。

(3) 混合式换热器

混合式换热器的特点是，冷热两种流体直接接触，彼此混合并换热，在热交换的同时存在着质量交换，如火力发电厂中的大型冷却水塔和空调系统中的中小型冷却水塔等。

11.2.2　平均传热温差

间壁式换热器中冷、热流体沿换热面流动时，沿途温度一般要发生变化，二者之间的温差也发生变化，且随着流体流动方式的不同而异。图 11-6 表示顺、逆流时冷、热流体的温度变化曲线(热流体的进、出口温度分别为 t_1'，t_1''，而冷流体的进、出口温度分别为 t_2'，t_2'')，因此在换热器的传热计算中必须引用平均传热温差 Δt_m 进行计算。

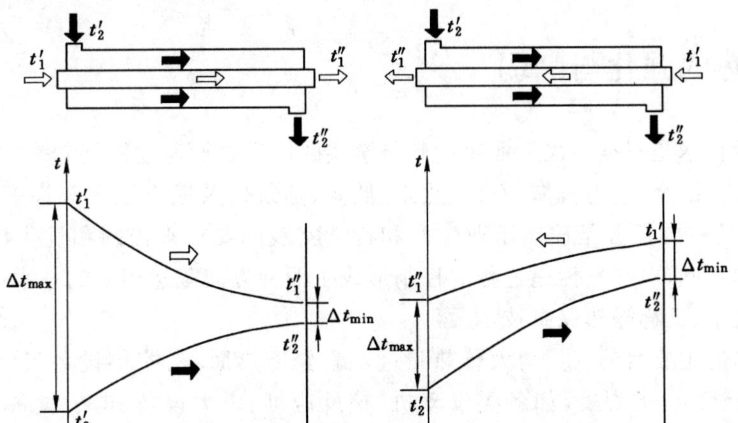

图 11-6　换热器中流体温度沿程变化

平均传热温差 Δt_m 的计算方法一般有两种：一是算术平均传热温差；二是对数平均传热

温差。下面就以套管式换热器为例分析其平均传热温差的计算方法。

（1）算术平均传热温差

算术平均传热温差是指热、冷流体间的温差沿整个换热面的算术平均值。如图 11-6 所示，不管是顺流还是逆流，换热器两端流体间的温差一定存在一个最大值 $\Delta t_{\max}$ 和一个最小值 $\Delta t_{\min}$，因而顺流或逆流的算术平均传热温差为：

$$\Delta t_{\mathrm{m}} = \frac{\Delta t_{\max} + \Delta t_{\min}}{2} \tag{11-7}$$

算术平均传热温差虽然算法简单，由于它是假设冷、热流体的温度均按直线变化时的平均温差，因而无法真实反映冷、热流体的实际温度变化情况，误差较大。一般当 $\Delta t_{\max}/\Delta t_{\min}<2$ 时，其计算误差小于 4%，这在工程计算中是允许的。当计算精度要求较高时，则采用对数平均传热温差，这也是换热器计算中通常采用的温差计算方法。

（2）对数平均传热温差

数学推导分析表明，无论是顺流还是逆流，都可以统一用式(11-8)计算换热器的平均传热温差：

$$\Delta t_{\mathrm{m}} = \frac{\Delta t_{\max} - \Delta t_{\min}}{\ln \dfrac{\Delta t_{\max}}{\Delta t_{\min}}} \tag{11-8}$$

由于式(11-8)中出现对数运算，因而通常称之为对数平均传热温差。

在各种流动形式中，顺流和逆流是两种极端情况。在冷、热流体进、出口温度相同的情况下，逆流的平均温差大于顺流的平均温差。由图 11-6 可以看出，顺流时冷流体的出口温度 t_2'' 总是低于热流体的出口温度 t_1''，而逆流时 t_2'' 却可以大于 t_1''，因此从强化传热的角度出发，换热器应当尽量布置成逆流。但逆流的缺点是热流体和冷流体的最高温度 t_1'，t_2'' 和最低温度 t_1''，t_2' 分别集中在换热器的两端，这使换热器的温度分布乃至热应力分布极不均匀，不利于换热器的安全运行，尤其对于高温换热器来说，这种情况应该避免。

11.3 传热的强化与削弱

传热工程技术是根据现代工业生产和科学实践的需要而蓬勃发展起来的先进科学与工程技术，在电力、冶金、动力机械、石油、化工、低温、建筑以及航空航天等许多领域发挥极其重要的作用，其主要任务是按照工业生产和科学实践的要求来控制和优化热量传递过程。根据应用目的不同，对热量传递过程的控制形成了两种方向截然相反的技术：强化传热技术与削弱传热技术（又称隔热保温技术）。

强化传热的主要目的：① 增大传热量；② 减少传热面积、缩小设备尺寸、降低材料消耗；③ 降低高温部件的温度，如各类发动机、核反应堆、电力设备、电子设备中元器件的冷却，保证设备安全运行；④ 降低载热流体的输送功率。削弱传热的主要目的：① 减少热力设备、载热流体的热损失，节约能源，如火力发电厂锅炉、汽轮机以及过热蒸汽输送管道的保温等；② 维护低温工程中的人工低温环境，防止外界热量的传入，如冷冻仓库、冷藏车、储液罐以及冰箱的隔热等；③ 保护工程技术人员的人身安全，避免遭受热或冷的伤害，创造温度适宜的工作和生活环境，如各类航天器在穿过大气层时，由于其表面和大气的摩擦，会产生几千摄氏度以上的高温，因此必须采取隔热措施，避免航天器烧毁；再如载人航天器在太空飞行时，面对太阳的高温辐射以及自身向温度约为 3 K 的低温太空环境的热辐射，如何保证宇航员座舱内近 20 ℃的工作、生活环境，是隔热保温技术必须解决的问题。

11.3.1　强化传热

由传热基本公式：

$$Q = kA\Delta t_m = \frac{\Delta t_m}{R_k} = \frac{\Delta t_m}{R_{\alpha 1} + R_\lambda + R_{\alpha 2}} \tag{11-9}$$

可知，强化传热有两种基本途径：

（1）加大传热温差

途径有两条：一是提高热流体温度和降低冷流体温度；二是换热面两侧流体的流动尽可能采用逆流方式。

（2）减小传热总热阻

传热总热阻由传热过程中各串联热阻所组成，分析各串联热阻的数值，并设法减小其中最大的热阻，就能降低传热总热阻。采取的措施可以有：① 减小导热热阻，其中包括换热面本身热阻和表面污垢热阻。一般换热器为金属壁，其导热系数较高，导热热阻可略去不计，而污垢层虽不厚，但会产生较大的热阻。例如，1 mm 厚水垢的导热热阻相当于 40 mm 厚普通钢板的导热热阻。所以，防止和及时清除污垢是保证换热设备正常高效运行的重要技术措施。② 减小对流换热热阻。工程上常用的方法有：在表面传热系数小的一侧加装肋片，并注意使肋基接触良好。这是减小对流换热热阻最有效的方法。另外，可以适当增加流速，采用小管径，增加流体的扰动和混合，以破坏边界层等措施来提高表面传热系数，如在管道内设置强化圈，则内壁面一侧的表面传热系数会显著提高。③ 减小辐射热阻，可采用增加系统黑度、增加物体间角系数等措施。

11.3.2　削弱传热

（1）采用隔热保温技术

与强化传热相反，除减小传热温差外，工程上削弱传热的主要措施是在壁面上敷设保温层以达到增加传热热阻的目的，即采用所谓的隔热保温技术。包括：① 选择导热系数低的材料进行保温，一般要低于 0.1 W/(m·K)，如石棉、泡沫塑料等；② 采用真空夹层减小对流与导热传热；③ 采用遮热板和减小材料的表面黑度（如铝箔）降低辐射传热；④ 在条件允许的情况下增加保温层厚度。

（2）采用临界热绝缘直径

在平壁上敷设保温层，热阻总是随保温层厚度的增加而增加，从而削弱传热。但在圆筒壁上敷设保温层，热阻并不总是随着厚度的增加而增加，可能会出现相反的由于外表面传热面积增大使总热阻减小的情况。这样就存在一个使传热效果发生转折的保温层外直径 d_c，称之为临界热绝缘直径。假设管壁材料的导热系数为 λ_1，保温材料的导热系数为 λ_x，如图 11-7 所示。

根据前面对通过圆管壁的传热过程的分析可知，两层圆管壁的传热热阻为 4 个热阻之和，即

$$R_k = R_{\alpha 1} + R_{\lambda 1} + R_{\lambda x} + R_{\alpha 2} = \frac{1}{\alpha_1 \pi d_1 l} + \frac{1}{2\pi\lambda_1 l}\ln\frac{d_2}{d_1} + \frac{1}{2\pi\lambda_x l}\ln\frac{d_x}{d_2} + \frac{1}{\alpha_2 \pi d_x l}$$

由上式可见，如 d_1，d_2，α_1，λ_1 和 λ_x 为定值，随着保温层厚度的增加，即 d_x 的增大，管内对

流换热热阻与管壁导热热阻之和$(R_{\alpha1}+R_{\lambda1})$保持不变，保温层的导热热阻$R_{\lambda x}$随之加大，但保温层外侧的对流换热热阻$R_{\alpha2}$却随之减小。当$d_2$较小时，总热阻$R_k$先随着$d_x$的增大而减小，然后随着$d_x$的增大而增大，中间出现极小值，相应热流量$Q$出现极大值，如图11-7(b)所示。当$dR_k/dd_x=0$时，总热阻$R_k$取得极小值，此时的保温层外直径即临界热绝缘直径$d_c$，可求得：

$$d_x=\frac{2\lambda_x}{\alpha_2}=d_c \tag{11-10}$$

图11-7(b)及式(11-10)表明，对于直径d_2小于d_c的小管子敷设保温层，必须考虑临界热绝缘直径的问题。若保温层外直径$d_x<d_c$，此时随着保温层厚度增加，总热阻减小，散热量增大；当保温层加厚到$d_x=d_c$时，总热阻最小，散热量最大；若再加厚，$d_x>d_c$，则随保温层厚度的增加，总热阻将增大，散热量减小，只有当保温层外直径$d_x>d_c$时，保温层才真正有减小散热损失的作用。工程上，绝大多数需要加保温层的管道外径都大于临界热绝缘直径，只有当管径很小，保温材料的导热系数又很大时，才会考虑临界热绝缘直径的问题。

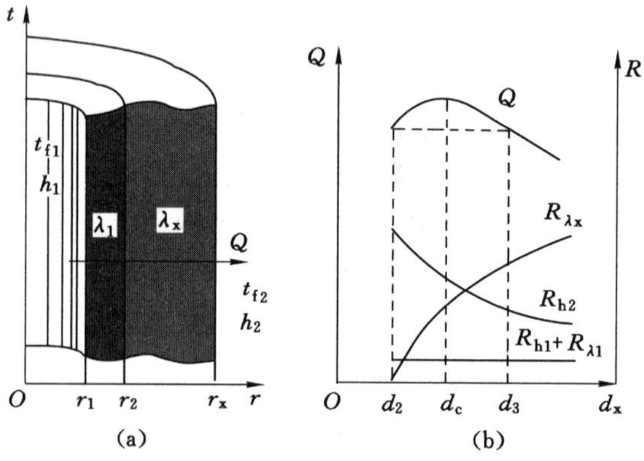

图11-7 临界热绝缘直径分析示意图

本章小结

实现冷、热流体间热量交换的设备称为换热器，流体在换热器中所进行的传热过程往往是导热、对流、辐射换热3种方式共同作用的复合传热过程。处理这类复合传热问题的有效方法是采用热阻分析方法，并从中找出主导换热方式，然后把次要的传热方式归纳到主导传热方式中加以考虑，或者忽略不计。换热器的种类繁多，按其工作原理，可分为混合式、回热式及间壁式等类型，其中间壁式换热器应用最为广泛。

思考与练习题

(1) 传热热阻由哪些分热阻组成？传热系数与传热热阻之间有什么关系？

(2) 平壁传热系数与圆管壁传热系数有什么不同？

(3) 何为换热器？它可以分为哪几种类型？

(4) 为什么逆流换热器比顺流换热器更为有效？

(5) 什么是对数平均传热温差？它与算术平均传热温差有什么区别和联系？

(6) 强化传热的目的是什么？采用哪些方法能强化传热？

(7) 保温层厚度越厚，隔热效果就越好吗？分别以平壁和圆管壁为例加以说明。

(8) 用热水散热器来采暖，如果不改变热水温度，只增加热水流速，能不能有效增加散热量，为什么？

(9) 野外工作者常用纸制容器来烧水。假设纸的厚度为 0.2 mm，纸的导热系数为 0.9 W/(m·K)，水在大气压力下沸腾，水侧的沸腾换热表面传热系数为 2 400 W/(m^2·K)。容器的加热火焰温度为 1 100 ℃，火焰与纸面的表面传热系数为 95 W/(m^2·K)。若纸的耐火温度为 200 ℃，试证明该纸制容器能够耐火。

(10) 有一气体冷却器，气侧表面传热系数 $\alpha_1 = 95$ W/(m^2·K)，壁厚 $\delta = 25$ mm，$\lambda = 46.5$ W/(m·K)。水侧表面传热系数 $\alpha_2 = 5\ 800$ W/(m^2·K)。计算每个环节的热阻以及传热热阻，并指出此时增强传热应从哪个环节着手？此设备运行一段时间后，如果在气侧积了一层 1 mm 厚的灰垢[$\lambda = 0.116$ W/(m·K)]，水侧结了一层厚为 0.5 mm 的水垢[$\lambda = 1.16$ W/(m·K)]，其他条件不变，问此时传热热阻为多少？

Chapter 11 Heat Transfer Process and Heat Exchangers

11.1 Heat Transfer Process

A heat exchanger typically involves two flowing fluids separated by a solid wall. Heat is first transferred from the hot fluid to the wall by convection, through the wall by conduction, and from the wall to the cold fluid again by convection. Any radiation effects are usually included in the convection heat transfer coefficients.

11.1.1 Heat Transfer through the Plane Wall

Consider a large plane wall of thickness δ, thermal conductivity λ, and surface area A. The two sides of the wall are maintained at constant temperatures of t_{w1} and t_{w2}, respectively, as shown in Fig. 11-1. Determine the variation of temperature within the wall and the rate of heat conduction through the wall under steady conditions. A plane wall with specified surface temperatures is given. The variation of temperature and the rate of heat transfer are to be determined. Heat conduction is steady. Heat conduction is one-dimensional since the wall is large relative to its thickness and the thermal conditions on both sides are uniform. Thermal conductivity is constant. There is no heat generation.

Taking the direction normal to the surface of the wall to be the x-direction, the differential equation for this problem can be expressed as:

$$Q_1 = \alpha_1 (t_{f1} - t_{w1}) A$$

The differential equation is linear and second order, and a quick inspection of it reveals that it has a single term involving derivatives and no terms involving the unknown function T as a factor. Thus, it can be solved by direct integration. Noting that an integration reduces the order of a derivative by one, the general solution of the differential equation above can be obtained by two simple successive integrations, each of which introduces an integration constant. As a check, if we take the derivative of this equation, we will obtain the original differential equation. This equation is not the solution yet since it involves a derivative.

Integrating one more time, we obtain:

$$Q = k(t_{f1} - t_{f2}) A$$

Which is the general solution of the differential equation. The general solution in this case resembles the general formula of a straight line whose slope is C_1 and whose value at

$x=0$ is C_2. This is not surprising since the second derivative represents the change in the slope of a function, and a zero second derivative indicates that the slope of the function remains constant. Therefore, any straight line is a solution of this differential equation.

The general solution contains two unknown constants C_1 and C_2, and thus we need two equations to determine them uniquely and obtain the specific solution. These equations are obtained by forcing the general solution to satisfy the specified boundary conditions. The application of each condition yields one equation, and thus we need to specify two conditions to determine the constants C_1 and C_2.

When applying a boundary condition to an equation, all occurrences of the dependent and independent variables and any derivatives are replaced by the specified values. Thus the only unknowns in the resulting equations are the arbitrary constants.

Substituting the C_1 and C_2 expressions into the general solution with boundary conditions: $x=0, t=t_{w1}$; $x=\delta, t=t_{w2}$, we obtain:

$$t_{w1}=t_{f1}-\frac{Q}{\alpha_1 A}$$

$$t_{w2}=t_{f2}+\frac{Q}{\alpha_2 A}$$

Which is the desired solution since it satisfies not only the differential equation but also the two specified boundary conditions.

11.1.2 Heat Transfer through Cylindrical Shell

Consider steady heat conduction through a hot water pipe. Heat is continuously lost to the outdoors through the wall of the pipe, and we intuitively feel that heat transfer through the pipe is in the normal direction to the pipe surface and no significant heat transfer takes place in the pipe in other directions (Fig. 11-2). The wall of the pipe, whose thickness is rather small, separates two fluids at different temperatures, and thus the temperature gradient in the radial direction will be relatively large. Further, if the fluid temperatures inside and outside the pipe remain constant, then heat transfer through the pipe is steady. Thus heat transfer through the pipe can be modeled as steady and one-dimensional. The temperature of the pipe in this case will depend on one direction only (the radial r-direction) and can be expressed as $T=T(r)$. The temperature is independent of the azimuthal angle or the axial distance. This situation is approximated in practice in long cylindrical pipes and spherical containers.

In steady operation, there is no change in the temperature of the pipe with time at any point. Therefore, the rate of heat transfer into the pipe must be equal to the rate of heat transfer out of it. In other words, heat transfer through the pipe must be constant, $Q_{cond,cyl}=$constant.

Consider a long cylindrical layer (such as a circular pipe) of inner radius r_1, outer radius r_2, length l, and average thermal conductivity λ. The two surfaces of the cylindrical layer are maintained at constant temperatures $T_1=t_{w1}$ and $T_2=t_{w2}$. There is no heat generation in the layer and the thermal conductivity is constant. For one-dimensional heat conduction through the cylindrical layer, we have $T(r)$. Then Fourier's law of heat conduction for heat transfer through the cylindrical layer can be expressed as:

$$Q_1 = \alpha_1 \pi d_1 l (t_{f1} - t_{w1})$$

Separating the variables in the above equation and integrating from $r = r_1$, where $T(r_1) = t_{w1}$, to $r = r_2$, where $T(r_2) = t_{w2}$, gives:

$$Q_2 = \frac{\pi l (t_{w1} - t_{w2})}{\frac{1}{2\lambda} \ln \frac{d_2}{d_1}}$$

$$Q_3 = \alpha_2 \pi d_2 l (t_{w2} - t_{f2})$$

$$Q = \frac{t_{f1} - t_{f2}}{\frac{1}{\alpha_1 \pi d_1 l} + \frac{1}{2\pi\lambda l} \ln \frac{d_2}{d_1} + \frac{1}{\alpha_2 \pi d_2 l}}$$

$$Q = k_1 \Delta t l = \frac{(t_{f1} - t_{f2}) l}{\frac{1}{\alpha_1 \pi d_1} + \frac{1}{2\pi\lambda} \ln \frac{d_2}{d_1} + \frac{1}{\alpha_2 \pi d_2}}$$

$$k_1 = \frac{1}{\frac{1}{\alpha_1 \pi d_1} + \frac{1}{2\pi\lambda} \ln \frac{d_2}{d_1} + \frac{1}{\alpha_2 \pi d_2}}$$

is the thermal resistance of the cylindrical layer against heat conduction, or simply the conduction resistance of the cylinder layer.

11.2 Heat Exchangers

Heat exchangers are devices that facilitate the exchange of heat between two fluids that are at different temperatures while keeping them from mixing with each other. Heat exchangers are commonly used in practice in a wide range of applications, from heating and air-conditioning systems in a household, to chemical processing and power production in large plants.

11.2.1 Classification of Heat Exchangers

The simplest type of heat exchanger consists of two concentric pipes of different diameters, as shown in Fig. 11-3, called the double-pipe heat exchanger. One fluid in a double-pipe heat exchanger flows through the smaller pipe while the other fluid flows through the annular space between the two pipes. Two types of flow arrangement are possible in a double-pipe heat exchanger: in parallel flow, both the hot and cold fluids enter the heat exchanger at the same end and move in the same direction. In counter flow, on the other hand, the hot and cold fluids enter the heat exchanger at opposite ends and flow in opposite directions.

Another type of heat exchanger, which is specifically designed to realize a large heat transfer surface area per unit volume, is the compact heat exchanger. Compact heat exchangers enable us to achieve high heat transfer rates between two fluids in a small volume, and they are commonly used in applications with strict limitations on the weight and volume of heat exchangers. The large surface area in compact heat exchangers is obtained by attaching closely spaced thin plate or corrugated fins to the walls separating the two fluids. Compact heat exchangers are commonly used in gas-to-gas and gas-to-liquid (or liquid-to-gas) heat exchangers to counteract the low heat transfer coefficient associated with gas flow with increased surface area. In a car radiator, which is a water-to-air compact heat exchanger, for example, it is no surprise that fins are attached to the air side of the tube surface. In compact heat exchangers, the two fluids usually move perpendicular to each other, and such flow configuration is called cross-flow. The cross-flow is further classified as unmixed and mixed flow, depending on the flow configuration.

The presence of mixing in the fluid can have a significant effect on the heat transfer characteristics of the heat exchanger. Perhaps the most common type of heat exchanger in industrial applications is the shell-and-tube heat exchanger, shown in Fig. 11-4. Shell-and-tube heat exchangers contain a large number of tubes (sometimes several hundred) packed in a shell with their axes parallel to that of the shell. Heat transfer takes place as one fluid flows inside the tubes while the other fluid flows outside the tubes through the shell. Baffles are commonly placed in the shell to force the shell-side fluid to flow across the shell to enhance heat transfer and to maintain uniform spacing between the tubes. Despite their widespread use, shell-and-tube heat exchangers are not suitable for use in automotive and aircraft applications because of their relatively large size and weight. Note that the tubes in a shell-and-tube heat exchanger open to some large flow areas called headers at both ends of the shell, where the tube-side fluid accumulates before entering the tubes and after leaving them. Shell-and-tube heat exchangers are further classified according to the number of shell and tube passes involved. Heat exchangers in which all the tubes make one U-turn in the shell, for example, are called one-shell-pass and two-tube- passes heat exchangers. Likewise, a heat exchanger that involves two passes in the shell and four passes in the tubes is called a two-shell-passes and four-tube-passes heat exchanger.

An innovative type of heat exchanger that has found widespread use is the plate and frame (or just plate) heat exchanger, which consists of a series of plates with corrugated flat flow passages (Fig. 11-5). The hot and cold fluids flow in alternate passages, and thus each cold fluid stream is surrounded by two hot fluid streams, resulting in very effective heat transfer. Also, plate heat exchangers can grow with increasing demand for heat transfer by simply mounting more plates. They are well suited for liquid-to-liquid heat exchanger applications, provided that the hot and cold fluid streams are at about the same pressure.

Heat exchangers are often given specific names to reflect the specific application for which they are used. For example, a condenser is a heat exchanger in which one of the fluids is cooled and condenses as it flows through the heat exchanger. A boiler is another heat exchanger in which one of the fluids absorbs heat and vaporizes. A space radiator is a heat exchanger that transfers heat from the hot fluid to the surrounding space by radiation.

11.2.2 Mean Temperature Difference

Earlier, we mentioned that the temperature difference between the hot and cold fluids varies along the heat exchanger, and it is convenient to have a mean temperature difference Δt_m for use in the relation $Q = kA\Delta t_m$.

In order to develop a relation for the equivalent average temperature difference between the two fluids, consider the parallel-flow double-pipe heat exchanger shown in Fig. 11-6. Note that the temperature difference Δt between the hot and cold fluids is large at the inlet of the heat exchanger but decreases exponentially toward the outlet. As you would expect, the temperature of the hot fluid decreases and the temperature of the cold fluid increases along the heat exchanger, but the temperature of the cold fluid can never exceed that of the hot fluid no matter how long the heat exchanger is.

$$\Delta t_m = \frac{\Delta t_{max} - \Delta t_{min}}{\ln \dfrac{\Delta t_{max}}{\Delta t_{min}}}$$

is the logarithmic mean temperature difference, which is the suitable form of the average temperature difference for use in the analysis of heat exchangers. Here Δt_{max} and Δt_{min} represent the temperature difference between the two fluids at the two ends (inlet and outlet) of the heat exchanger. It makes no difference which end of the heat exchanger is designated as the inlet or the outlet (Fig. 11-6). The temperature difference between the two fluids decreases from Δt_{max} at the inlet to Δt_{min} at the outlet. Thus, it is tempting to use the arithmetic mean temperature $\Delta t_m = (\Delta t_{max} + \Delta t_{min})/2$ as the average temperature difference. The logarithmic mean temperature difference is obtained by tracing the actual temperature profile of the fluids along the heat exchanger and is an exact representation of the average temperature difference between the hot and cold fluids. It truly reflects the exponential decay of the local temperature difference.

Note that the logarithmic mean temperature difference is always less than the arithmetic mean temperature difference. Therefore, using the logarithmic mean temperature difference in calculations instead of the arithmetic mean temperature difference will overestimate the rate of heat transfer in a heat exchanger between the two fluids. When Δt_{max} differs from Δt_{min} by no more than 40 percent, the error in using the arithmetic mean temperature difference is less than 1 percent. But the error increases to undesirable levels when Δt_{max} differs from Δt_{min} by greater amounts. Therefore, we should always use the logarithmic mean temperature difference when determining the rate of heat transfer in a heat exchanger.

附 录

附录 1 常用单位换算表

Appendix 1 Common Units Conversion Table

附表 1 压力单位换算表

Pa	bar	at(kgf/cm^2)	atm	mmHg	mmH$_2$O(kgf/m^2)
帕	巴	工程大气压	标准大气压	毫米汞柱	毫米水柱
1×10^{5}	1	1.019 7	$9.869\ 2\times10^{-1}$	$7.500\ 6\times10^{2}$	$1.019\ 7\times10^{4}$
1	1×10^{-5}	$1.019\ 7\times10^{-5}$	$9.869\ 2\times10^{-6}$	$7.500\ 6\times10^{-3}$	$1.019\ 7\times10^{-1}$
$9.806\ 7\times10^{4}$	$9.806\ 7\times10^{-1}$	1	$9.678\ 4\times10^{-1}$	$7.356\ 6\times10^{2}$	1×10^{4}
$1.013\ 3\times10^{5}$	1.013 3	1.033 2	1	$7.600\ 0\times10^{2}$	$1.033\ 2\times10^{4}$
$1.333\ 2\times10^{2}$	$1.333\ 2\times10^{-3}$	$1.359\ 5\times10^{-3}$	$1.315\ 8\times10^{-3}$	1	$1.359\ 5\times10^{1}$
9.806 7	$9.806\ 7\times10^{-5}$	1×10^{-4}	$9.678\ 4\times10^{-5}$	$7.356\ 6\times10^{-2}$	1

附表 2 功、热量、能量单位换算表

kJ	kgf·m	kcal	kW·h	
千焦	千克力·米	千卡	千瓦·时	马力·时
1	$1.019\ 7\times10^{2}$	$2.388\ 5\times10^{-1}$	$2.777\ 8\times10^{-4}$	$3.776\ 7\times10^{-4}$
$9.806\ 7\times10^{-3}$	1	$2.342\ 3\times10^{-3}$	$2.724\ 1\times10^{-6}$	$3.703\ 7\times10^{-6}$
4.186 8	$4.269\ 4\times10^{2}$	1	$1.163\ 0\times10^{-3}$	$1.581\ 2\times10^{-3}$
$3.600\ 7\times10^{3}$	$3.671\ 0\times10^{5}$	$8.598\ 5\times10^{2}$	1	1.359 6
$2.647\ 8\times10^{3}$	$2.700\ 5\times10^{5}$	$6.324\ 2\times10^{2}$	$7.355\ 0\times10^{-1}$	1

附表 3 其他单位换算表

热流密度	W/m^2	kcal/(m^2·h)	导热系数	W/(m·K)	kcal/(m·h·℃)
	1	$8.598\ 5\times10^{-1}$		1	$8.598\ 5\times10^{-1}$
	1.163	1		1.163	1
表面传热系数	W/(m^2·K)	kcal/(m^2·h·℃)	比热	kJ/(kg·K)	kcal/(kg·℃)
	1	$8.598\ 5\times10^{-1}$		1	$2.388\ 5\times10^{-1}$
	1.163	1		4.186 8	1
动力黏度	kg/(m·s)	kgf·s/m^2	运动黏度、导温系数	m^2/s	m^2/h
	1	$1.019\ 7\times10^{-1}$		1	3 600
	9.806 7	1		$2.777\ 8\times10^{-4}$	1

附录 2　气体的热力参数

Appendix 2　Thermodynamic Parameters of Gases

附表 4　常用气体的热力特性

物质	M /(g/mol)	c_p /[kJ/(kg·K)]	$c_{p\cdot m}$ /[J/(mol·K)]	c_v /[kJ/(kg·K)]	$c_{v\cdot m}$ /[J/(mol·K)]	R /[kJ/(kg·K)]	κ
氩气 Ar	39.940	0.523	20.89	0.315	12.57	0.208	1.66
氦气 He	4.003	5.200	20.81	3.123	12.50	2.077	1.67
氢气 H_2	2.016	14.320	28.86	10.190	20.55	4.124	1.41
氮气 N_2	28.020	1.038	29.08	0.742	20.77	0.297	1.40
氧气 O_2	32.000	0.917	29.34	0.657	21.03	0.260	1.40
一氧化碳 CO	28.010	1.042	29.19	0.745	20.88	0.297	1.40
空气	28.970	1.004	29.09	0.718	20.78	0.287	1.40
水蒸气	18.016	1.867	33.64	1.406	25.33	0.461	1.33
二氧化碳 CO_2	44.010	0.845	37.19	0.656	28.88	0.189	1.29
二氧化硫 SO_2	64.070	0.644	41.26	0.514	32.94	0.130	1.25
甲烷 CH_4	16.040	2.227	35.72	1.709	27.41	0.518	1.30
丙烷 C_3H_8	44.090	1.691	74.56	1.502	66.25	0.189	1.13

附表 5　常见气体的定压摩尔比热随温度的变化(真实比热)

关系式 $c_{p\cdot m}=a_0+a_1T+a_2T^2+a_3T^3$, kJ/(kmol·K)

气体名称	分子式	a_0	$a_1/\times10^3$	$a_2/\times10^6$	$a_3/\times10^9$	温度范围/K	最大误差/%
空气		28.106	1.966 5	4.802 3	−1.966 1	273～1 800	0.72
氢气	H_2	19.107	−1.915 9	−4.003 8	−0.870 4	273～1 800	1.01
氧气	O_2	25.477	15.202 2	−5.061 8	1.311 7	273～1 800	1.19
氮气	N_2	28.901	−1.571 3	8.080 5	−28.725 6	273～1 800	0.59
一氧化碳	CO	28.160	1.675 1	5.371 7	−2.221 9	273～1 800	0.89
二氧化碳	CO_2	22.257	59.808 4	−35.010 0	7.469 3	273～1 800	0.647
水蒸气	H_2O	32.238	1.923 4	10.554 9	−3.595 2	273～1 800	0.53
乙烯	C_2H_2	4.1261	155.021 3	−81.545 5	16.975 5	298～1 500	0.30
丙烯	C_3H_6	3.745 7	234.010 7	−115.127 8	21.735 3	298～1 500	0.44
甲烷	CH_4	19.887	50.241 6	12.686 0	−11.011 3	273～1 800	1.33
乙烷	C_2H_6	5.413	178.087 2	−69.374 9	8.714 7	298～1 500	0.70
丙烷	C_3H_8	−4.233	306.264 0	−158.631 6	32.145 5	298～1 500	0.28

附表 6　气体的平均比热容(直线关系式)

$$c|_{t_1}^{t_2}=a+bt,\mathrm{kJ/(kg\cdot K)}(0\sim 1\ 500\ ℃)$$

气体名称	平均定容比热	平均定压比热
空气	0.708 8+0.000 093t	0.995 6+0.000 093t
氢气 H_2	10.12+0.000 594 5t	14.33+0.000 594 5t
氮气 N_2	0.730 4+0.000 089 55t	1.032+0.000 089 55t
氧气 O_2	0.659 4+0.000 106 5t	0.919+0.000 106 5t
一氧化碳 CO	0.733 1+0.000 096 81t	1.035+0.000 096 81t
水蒸气 H_2O	1.372+0.000 311 1t	1.833 +0.000 311 1t
二氧化碳 CO_2	0.683 7+0.000 240 6t	0.872 5+0.000 240 6t

附录 3 饱和水与饱和水蒸气的热力性质表(以温度为序)

Appendix 3 Table for Properties of Saturated Water and Vapor(by Temperature)

附表 7 饱和水与饱和水蒸气的热力性质表(以温度为序)

温度	压力	比容		比焓		汽化潜热	比熵	
		饱和水	饱和蒸汽	饱和水	饱和蒸汽		饱和水	饱和蒸汽
t/℃	p/MPa	v'/(m³/kg)	v''/(m³/kg)	h'/(kJ/kg)	h''/(kJ/kg)	r/(kJ/kg)	s'/[kJ/(kg·K)]	s''/[kJ/(kg·K)]
0	0.000 611 2	0.001 000 22	206.154	−0.05	2 500.51	2 500.6	−0.000 2	9.154 4
0.01	0.000 611 7	0.001 000 21	206.012	0	2 500.53	2 500.5	0	9.154 1
1	0.000 657 1	0.001 000 18	192.464	4.18	2 502.35	2 498.2	0.015 3	9.127 8
4	0.000 813 5	0.001 000 08	157.151	16.82	2 507.87	2 491.1	0.061 1	9.049 3
6	0.000 935 2	0.001 000 10	137.670	25.22	2 511.55	2 486.3	0.091 3	8.998 2
8	0.001 072 8	0.001 000 19	120.868	33.62	2 515.33	2 481.6	0.121 3	8.948 0
10	0.001 224 9	0.001 000 34	106.341	42.00	2 519.90	2 476.9	0.151 0	8.898 8
14	0.001 598 5	0.001 000 80	82.828	58.76	2 526.24	2 467.5	0.209 8	8.820 9
18	0.002 064 0	0.001 001 45	65.029	75.50	2 533.55	2 458.1	0.267 7	8.710 3
22	0.002 644 4	0.001 002 99	51.445	92.23	2 540.84	2 448.6	0.324 7	8.621 0
26	0.003 362 5	0.001 003 28	40.997	108.95	2 548.10	2 439.2	0.381 0	8.534 7
30	0.004 245 1	0.001 004 42	32.899	125.68	2 555.35	2 429.7	0.436 6	8.451 4
35	0.005 626 3	0.001 006 05	25.222	146.59	2 564.38	2 417.8	0.505 0	8.351 1
40	0.007 381 1	0.001 007 89	19.529	167.50	2 573.36	2 405.9	0.572 3	8.255 1
45	0.009 589 7	0.001 009 93	15.263 6	188.42	2 582.30	2 393.9	0.638 6	8.163 0
50	0.012 344 6	0.001 012 16	12.036 5	209.33	2 591.19	2 381.9	0.703 8	8.074 5
60	0.019 933	0.001 017 13	7.674 0	251.15	2 608.79	2 357.6	0.831 2	7.908 0
70	0.031 178	0.001 022 76	5.044 3	293.01	2 626.10	2 333.1	0.955 0	7.754 0
80	0.047 376	0.001 029 03	3.408 6	334.93	2 643.06	2 308.1	1.075 3	7.611 2
100	0.101 325	0.001 043 44	1.673 6	419.06	2 675.71	2 256.6	1.306 9	7.354 5
120	0.198 483	0.001 060 31	0.892 19	503.76	2 706.18	2 202.4	1.527 7	7.129 7
140	0.361 190	0.001 079 72	0.509 00	589.21	2 733.81	2 144.6	1.739 3	6.930 2
160	0.617 66	0.001 101 93	0.307 09	657.62	2 757.92	2 082.3	1.942 9	6.750 2
180	1.001 93	0.001 127 32	0.194 03	763.22	2 777.74	2 014.5	2.139 6	6.585 2
200	1.553 66	0.001 156 40	0.127 32	852.34	2 792.47	1 940.1	2.330 7	6.431 2
220	2.317 83	0.001 190 00	0.086 157	943.46	2 801.20	1 857.7	2.517 5	6.284 6

附表 7(续)

温度	压力	比容		比焓		汽化潜热	比熵	
		饱和水	饱和蒸汽	饱和水	饱和蒸汽		饱和水	饱和蒸汽
t/℃	p/MPa	v'/(m^3/kg)	v''/(m^3/kg)	h'/(kJ/kg)	h''/(kJ/kg)	r/(kJ/kg)	s'/[kJ/(kg·K)]	s''/[kJ/(kg·K)]
240	3.344 59	0.001 229 22	0.059 743	1 037.2	2 802.88	1 765.7	2.701 3	6.142 2
260	4.689 23	0.001 275 79	0.042 195	1 134.3	2 796.14	1 661.8	2.883 7	6.000 7
280	6.412 73	0.001 332 42	0.030 165	1 236.0	2 779.08	1 543.1	3.066 8	5.856 4
300	8.583 08	0.001 403 69	0.021 669	1 344.0	2 748.71	1 404.7	3.253 3	5.704 2
320	11.278	0.001 498 44	0.015 479	1 461.2	2 699.72	1 238.5	3.447 5	5.535 6
340	14.593	0.001 637 28	0.010 790	1 593.7	2 621.32	1 027.6	3.658 6	5.334 5
360	18.657	0.001 894 23	0.006 958	1 761.1	2 481.68	720.6	3.915 5	5.053 6
370	21.033	0.002 214 80	0.444 982	1 891.7	2 338.79	447.1	4.112 5	4.807 6
373.99	22.064	0.003 106	0.003 106	2 085.9	2 085.87	0	4.409 2	4.409 2

附录 4 饱和水与饱和水蒸气的热力性质表(以压力为序)

Appendix 4 Table for Properties of Saturated Water and Vapor(by Pressure)

附表 8 饱和水与饱和水蒸气的热力性质表(以压力为序)

压力	温度	比容		比焓		汽化潜热	比熵	
		饱和水	饱和蒸汽	饱和水	饱和蒸汽		饱和水	饱和蒸汽
p/MPa	t/℃	v'/(m³/kg)	v''/(m³/kg)	h'/(kJ/kg)	h''/(kJ/kg)	r/(kJ/kg)	s'/[kJ/(kg·K)]	s''/[kJ/(kg·K)]
0.001	6.949 1	0.001 000 1	129.185	29.21	2 513.29	2 484.1	0.105 6	8.973 5
0.003	24.114 2	0.001 002 8	45.666	101.07	2 544.68	2 443.6	0.354 6	8.575 8
0.005	32.879 3	0.001 005 3	28.191	137.72	2 560.55	2 422.8	0.476 1	8.393 0
0.007	38.996 7	0.001 007 5	20.528	163.31	2 571.56	2 408.3	0.558 9	8.273 7
0.009	43.790 1	0.001 009 4	16.204	183.36	2 580.15	2 396.8	0.622 6	8.185 4
0.010	43.790 1	0.001 010 3	14.673	191.76	2 583.72	2 392.0	0.649 0	8.148 1
0.015	53.970 5	0.001 014 0	10.022	225.93	2 598.21	2 372.3	0.754 8	8.006 5
0.020	60.065 0	0.001 017 2	7.649 7	251.43	2 608.90	2 357.5	0.832 0	7.906 8
0.025	64.972 6	0.001 019 8	6.204 7	271.96	2 617.43	2 345.5	0.893 2	7.829 8
0.030	69.104 1	0.001 022 2	5.229 6	289.26	2 624.56	2 335.3	0.944 0	7.767 1
0.040	75.872 0	0.001 026 4	3.993 9	317.61	2 636.10	2 318.5	1.026 0	7.668 8
0.060	85.949 6	0.001 033 1	2.732 4	359.91	2 652.97	2 293.1	1.145 4	7.531 0
0.080	93.510 7	0.001 038 5	2.087 6	391.71	2 665.33	2 273.6	1.233 0	7.433 9
0.100	99.634	0.001 043 2	1.694 3	417.52	2 675.14	2 257.6	1.302 8	7.358 9
0.120	104.810	0.001 047 3	1.428 7	439.37	2 683.26	2 243.9	1.360 9	7.297 8
0.140	109.318	0.001 051 0	1.236 8	458.44	2 690.22	2 231.8	1.411 0	7.246 2
0.160	113.326	0.001 054 4	1.091 59	475.42	2 696.29	2 220.9	1.494 6	7.201 6
0.180	116.941	0.001 057 6	0.977 67	490.76	2 701.69	2 210.9	1.530 3	7.162 3
0.200	120.240	0.001 060 5	0.885 85	504.78	2 706.53	2 201.7	1.607 5	7.127 2
0.300	133.556	0.001 073 2	0.605 87	561.58	2 725.26	2 163.7	1.672 1	6.992 1
0.400	143.642	0.001 083 5	0.462 46	604.88	2 738.49	2 133.6	1.776 9	6.896 1
0.500	151.867	0.001 092 5	0.374 86	640.35	2 748.59	2 108.2	1.861 0	6.821 4
0.600	158.863	0.001 100 6	0.315 63	670.67	2 756.66	2 086.0	1.931 5	6.760 0
0.700	164.983	0.001 107 9	0.272 81	697.32	2 763.29	2 066.0	1.992 5	6.707 9
0.800	170.444	0.001 114 8	0.240 37	721.20	2 768.86	2 047.7	2.046 4	6.662 5
0.900	175.389	0.001 121 2	0.214 91	742.90	2 773.59	2 030.7	2.094 8	6.622 2

附表 8(续)

压力	温度	比容		比焓		汽化潜热	比熵	
		饱和水	饱和蒸汽	饱和水	饱和蒸汽		饱和水	饱和蒸汽
p/MPa	t/℃	v'/(m^3/kg)	v''/(m^3/kg)	h'/(kJ/kg)	h''/(kJ/kg)	r/(kJ/kg)	s'/[kJ/(kg·K)]	s''/[kJ/(kg·K)]
1.0	179.916	0.001 127 2	0.194 38	762.84	2 777.67	2 014.8	2.138 8	6.585 9
2.0	212.417	0.001 176 7	0.099 588	908.64	2 798.66	1 890.0	2.447 1	6.339 5
4.0	250.394	0.001 252 4	0.049 771	1 087.2	2 800.53	1 713.4	2.796 2	6.068 8
8.0	295.048	0.001 384 3	0.023 520	1 316.5	2 757.70	1 441.2	3.206 6	5.743 0
12.0	324.715	0.001 526 0	0.014 263	1 490.7	2 684.50	1 193.8	3.495 2	5.492 0
16.0	347.396	0.001 709 9	0.009 311	1 649.4	2 580.21	930.8	3.745 1	5.245 0
20.0	365.789	0.002 037 9	0.005 870	1 827.2	2 413.05	585.9	4.015 3	4.932 2
22.0	373.752	0.002 704 0	0.003 684	2 013.0	2 084.02	71.0	4.296 9	4.406 6
22.064	373.990	0.003 106 0	0.003 106	2 085.9	2 085.87	0	4.409 2	4.409 2

附录 5　材料的导热系数
Appendix 5　Thermal Conductivities of the Materials

附表 9　金属材料的密度、比热及导热系数

材料名称	密度 ρ/(kg/m³)	比热 c/[J/(kg・K)]	导热系数 λ/[W/(m・K)]		
	20 ℃	20 ℃	0 ℃	20 ℃	100 ℃
纯铝	2 710	902	236	236	240
铝合金(92Al-8Mg)	2 610	904	102	107	123
铝合金(87Al-13Si)	2 660	871	158	162	173
铍	1 850	1 758	218	219	170
纯铜	8 930	386	401	398	393
青铜(89Cu-11Sn)	8 800	343	24	24.8	28.4
黄铜(70Cu-30Zn)	8 440	377	106	109	131
铜合金(60Cu-40Ni)	8 920	410	22.2	22.2	23.4
黄金	19 300	127	318	315	313
纯铁	7 870	455	83.5	81.1	72.1
阿姆可铁	7 860	455	74.7	73.2	67.5
灰铸铁($\omega_C \approx 3\%$)	7 570	470	28.5	39.2	32.4
碳钢($\omega_C \approx 1.0\%$)	7 790	470	43.0	43.2	42.8
铬钢($\omega_{Cr} \approx 13\%$)	7 740	460	26.5	26.8	27.0
镍钢($\omega_{Ni} \approx 1\%$)	7 900	460	45.2	45.5	46.8
镍钢($\omega_{Ni} \approx 35\%$)	8 110	460	13.4	13.8	15.4
镍钢($\omega_{Ni} \approx 50\%$)	8 260	460	19.4	19.6	20.5
锰钢($\omega_{Mn} \approx 0.4\%$)	7 860	440		51.2	51.0
铅	11 340	128	35.5	35.3	34.3
镁	1 730	1 020	157	156	154
钼	9 590	255	139	138	135
镍	8 900	444	94	91.4	82.8
铂	21 450	133	71.5	71.4	71.6
银	10 500	234	428	427	422
锡	7 310	228	68.2	67	63.2
钛	4 500	520	22.4	22	20.7
铀	19 070	116	27	27.4	29.1
锌	7 140	388	122	121	117
锆	6 570	276	23.2	22.9	21.8
钨	19 350	134	182	179	166

附表 10　保温和建筑及其他材料的密度和导热系数

材料名称	温度 t/℃	密度 ρ/(kg/m³)	导热系数 λ/[W/(m·K)]
膨胀珍珠岩散料	25	60～300	0.021～0.062
沥青膨胀珍珠岩	31	233～282	0.069～0.076
磷酸盐膨胀珍珠岩制品	20	200～250	0.044～0.052
水玻璃膨胀珍珠岩制品	20	200～300	0.056～0.065
岩棉制品	20	80～150	0.035～0.038
膨胀蛭石	20	100～130	0.051～0.07
沥青蛭石板管	20	350～400	0.081～0.10
石棉粉	22	744～1 400	0.099～0.19
石棉砖	21	384	0.099
石棉绳		590～730	0.10～0.21
石棉绒		35～230	0.055～0.077
石棉板	30	770～1 045	0.10～0.14
碳酸镁石棉灰		240～490	0.077～0.086
硅藻土石棉灰		280～380	0.085～0.11
粉煤灰砖	27	458～589	0.12～0.22
矿渣棉	30	207	0.058
玻璃丝	35	120～492	0.058～0.07
玻璃棉毡	28	18.4～38.3	0.043
软木板	20	105～437	0.044～0.079
木丝纤维板	25	245	0.048
稻草浆板	20	325～365	0.068～0.084
麻秆板	25	108～147	0.056～0.11
甘蔗板	20	282	0.067～0.072
葵芯板	20	95.5	0.05
玉米梗板	22	25.2	0.065
棉花	20	117	0.049
丝	20	57.7	0.036
锯木屑	20	179	0.083
硬泡沫塑料	30	29.5～56.3	0.041～0.048
软泡沫塑料	30	41～162	0.043～0.056
铝箔间隔层(5层)	21		0.042
红砖(营造状态)	25	1 860	0.87
红砖	35	1 560	0.49
松木(垂直木纹)	15	496	0.15
松木(平行木纹)	21	527	0.35
水泥	30	1 900	0.30

附表 10(续)

材料名称	温度 t/℃	密度 ρ/(kg/m³)	导热系数 λ/[W/(m·K)]
混凝土板	35	1 930	0.79
耐酸混凝土板	30	2 250	1.50～1.60
黄沙	30	1 580～1 700	0.28～0.34
泥土	20		0.83
瓷砖	37	2 090	1.10
玻璃	45	2 500	0.65～0.71
聚苯乙烯	30	24.7～37.8	0.04～0.043
花岗石		2 643	1.73～3.98
大理石		2 499～2 707	2.70
云母		290	0.58
水垢	65		1.31～3.14
冰	0	913	2.22
黏土	27	1 460	1.30

附录 6　常见材料的表面(法向)发射率
Appendix 6　Surface (Normal) Emissivity of Common Materials

附表 11　常见材料的表面(法向)发射率

材料类别和表面状况	温度/℃	发射率(黑度)ε
磨光的钢铸体	770～1 035	0.52～0.56
碾压的钢板	21	0.657
具有非常粗糙的氧化层的钢板	24	0.80
磨光的铬	150	0.058
粗糙的铝板	20～25	0.06～0.07
基体为铜的镀铝表面	190～600	0.18～0.19
在磨光的铁上电镀一层镍,但不再磨光	38	0.11
铬镍合金	52～1 034	0.64～0.76
粗糙的铅	38	0.43
灰色、氧化的铝	38	0.28
磨光的铸铁	200	0.21
生锈的铁板	20	0.685
粗糙的铁锭	926～1 120	0.87～0.95
经过车磨加工的铸铁	882～987	0.60～0.70
稍加磨光的黄铜	38～260	0.12
无光泽的黄铜	38	0.22
粗糙的黄铜	38	0.74
磨光的纯铜	20	0.03
氧化了的纯铜	20	0.78
镀有锡且发亮的铁片	25	0.043～0.064
镀锌的铁皮	38	0.23
镀锌的铁片,被氧化呈灰色	24	0.276

附录 7　水蒸气的焓熵(*h*-*s*)图

Appendix 7　Diagram of Enthalpy and Entropy (*h*-*s*) of Water Vapor

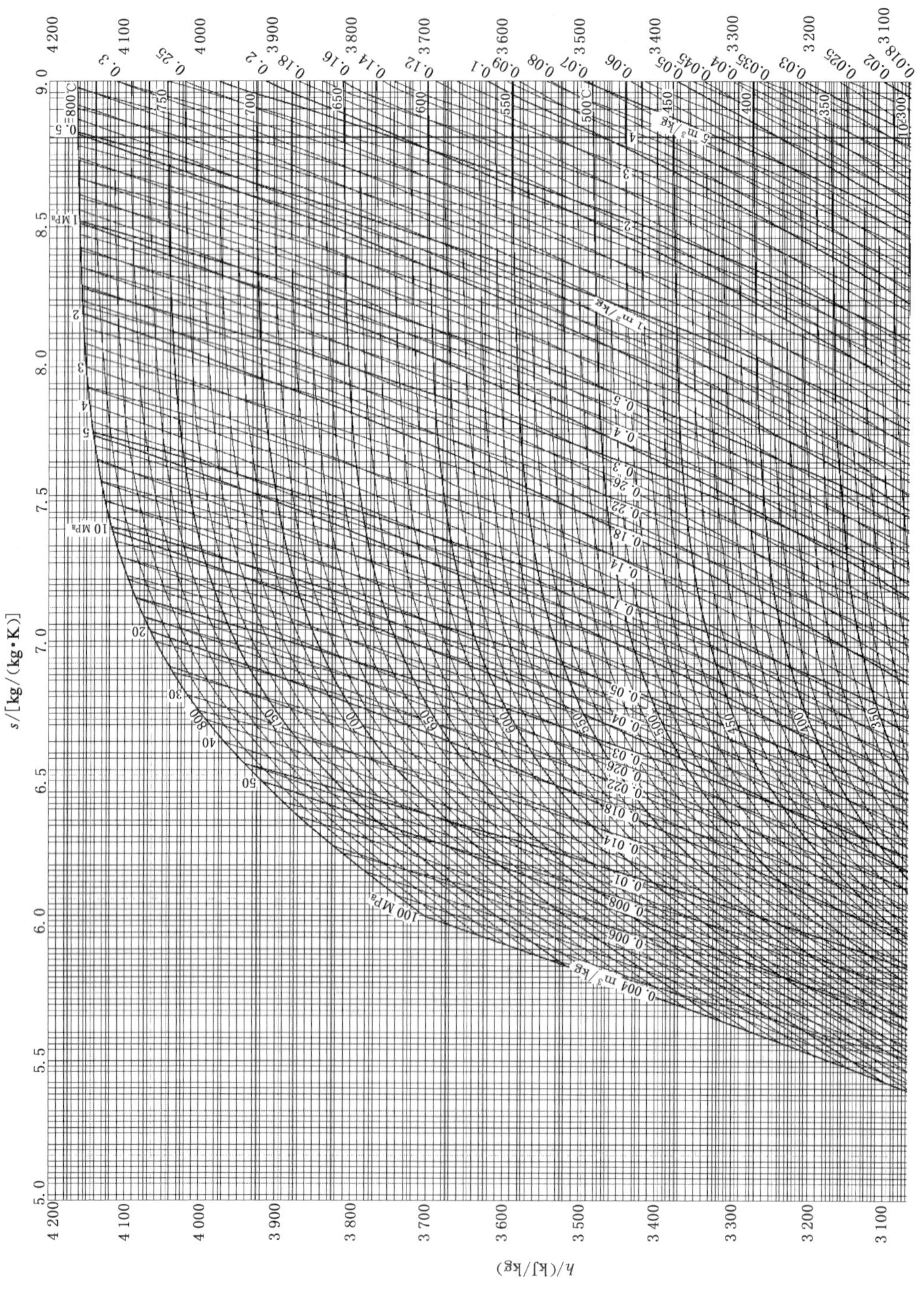

附图 1　水蒸气的焓熵 (*h*-*s*) 图

附录 8　湿空气的焓湿(h-d)图

Appendix 8　Diagram of Enthalpy and Humidity Ration (h-d) of Moist Air

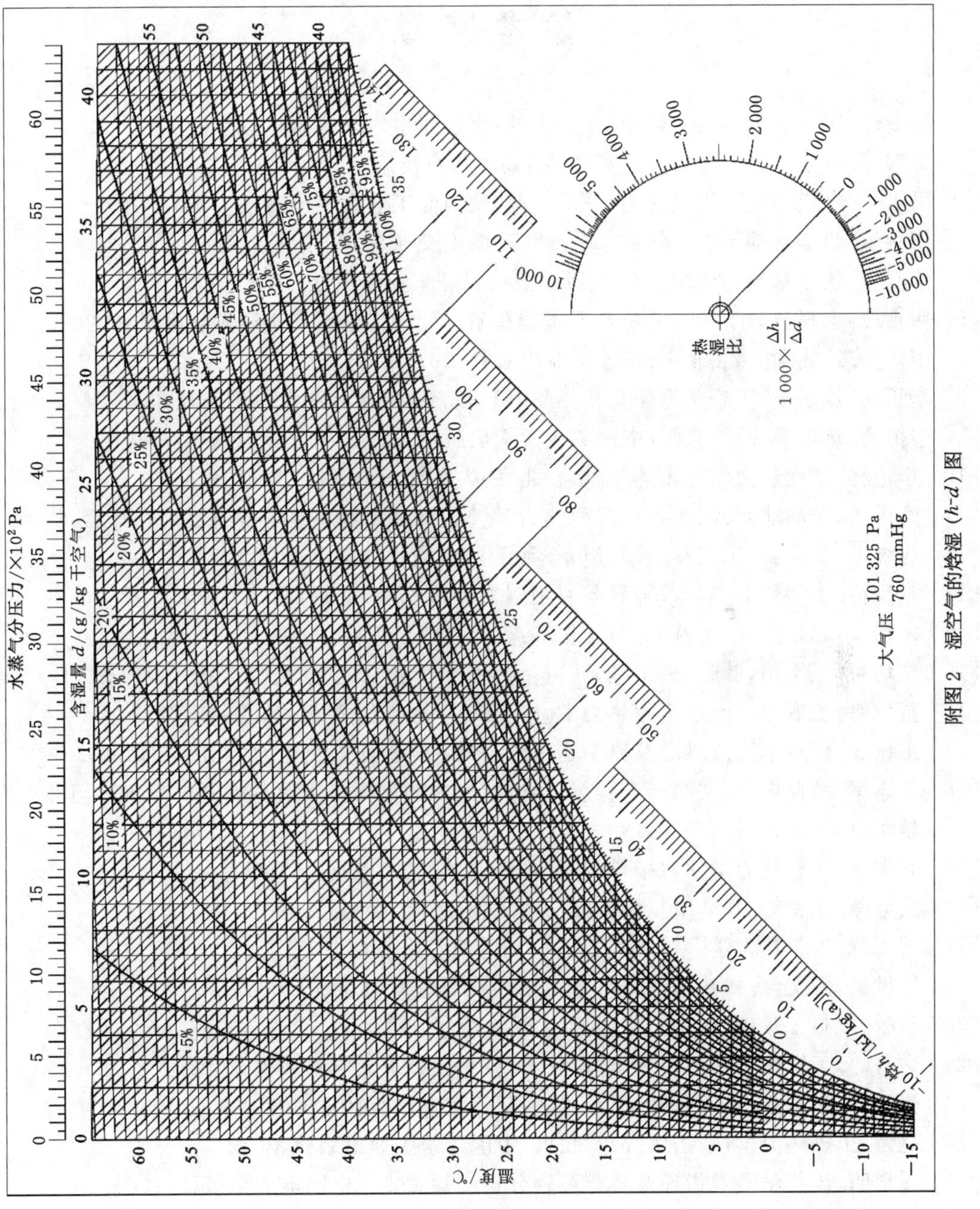

附图 2　湿空气的焓湿(h-d)图

参考文献

[1] 陈爱玲. 工程热力学与传热学[M]. 大连:大连海事大学出版社,2005.

[2] 陈黟,吴味隆. 热工学[M]. 3 版. 北京:高等教育出版社,2004.

[3] 冯青,李世武,张丽. 工程热力学[M]. 西安:西北工业大学出版社,2006.

[4] 傅秦生. 热工基础要点与解题[M]. 西安:西安交通大学出版社,2006.

[5] 傅秦生. 热工基础与应用[M]. 3 版. 北京:机械工业出版社,2016.

[6] 何雅玲. 工程热力学精要分析及典型题精解[M]. 西安:西安交通大学出版社,2000.

[7] 霍尔曼. 传热学[M]. 北京:机械工业出版社,2011.

[8] 李长友,钱东平. 工程热力学与传热学[M]. 北京:中国农业大学出版社,2004.

[9] 李传统. 热工学[M]. 徐州:中国矿业大学出版社,1994.

[10] 李岳林. 工程热力学与传热学[M]. 北京:人民交通出版社,2007.

[11] 廉乐明,谭羽非,吴家正,等. 工程热力学[M]. 5 版. 北京:中国建筑工业出版社,2007.

[12] 沈维道,童钧耕. 工程热力学[M]. 5 版. 北京:高等教育出版社,2016.

[13] 童钧耕. 工程热力学学习辅导与习题解答[M]. 2 版. 北京:高等教育出版社,2008.

[14] 童均耕,赵镇南. 热工基础[M]. 北京:高等教育出版社,2009.

[15] 童均耕,王平阳,叶强. 热工基础[M]. 3 版. 上海:上海交通大学出版社,2016.

[16] 童均耕,王丽伟,叶强. 工程热力学[M]. 6 版. 北京:高等教育出版社,2022.

[17] 王秋旺. 传热学重点难点及典型题精解[M]. 西安:西安交通大学出版社,2001.

[18] 王志军,高保彬,宋文婷.《工程热力学》绪论课的重要作用及其课堂教学设计[J]. 教育教学论坛,2012(9):237-238.

[19] 武淑萍. 工程热力学学习指导[M]. 北京:中国电力出版社,2004.

[20] 武淑萍. 工程热力学[M]. 重庆:重庆大学出版社,2006.

[21] 严家騄. 工程热力学[M]. 6 版. 北京:高等教育出版社,2021.

[22] 杨世铭,陶文铨. 传热学[M]. 4 版. 北京:高等教育出版社,2006.

[23] 于秋红. 热工基础[M]. 北京:北京大学出版社,2009.

[24] 岳丹婷. 工程热力学和传热学[M]. 大连:大连海事大学出版社,2002.

[25] 张学学,李桂馥. 热工基础[M]. 北京:高等教育出版社,2000.

[26] 赵蕾. 工程热力学:双语版[M]. 北京:中国建筑工业出版社,2012.

[27] 周根明. 传热学学习指导及典型习题分析[M]. 北京:中国电力出版社,2004.

[28] 朱明善,刘颖,林兆庄,等. 工程热力学[M]. 2 版. 北京:清华大学出版社,2011.

[29] ENGEL Y A. Heat transfer:a practical approach[M]. [S. l. :s. n.],2003.

[30] ENGEL Y A,BOLES M A. Thermodynamics: an engineering approach[M]. 6th ed. [S. l. :s. n.],2007.